New Trends in Mechanical and Tribological Properties of Materials and Components

New Trends in Mechanical and Tribological Properties of Materials and Components

Guest Editors

Andrea Mura
Luigi Mazza
Edoardo Goti

MDPI

Basel • Beijing • Wuhan • Barcelona • Belgrade • Novi Sad • Cluj • Manchester

Guest Editors

Andrea Mura
Department of Mechanical and Aerospace Engineering
Politecnico di Torino
Torino
Italy

Luigi Mazza
Department of Mechanical and Aerospace Engineering
Politecnico di Torino
Torino
Italy

Edoardo Goti
Department of Mechanical and Aerospace Engineering
Politecnico di Torino
Torino
Italy

Editorial Office
MDPI AG
Grosspeteranlage 5
4052 Basel, Switzerland

This is a reprint of the Special Issue, published open access by the journal *Materials* (ISSN 1996-1944), freely accessible at: https://www.mdpi.com/journal/materials/special_issues/5JFSUGFT4T.

For citation purposes, cite each article independently as indicated on the article page online and as indicated below:

Lastname, A.A.; Lastname, B.B. Article Title. *Journal Name* **Year**, *Volume Number*, Page Range.

ISBN 978-3-7258-6624-3 (Hbk)
ISBN 978-3-7258-6625-0 (PDF)
https://doi.org/10.3390/books978-3-7258-6625-0

Contents

Preface

It is our great pleasure to introduce this Special Issue of *Materials*, titled "New Trends in Mechanical and Tribological Properties of Materials and Components." The rapid advancement of modern engineering systems, from transportation and energy to manufacturing, biomedical devices, and robotics, has intensified the demand for materials and components capable of delivering high mechanical reliability, a superior wear performance, and an extended service life. As industries strive for higher efficiency, lower environmental impacts, and improved sustainability, research on tribology and fatigue behavior has become crucial for guiding next-generation material design.

The characterization and optimization of advanced materials remain central themes. Many contributions explore how tailored material design, surface engineering, and coating technologies can fundamentally alter wear mechanisms, resistance to cyclic loading, and fatigue behavior under diverse service conditions. By leveraging advanced characterization techniques, including high-resolution microscopy, surface topography analysis, and mechanical testing across scales, authors dissect the complex interactions between microstructure, surface state, loading conditions, and material response that dictate performance limits and failure modes.

Enhancing the tribological performance and fatigue resistance of materials through surface treatments and engineered coatings is another pivotal focus. These strategies, ranging from hard ceramic layers to tailored tribo-films, not only reduce friction and wear but also significantly extend component lifespan in challenging environments. The work collected here advances our understanding of how specific surface modifications interact with base materials and external influences to mitigate surface damage and delay crack initiation, thereby improving service life and operational reliability.

We are delighted to have received contributions that address these diverse, yet interconnected topics, fostering a comprehensive perspective on what constitutes high performance in mechanical and tribological behavior. The insights provided in this Special Issue, from fundamental mechanisms to applied system performance, are highly relevant to both academic and industrial communities striving to push the boundaries of material capabilities.

Finally, we extend our sincere gratitude to all authors for submitting their high-quality research, to the reviewers for their thoughtful evaluations, and to the *Materials* Editorial Office for their unwavering support. It is our hope that this collection not only showcases the latest scientific advances but also inspires future research efforts that continue to elevate the understanding and application of materials under mechanical and tribological demands.

Andrea Mura, Luigi Mazza, and Edoardo Goti
Guest Editors

materials

MDPI

Article

Dynamic Testing of Materials for Galvanising Pot Roll Bearings with Improved Performance

Giovanni Paolo Alparone [1,*], James Sullivan [1], Christopher Mills [2], James Edy [2] and David Penney [1]

[1] Department of Materials Engineering, Swansea University, Swansea SA2 8PP, UK
[2] Tata Steel, Research and Development, Swansea Technology Centre, Swansea SA2 8PP, UK
* Correspondence: 2043556@swansea.ac.uk

Abstract: Galvanising pot roll bearings are subjected to severe deterioration due to the corrosion of the bearing materials in liquid Zn, resulting in maintenance stops that can cost thousands of pounds per hour in downtime. Dynamic wear testing in molten Zn-Al and Zn-Al-Mg was conducted to assess the corrosion and wear resistance of three material pairs using a bespoke testing rig. The materials investigated in this study were Wallex6™ coated with WC-Co, stainless steel 316L coated with Al_2O_3, and as-received Wallex6™ and Wallex4™ alloys. It was found that only the Al_2O_3 coating remained unreactive in Zn alloy, whereas the materials containing Co were corroded, as evidenced by the formation of intermetallic compounds containing Al-Co-Zn-Fe. The results also highlighted that the dissolution of the Co matrix and diffusion of Zn and Al from the bath occurred in Wallex6™ and Wallex4™. However, the diffusion of Zn into the Wallex™ alloys was reduced by approximately 60% in the Zn-Al-Mg bath compared to Zn-Al. The wear scars were analysed to determine the wear coefficient of the worn specimens. Out of the three material couplings investigated in this study, minimal wear damage in both Zn-Al and Zn-Al-Mg was only obtained by pairing Wallex6™ with Al_2O_3 coatings.

Keywords: ceramics; corrosion; wear testing; galvanising pot journal bearings; continuous galvanising; galvanising pot hardware

1. Introduction

Continuous galvanising is a hot-dip process in which the strip steel is immersed in a bath containing liquid Zn alloy of the desired composition. Automotive customers require a 'full finish', high surface quality, which is important for visible parts that must be free of imperfections [1]. As a result, the galvanised steel sheet demanded by the automotive industry must have an excellent surface finish free from defects and with a controlled surface roughness [2]. Recently, the use of Zn-Al-Mg coatings has become popular in the automotive industry due to their potential to reduce fuel consumption via lightweighting, as they require a lower thickness compared to Zn-Al coatings [3]. In addition to this, automotive industries use Zn-Al-Mg coatings due to their excellent galling resistance [4,5].

During the immersion process in the hot-dip galvanising bath, the strip steel is guided by the pot hardware, which includes the pot rolls and the roll journal bearings. In order to produce galvanised products that meet the requirements of automotive customers, the stability of the strip steel through the pot of liquid Zn alloy is important. Vibrations in the strip, especially after exiting from the Zn pot and around the gas knives section of the line, must be minimised to produce a high-quality product, as the gas knives control the thickness of the coating. However, the materials of the pot roll bearing components, namely the journal sleeve and bushing, react with the molten Zn alloy at temperatures greater than 400 °C and the deterioration of the bearings induces vibrations in the strip as it passes through the gas knives section of the line [6–9]. Therefore, to avoid quality issues, the pot hardware must be changed and reconditioned, leading to downtime and loss of yield [10,11]. For this reason, increasing the lifetime of the bearings by suppressing the reactions of the bearing materials with the Zn bath will enable the production window of automotive-grade galvanised products to be extended, resulting in cost benefits for the galvanising industry.

The properties of ceramic materials have been studied for developing bearings with extended durability, due to their potential to resist the attacks of many molten metals at high temperatures [12–16]. In previous work, the corrosion behaviour of Al_2O_3 coatings was examined by conducting static immersion tests in liquid Zn-Al and Zn-Al-Mg [17]. The findings obtained in this study suggested that Al_2O_3 showed superior performance in Zn-Al, compared to materials that have been traditionally used to make hardware components, such as steel and Co-based alloys or coatings, which were found to be severely corroded by the molten metal [7,18,19].

However, the corrosion behaviour of these materials was assessed under static conditions. The behaviour of materials under dynamic conditions could differ from that observed during static testing. In the dynamic situation, the pot hardware was exposed to chemical attacks from the liquid metal and dynamic wear due to the sliding of the bearing materials under load. It was demonstrated that the corrosion rates of materials in molten metal can differ from those observed in static baths [20]. Previous studies attempted to carry out dynamic tests on materials with potential use as pot-bearing materials in continuous galvanising [21,22]. However, a limited selection of materials has been tested under dynamic conditions; in addition to this, there is no evidence of dynamic tests conducted in Zn-Al-Mg baths in the literature.

For this reason, the present work investigates the performance of three material couplings tested under dynamic conditions in Zn-Al and Zn-Al-Mg using a bespoke dynamic testing rig. These experiments were performed on a current configuration, which coupled WC-Co coatings with Wallex6^{TM} and two potential upgrades to the benchmark configuration. The first of these potential upgrades involved replacing Wallex6^{TM} with Wallex4^{TM} and the second potential upgrade consisted of pairing stainless steel (SS) 316L coated with Al_2O_3 and Wallex6^{TM}.

2. Materials and Methods

2.1. Materials

The specimens used in this study consisted of 20 mm diameter × 150 mm-coated cylinders, which were tested against 25 × 50 × 10 mm pads. The three material couplings chosen for dynamic testing are summarised in Table 1. Wallex6^{TM} and Wallex4^{TM} were procured from Wall Colmonoy (Swansea, UK). WC-Co coatings were applied with a thickness of 150 μm. Al_2O_3 coatings were applied to SS 316L bars with a thickness of 250 μm via a high-velocity oxygen fuel (HVOF) thermal spray process, carried out externally by Engineered Performance Coatings (Cardiff, UK). The reason for choosing thermal spraying over other coating fabrication methods is that previous studies have shown that Al_2O_3 could be successfully deposited onto stainless steel [12]. The compositions of Wallex6^{TM} and Wallex4^{TM} were provided by the supplier and are found in Table 2.

Table 1. Material couplings chosen for dynamic testing.

Material Pair No.	Bar Material	Pad Material
1	Wallex6^{TM} coated with HVOF WC-Co	Wallex6^{TM}
2	Wallex6^{TM} coated with HVOF WC-Co	Wallex4^{TM}
3	SS 316L coated with HVOF Al_2O_3	Wallex6^{TM}

Table 2. Composition of Wallex6™ and Wallex4™.

Alloy	%Fe	%Cr	%Ni	%Mn	%Si	%Co	%C	%W	%Mo
Wallex6	2.3	28.4	2.6	0.8	1	BAL	1.2	3.8	0.8
Wallex4	1.7	30.4	2.6	0.6	0.8	BAL	0.9	14.3	1.5

2.2. Experimental Procedure

The material pairs listed in Table 1 were mounted on a bespoke dynamic testing rig, as shown in Figure 1. The rig enabled the testing of a round bar specimen that was mounted on a shaft connected to an electric motor. The bar was capable of rotating up to a speed of 300 RPM and it was allowed to slide against a static pad specimen with a force of 60 N. The rig was equipped with a heated Zn pot so that the two specimens could be submerged in molten metal. The pot was loaded with approximately 40 kg of Zn alloy ingots. Each material coupling was tested in Zn-0.3wt%Al, also known as GI, and Zn-1.5wt%Al-1.5wt%Mg bath compositions. The pot was set at a temperature of 465 °C, which was monitored using thermocouples installed inside the pot. As the Zn melted, dross formed on top of the bath, which was removed before starting the test. Once the Zn was fully molten, the pot was raised so that the bar was just above the Zn level and was held for 24 h to preheat the specimens and to minimise the effects of thermal shock. Subsequently, the pot was raised to submerge the bar and the pad in liquid Zn and the lid was closed. The bar was rotated in a clockwise direction at a speed of 300 RPM for 48 h. After the test, the pot was lowered, and the specimens were removed from the holders. Samples from the bar and the pad were cross-sectioned for characterisation. Images were captured using a ZEISS (Oberkochen, Germany) EVO scanning electron microscope (SEM) and a Hitachi (Tokyo, Japan) TM4000 desktop SEM, both equipped with backscattered electron detectors (BSD). Energy Dispersive X-ray analysis (EDS) was performed using an Oxford Instruments (Abingdon, UK) EDS detector with Aztec 6.1 software. The solidified Zn was removed with 35% HCl on the pad specimens, which required measurement of the volume loss. The displacement of the bar after 48 h of sliding was determined by measuring the depth of the wear scar left on the pad specimen. The 3D maps of the wear scar were generated using the Keyence VHX-7000 (Osaka, Japan) digital microscope. The volume loss 'V' was obtained from the cross-sectional area and the length of the scar according to Equation (1):

$$V = A \times L, \quad (1)$$

where 'A' is the measured cross-sectional area of the wear scar and 'L' is the length of the specimen. The worn track section 'A' was determined using the digital microscope. The total sliding distance 'S' was obtained by multiplying the sliding speed 'v' in [m/min] and the duration of sliding 't' in [min] (Equation (2)):

$$S = v \times t, \quad (2)$$

This value was subsequently used to calculate the wear coefficient 'k', as shown by Equation (3) [23]:

$$k = \frac{V}{F \times S} \left[\mathrm{mm}^3/\mathrm{Nm}\right], \quad (3)$$

where 'F' is the contact load and 'S' is the sliding distance.

(a) Frame
(b) Motor
(c) Bar and pad system
(d) Controller
(e) Zinc pot
(f) Scissor lift trolley
(g) Chuck
(h) Pad holder
(i) Bar sample
(j) Pad sample

Figure 1. Bespoke dynamic testing rig showing the main components.

3. Results and Discussion

3.1. Imaging of As-Received Samples

Images were taken on the as-received samples before exposure to Zn-Al and Zn-Al-Mg. Cross-sections were captured on as-received WC-Co/Wallex6^{TM} (Figure 2a,b) and Al_2O_3/SS 316L bar specimens (Figure 3). Figure 2b shows the microstructure of WC-Co coatings, which is characterised by WC grains (bright contrast) surrounded by the Co binder phase (dark contrast) [24]. The microstructure of Al_2O_3 coatings was described in a previous work, where the same coatings were tested under static conditions [17]. The as-received pad specimens, namely Wallex6^{TM} and Wallex4^{TM}, are shown in Figure 4a,b and Figure 5a,b respectively.

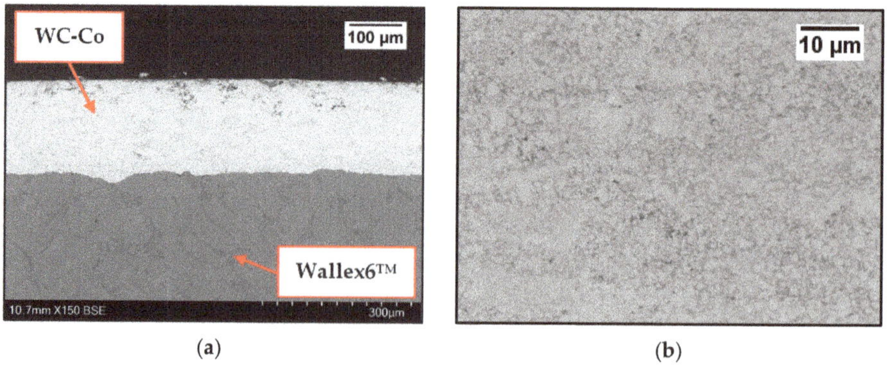

Figure 2. Cross-section of the as-received Wallex6^{TM} bar coated with WC-Co (**a**); details of the WC-Co coating (**b**). Please note the different magnifications.

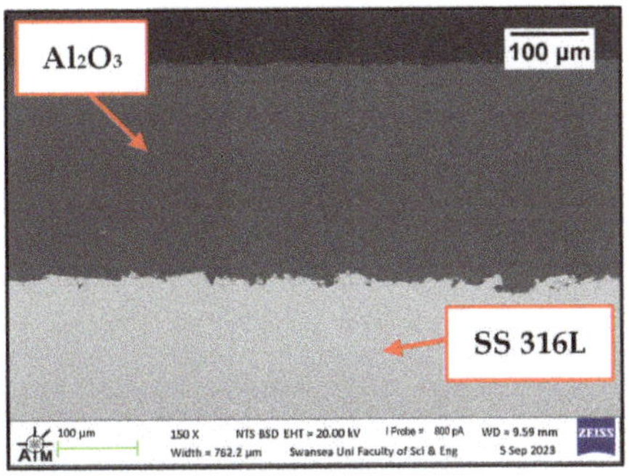

Figure 3. SEM image of the cross-section of high-velocity oxygen fuel (HVOF) Al_2O_3 coated onto the surface of stainless steel (SS) 316 L.

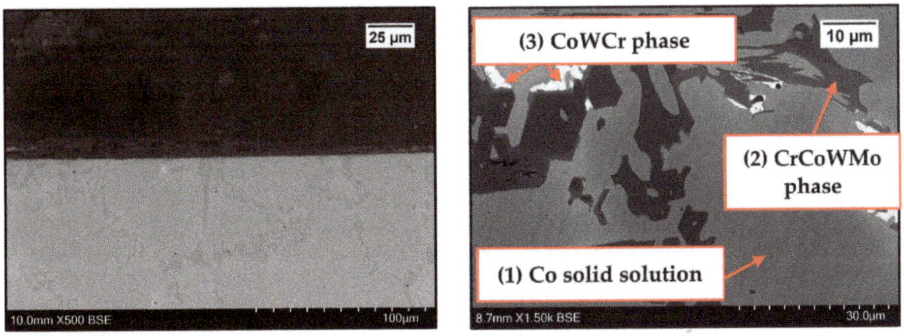

(**a**) (**b**)

Figure 4. SEM images of the cross-sections of Wallex6^{TM} captured at the interface (**a**) and in the bulk of the material (**b**). Please note the different magnifications.

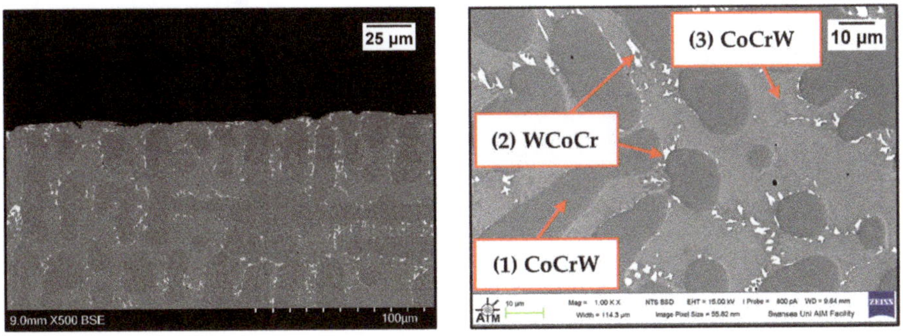

(**a**) (**b**)

Figure 5. SEM images of the cross-sections of Wallex4^{TM} captured at the interface (**a**) and in the bulk of the material (**b**). Please note the different magnifications.

Wallex6^{TM} is the equivalent of Stellite 6 and is characterised by a Co solid solution matrix phase (1), with Co, Cr, and W as the main constituents. EDS found that the average element composition of this phase was Co (61.3 ± 0.8 wt.%), Cr (23.8 ± 0.6 wt.%), and W (9.3 ± 1.5 wt.%), as shown in Figure 6. Two eutectic phases were observed in the Co matrix phase: a primary CrCoWMo (2) phase and a secondary eutectic phase of CoWCr (3)

(Figure 4b). The results of EDS elemental analysis showed that the primary eutectic phase was composed of Cr (69.6 ± 7.8 wt.%), Co (17.3 ± 2.8 wt.%), W (10.6 ± 4.6 wt.%), and Mo (1.4 ± Mo wt.%), with traces of Fe and Ni. The secondary phase mainly contained Co (49.2 ± 8.1 wt.%), W (23.2 ± 8.4 wt.%) and Cr (21.6 ± 2.5 wt.%).

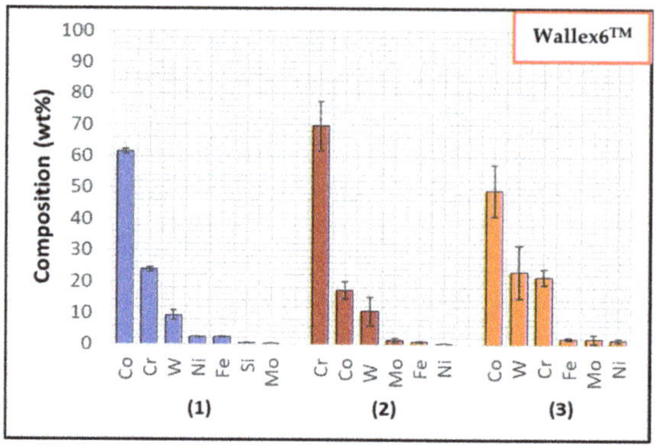

Figure 6. EDS phase elemental analysis of Wallex6™. The numbers refer to the phases in Figure 4, where (1) is the Co solid solution phase, (2) is the CrCoWMo phase and (3) the CoWCr phase.

The composition of Wallex4™ is similar to that of Wallex6™. The SEM image showed the presence of three different phases (Figure 5b). EDS analysis was conducted to determine the composition of these phases, as shown in Figure 7. EDS revealed the presence of a dark grey phase (1), that is a solid solution matrix phase containing Co (56.1 ± 0.4 wt.%), Cr (27.8 ± 0.2 wt.%), W (11.8 ± 0.2 wt.%), Ni (2.7 ± 0.2 wt.%), and Fe (1.6 ± 0.1 wt.%). In addition to this, two eutectic phases were present. The bright phase (2) was made of W (57.0 ± 0.7 wt.%), Co (23.0 ± 0.7 wt.%), Cr (18.8 ± 0.2 wt.%), Ni (0.8 ± 0.0 wt.%), and Fe (1.3 ± 0.1 wt.%), whereas the light grey phase (3) contained Co (44.9 ± 3.6 wt.%), Cr (34.9 ± 2.7 wt.%), W (17.1 ± 1.1 wt.%), Ni (1.8 ± 0.1 wt.%), and Fe (1.3 ± 0.1 wt.%). It was observed that the bright eutectic phase had a greater portion of W compared to the Co-rich matrix; in addition to this, the composition of the dark grey phase was similar to that of the matrix, with greater Cr and W content. The composition does not significantly differ from Wallex6™, as shown in Table 2; however, Wallex4™ was found to have lower Co content (47.2 wt.%) and higher W additions (14.3 wt.%) compared to Wallex6™, which contained Co (59.1 wt.%) and W (3.8 wt.%).

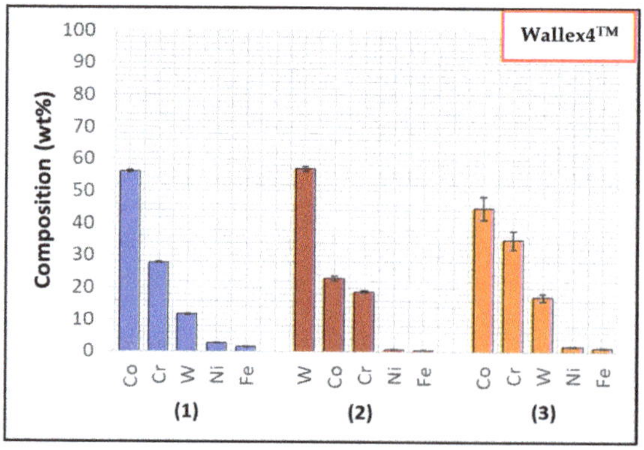

Figure 7. EDS phase elemental analysis of Wallex4™. The numbers refer to the phases in Figure 5, where (1) is the CoCrW phase, (2) is the WCoCr phase and (3) the CoCrW phase.

3.2. *Dynamic Corrosion Testing*

3.2.1. Wallex6™ with HVOF WC-Co and Wallex6™

Figure 8 shows a cross-section of the WC-Co/Wallex6™ bar after 48 h of testing in Zn-Al. The SEM image showed that the coating was not prone to damage after contact with the static pad specimen. Cracks, as well as pores, can provide pathways for liquid Zn alloy to penetrate the coating and corrode the base metal [16]. However, no reaction products accumulated below the coating, highlighting that Zn did not diffuse into the base metal. The integrity of the coating upon sliding with the Wallex6™ counterpart was linked with the hardness of the material. WC is harder than Wallex6™; it is known that the hardness of spherical WC lies between 2100–2500 HV, which is significantly higher than the hardness of Wallex6™ (400 HV) and Wallex4™ (570 HV) [25], suggesting that the Wallex™ pads could not inflict severe wear damage to the bar specimens.

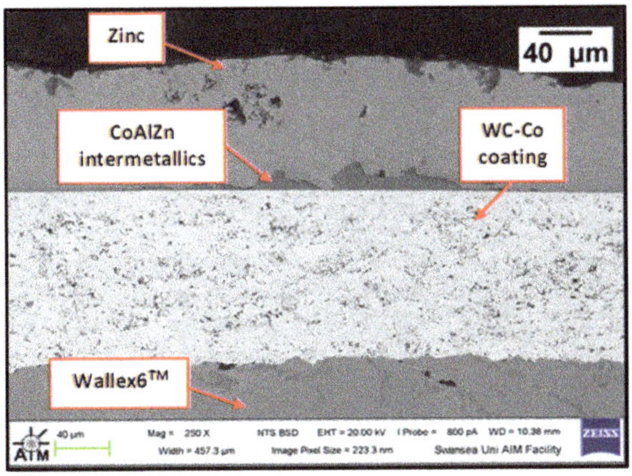

Figure 8. SEM image of the WC-Co/Wallex6™ bar after testing in Zn-Al.

Although the WC-Co coating protected Wallex6™ from the attack of liquid Zn-Al, the analysis of the cross-section revealed that intermetallic compounds formed on the contact surface, as illustrated in the high magnification image (Figure 9). These compounds were absent in the as-received specimen (Figure 2a). The results of the EDS phase elemental analysis conducted in this region (Figure 10) showed that the intermetallic compounds are Zn-based phases, which also contain Al (42.6 $\pm$ 0.9 wt.%) and Co (29.3 $\pm$ 0.9 wt.%), with traces of Fe, W, and Ni. Previous studies reported the presence of intermetallic phases on the surface of WC-Co coatings [26,27].

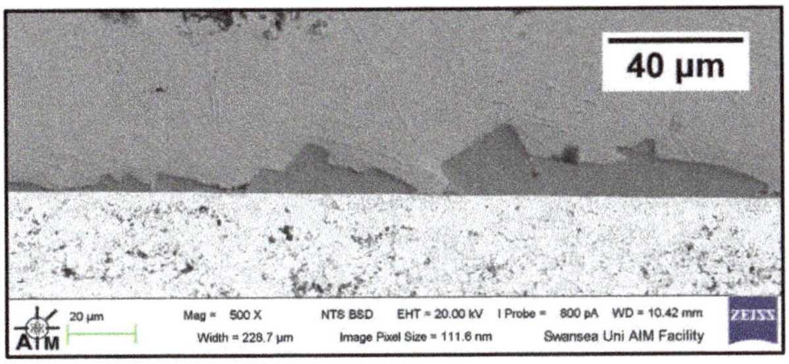

Figure 9. Higher magnification image of the intermetallic phases observed on the WC-Co coating after testing in Zn-Al.

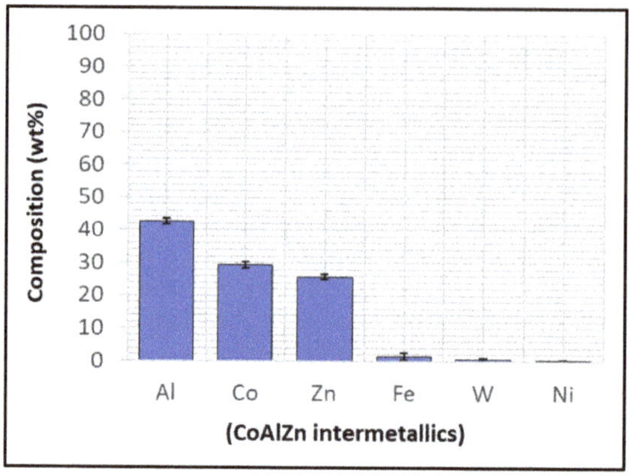

Figure 10. EDS phase elemental analysis of the intermetallic compounds on the WC-Co coating shown in Figure 9.

Co is known to have a strong affinity for Al present in the melt and, for this reason, Co-Al particles can develop in liquid Zn baths containing Al [28]. In the present study, Co is contained in the WC-Co coating of the bar specimen, as well as in the Wallex6^{TM} pad. Co-rich surfaces are known for being ideal sites for the attachment of intermetallic compounds existing in the liquid metal bath [28,29]. Experiments were performed by Zhang [29] on Stellite 6 bearings in GI, and it was found that Co-based aluminides were present on the bushing surface as a result of the reaction between the wear particles generated during testing and the Zn alloy bath. Therefore, the presence of aluminides illustrated in Figure 9 could be related to the adhesion of these particles to the coating surface. However, the formation of Al-Co-Zn intermetallic compounds in WC-Co coatings is often the result of the corrosion of the Co-rich matrix phase in the coatings by the Zn bath. As discussed by Seong et al. [26] and Tani et al. [27], Co reacted with the Al present in the Zn bath due to Co dissolution into the melt, leading to the formation of Al-rich compounds containing Co and Zn.

The Wallex6^{TM} pad was found to react with the molten metal bath, after exposure to Zn-Al. Cross-sections of the specimens were analysed in the unworn region (Figure 11) and they were compared to the as-received specimen (Figure 4). EDS point spectrum analysis was performed to reveal the composition of the phases observed in this area (Figure 12). The results clearly showed the presence of dross phases (2) within the top Zn phase (1) and on top of the surface in contact with molten Zn-Al. These Zn-based intermetallic compounds were found to mainly contain Al (45.9 ± 1.0 wt.%), Co (17.1 ± 1.2 wt.%), Fe (14.5 ± 1.0 wt.%) and small amounts of Cr and W. Co has a strong affinity for Al and there is evidence that the Co solid solution phase of Co-Cr-W alloys reacts with the Al contained in liquid Zn baths [30]. The products of this reaction are CoAl particles forming on the sample surface. The information obtained in the present study showed that Fe was contained in the CoAl intermetallic compounds after testing in Zn-Al. This finding aligns with previous studies, which reported that CoAl particles transformed to FeAl complexes, as the exposure to liquid Zn alloy increased because of the presence of dissolved Fe in the Zn alloy [26,28].

A reaction layer developed beneath the surface of the sample that was found to mainly contain Al-Zn-Co-Fe-Cr-W. Similar findings were obtained by Zhang [29] who reported the formation of a layer of cobalt-based aluminides of Al-Co-Zn-Cr-Fe-W on the contact surface of a Stellite 6 bearing after exposure to GI. In this study, the reaction layer was interrupted by the CoCrWMo eutectic phases present in the alloy, meaning that the diffusion layer did not form at locations where the eutectic phases were present. In Figure 11, (3) and (4), respectively, denote the region within the subsurface reaction layer and the interface between the reaction layer and the bulk of the alloy. It was observed that Al was present in the reaction layer, whereas the Co content diminished (20.6 ± 3.8 wt.%) compared to the unreacted Co-solid solution phase, as shown in Figure 6 (56.1 ± 0.4 wt.%). The Co content was found to increase (37.7 ± 2.4 wt.%) in the vicinity of the interface with the Co-solid solution phase, indicating that Co was depleted from the reaction layer below the surface.

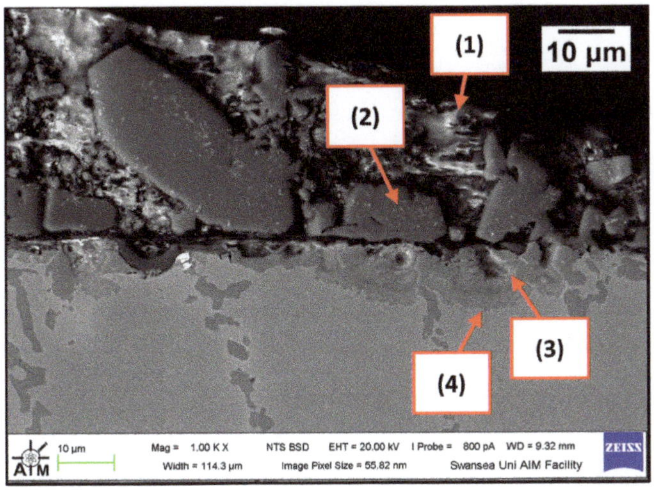

Figure 11. Cross-section of Wallex6^{TM} pad after testing in Zn-Al: (1) Zn phase, (2) intermetallic particles, (3,4) diffusion layer.

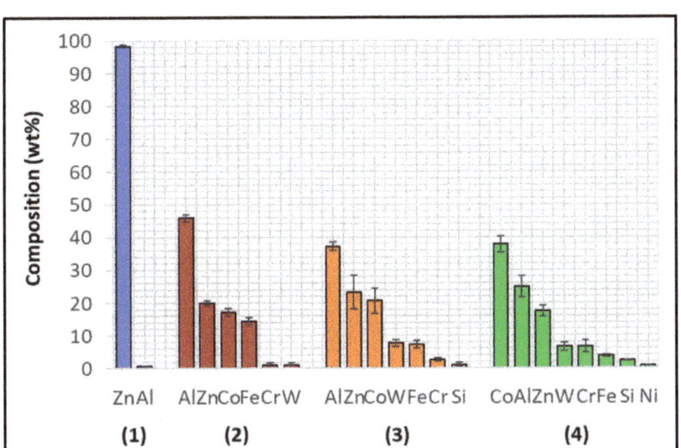

Figure 12. EDS point spectrum analysis of the phases present in the Wallex6^{TM} pad after testing in Zn-Al. The numbers refer to the phases shown in Figure 11, namely the (1) Zn phase, (2) intermetallic particles and (3,4) diffusion layer.

The experiment was repeated on as-received specimens in Zn-Al-Mg. The cross-section of the WC-Co/Wallex6^{TM} bar specimen was imaged (Figure 13) and it was observed that particles of intermetallic compounds were present (Figure 14), showing similar behaviour to Zn-Al. However, the results of EDS analysis (Figure 15) revealed that these particles contained more Fe (18.5 ± 1.2 wt.%) than Co (12.9 ± 1.6 wt.%), whereas the intermetallic particles grown in Zn-Al were richer in Co.

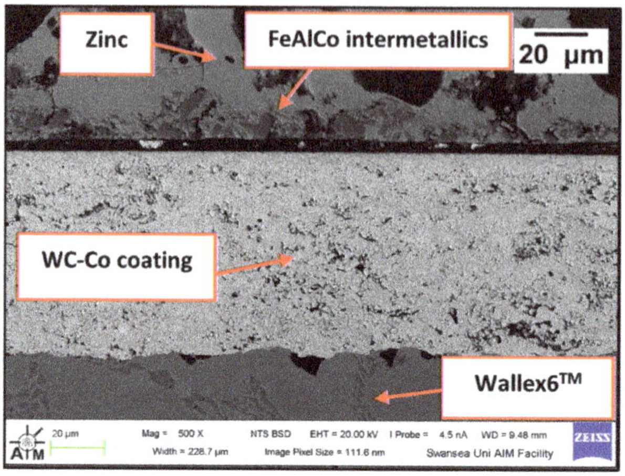

Figure 13. Cross-section of the WC-Co/Wallex6^{TM} bar specimen after exposure to Zn-Al-Mg.

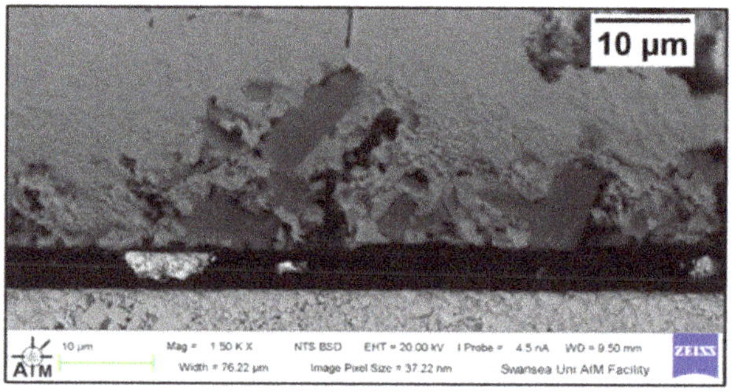

Figure 14. Higher magnification image of the intermetallic phases observed on the WC-Co coating after testing in Zn-Al-Mg.

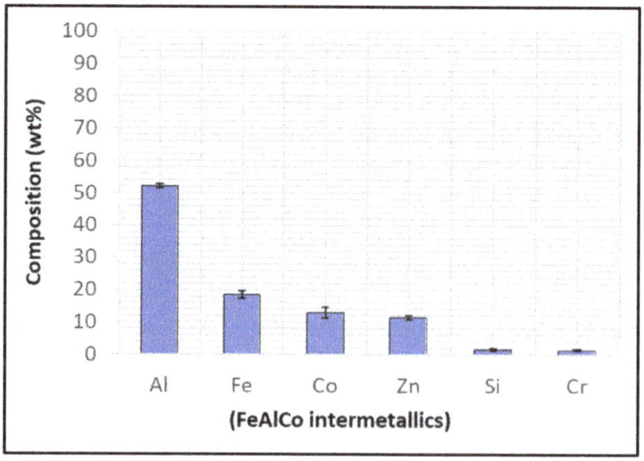

Figure 15. EDS phase elemental analysis of the intermetallic compounds observed on the WC-Co coating after exposure to Zn-Al-Mg.

The present literature highlights that the solubility of Fe in liquid Zn alloy depends on the Al content. In this experiment, the solubility limit corresponds to the maximum quantity of Fe that can dissolve in the Zn bath with the addition of Al or Al and Mg. The solubility limit of Fe in Zn-Al was determined from the experiments conducted by Tang [31] and it is a function of temperature [32]. Figure 16 shows a phase diagram constructed for the Zn-rich corner of the Zn-Al-Fe system, based on the more recent experiments conducted by McDermid, et al. [33].

The diagram shows the Fe solubility limits for Al concentrations up to approximately 0.3 wt.% at 460 °C. The area under the Fe/Al solubility curve indicates the region where Fe and Al remain in solution. It can be observed that this area decreases as the Al content in the bath is increased. Intermetallic dross compounds between Fe and Al do not form below the solubility limit. However, FeAl intermetallic phases form when the Fe solubility is exceeded [9,34]. Al tends to react with Co dissolved in the melt at low Al contents [18]. Therefore, the strong affinity of Co for Al could explain the higher Co content in the intermetallic phases observed for the Zn-Al bath with 0.3 wt.% Al.

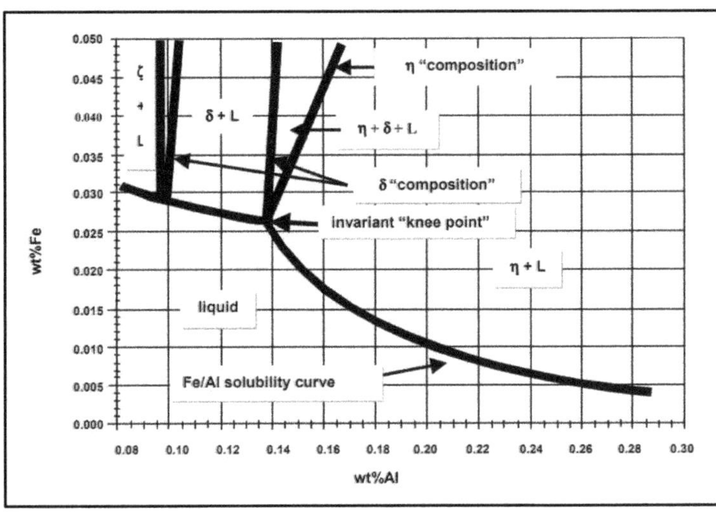

Figure 16. Phase diagram showing the Zn-rich corner of the Zn-Al-Fe system at 460 °C. The diagram illustrates the Fe solubility limit for Al concentrations up to ~0.3 wt.% Al [35].

The Zn-Al-Mg bath contains 1.5 wt.% Al and 1.5 wt.% Mg and it is reported that the presence of Mg does not significantly alter the Fe/Al solubility in the phase diagram. The area under the solubility line remains small as the Al content is increased from the Zn-0.3wt.%Al composition to the levels encountered in the Zn-Al-Mg bath and the Fe solubility is even smaller at higher Al content [34]. This information is supported by the results reported in the present study, which showed that the Fe content in the intermetallic phases detected after immersion in Zn-Al-Mg was higher by 17.1 wt.%.

The Wallex6^{TM} pad was corroded in a similar fashion to the specimen immersed in Zn-Al (Figure 17). CoAlZn intermetallic compounds grew on the surface of the material (1) and a diffusion layer formed beneath the surface of the sample (2). The results of the EDS analysis (Figure 18) showed that this reaction layer contained Al (35.1 $\pm$ 1.5 wt.%) and Co (30.9 $\pm$ 5.3 wt.%), whereas, outside this layer, the composition changed to Co (39.9 $\pm$ 2.4 wt.%) and Al (26.6 $\pm$ 2.9 wt.%), indicating Co depletion from the reaction layer, as observed after exposure to the Zn-Al bath, confirming that the introduction of Mg into the bath did not alter the corrosion behaviour of the materials present in the current configuration.

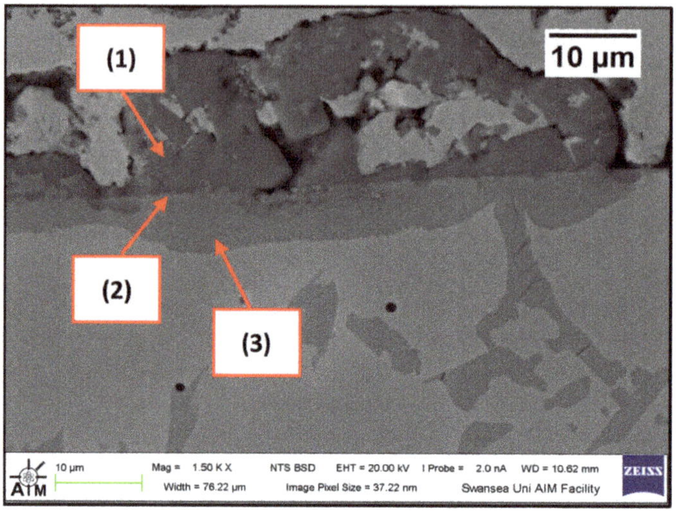

Figure 17. Cross-section of the Wallex6^{TM} pad specimen after exposure to Zn-Al-Mg: (1) intermetallic particles, (2,3) diffusion layer.

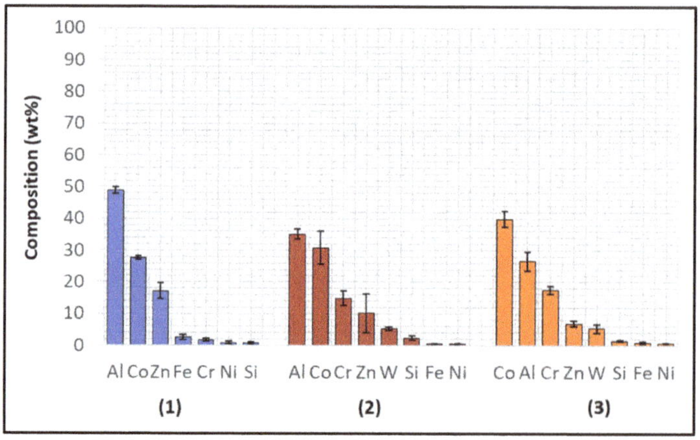

Figure 18. EDS point spectrum analysis of the phases present in Wallex6^{TM} after exposure to Zn-Al-Mg. The numbers refer to the phases shown in Figure 17, namely the (1) intermetallic particles and (2,3) diffusion layer.

3.2.2. Wallex6^{TM} with HVOF WC-Co and Wallex4^{TM}

The first of the two potential upgrades to the benchmark configuration involved replacing Wallex6^{TM} with Wallex4^{TM}. The bar specimen used in the experiment was again coated with WC-Co applied via HVOF, as described in Section 2, Materials and Methods. Figure 19 was captured on the WC-Co/Wallex6^{TM} bar after sliding against the Wallex4^{TM} pad in Zn-Al. Similar behaviour to the bar specimens analysed previously can be observed. The Wallex6^{TM} base metal remained unreactive in Zn-Al, as the WC-Co coating prevented exposure to the molten metal bath. No cracks or other signs of damage to the coating were observed after the experiment. In a similar fashion to the coatings discussed in the previous section, the image was compared to the as-received specimen (Figure 2) and it was observed that a layer of intermetallic particles formed on the surface. The EDS phase elemental analysis (Figure 20) revealed the composition of the intermetallic particles, which contained Al (48.6 ± 0.2 wt.%), Fe (18.6 ± 0.4 wt.%), Zn (17.6 ± 0.3 wt.%), and Co (13.5 ± 0.4 wt.%).

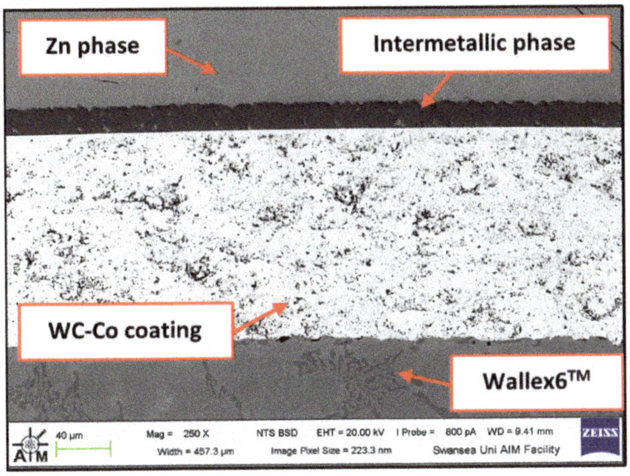

Figure 19. Cross-section of the WC-Co/Wallex6^{TM} bar after dynamic testing against Wallex4^{TM} in Zn-Al.

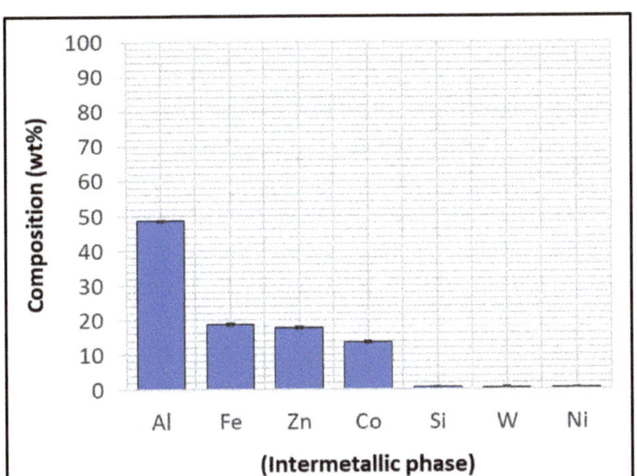

Figure 20. Composition of the intermetallic phase present in the WC-Co coating after dynamic testing with Wallex4^{TM} in Zn-Al.

Several corrosion products were identified on the Wallex4^{TM} pad specimen (Figure 21), and the composition of the new phases was investigated (Figure 22). Intermetallic dross phases were found on the surface of the alloy (1), which contained Al (44.2 ± 2.6 wt.%), Co (23.4 ± 4.0 wt.%), Zn (22.3 ± 3.1 wt.%), and Fe (3.4 ± 1.0 wt.%). The results of EDS showed that the composition of the dross was similar to that of the intermetallic particles found on the bar specimen (Figure 19), although it contained 9.9 wt.% more Co and 15.2 wt.% less Fe. A subsurface reaction layer was still present (2), where both Al and Zn were detected, and the Zn content was lower compared to the dross phase. The Co and Cr content measured in this layer was found to be lower compared to the as-received specimen (Figure 7). The results of EDS analysis showed that the reaction layer contained Co (36.3 ± 7.0 wt.%) and Cr (2.9 ± 0.7 wt.%), whereas the solid solution phase of the as-received specimen contained Co (56.1 ± 0.4 wt.%) and Cr (27.8 ± 0.2 wt.%). Therefore, the analysis suggested that the depletion of these elements from the reaction layer occurred. Moreover, it was observed that the CoCrW eutectic phase present in Wallex4™ (Figure 5b) reacted with the molten metal bath as highlighted in the EDS maps (Figure 23), which show evidence of Al diffusion from the bath.

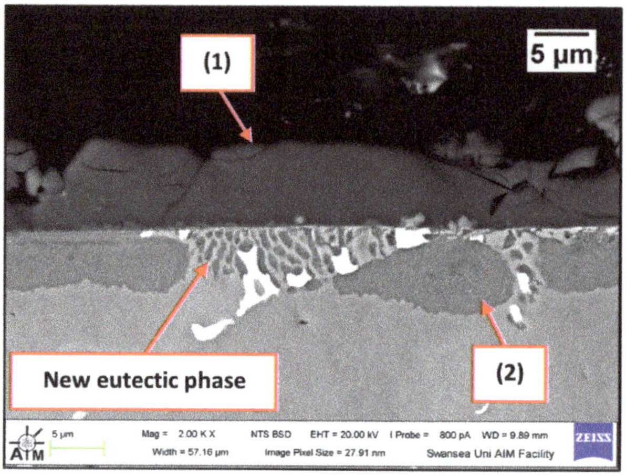

Figure 21. Cross-section of the Wallex4™ pad specimen after exposure to Zn-Al: (1) intermetallic particles, (2) diffusion layer.

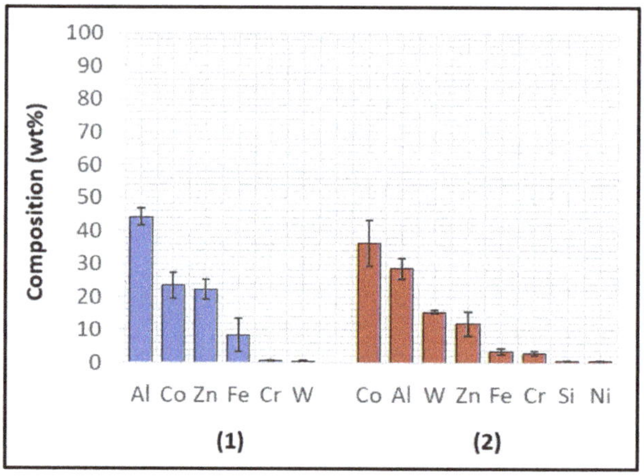

Figure 22. EDS analysis of the corrosion products in Wallex4™ after exposure to Zn-Al. The numbers refer to the phases shown in Figure 21: (1) intermetallic compounds and (2) diffusion layer.

A similar corrosion behaviour was observed in the samples immersed in Zn-Al-Mg. Figure 24 illustrates that reaction products developed onto the WC-Co coating (1), which were found to mainly contain Al (45.3 ± 3.7 wt.%), Zn (19.4 ± 5.8 wt.%), Co (11.7 ± 1.7 wt.%) and Fe (10.3 ± 1.2 wt.%), as shown in Figure 25. The reaction products of this specimen formed a distinct layer, and their presence confirms that the Co matrix of the thermal sprayed coating reacts with the elements present in the bath. No reaction with Mg was detected, as observed in the benchmark configuration.

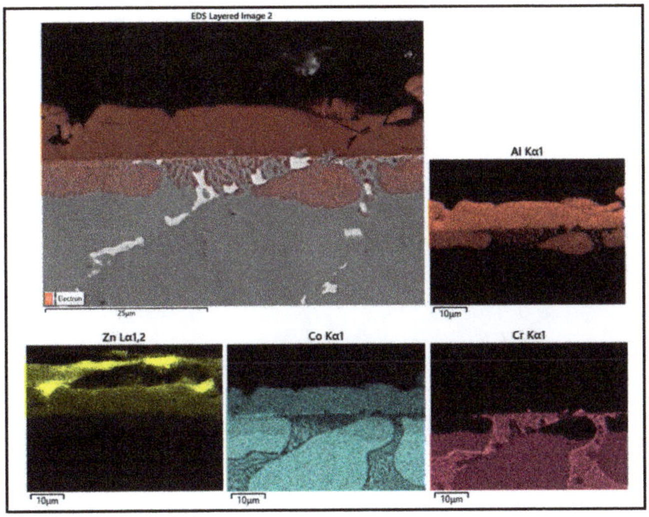

Figure 23. EDS mapping of the elements present in Wallex4^{TM} after testing in Zn-Al.

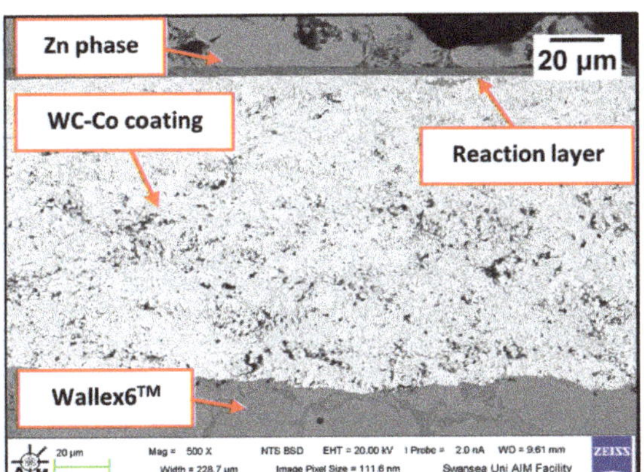

Figure 24. Cross-section of WC-Co/Wallex6^{TM} after dynamic testing with Wallex4^{TM} in Zn-Al-Mg.

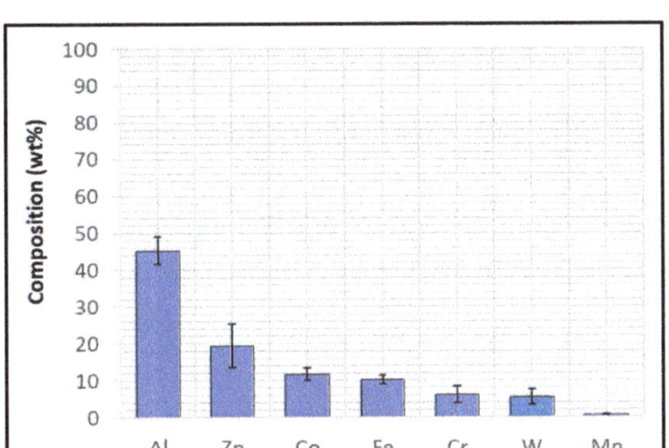

Figure 25. EDS analysis on the reaction layer shown in Figure 24.

Figures 26 and 27 show that AlCoZn dross was again present on the Wallex4^{TM} pad specimen (1); in addition to this, CoCrW particles were present within this layer (2), which are likely wear debris from the alloy due to their composition. Al diffused into the Co solid solution phase forming a subsurface reaction layer (3), as well as in the CoCrW eutectic phases, in a similar fashion to the specimen tested in Zn-Al. The reaction layer was found to contain Al (8.22 ± 4.5 wt.%) and only small amounts of Mg (~0.1 wt.%).

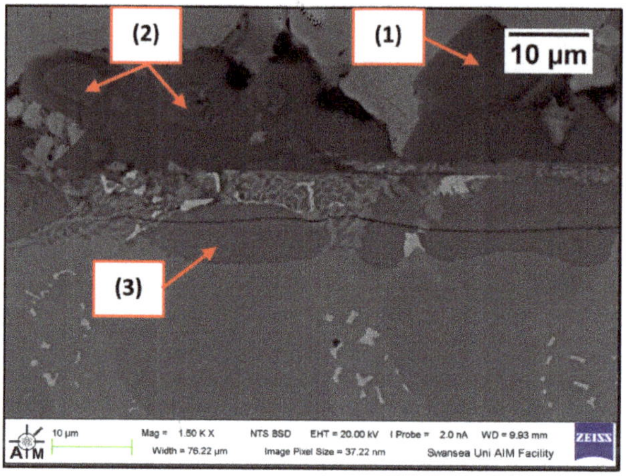

Figure 26. SEM image of the Wallex4^{TM} pad after dynamic testing in Zn-Al-Mg: (1) intermetallic particles, (2) possible wear debris, (3) diffusion layer.

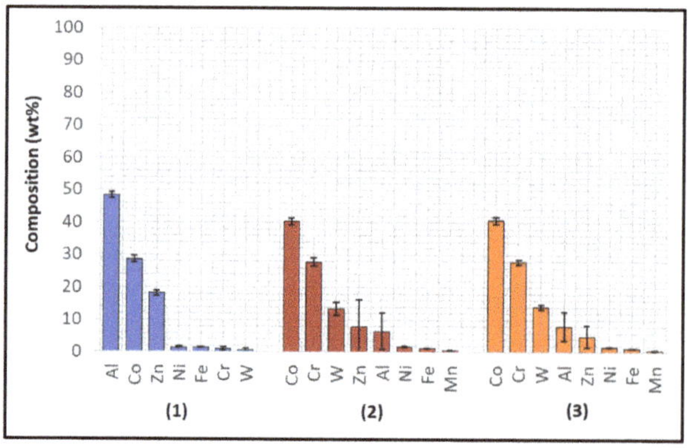

Figure 27. EDS analysis conducted on Wallex4^{TM} after exposure to Zn-Al-Mg. The numbers refer to the phases shown in Figure 26: (1) intermetallic particles, (2) possible wear debris, (3) diffusion layer.

Similarly, EDS elemental mapping showed that mostly Al diffused into the eutectic phases from the molten metal bath (Figure 28). Overall, changing the bath from Zn-Al to Zn-Al-Mg did not alter the interactions between the materials and the liquid metal significantly. Wallex4^{TM} was corroded in both baths and WC-Co reacted with Al. Moreover, the corrosion behaviour of Wallex4^{TM} showed similarities with Wallex6^{TM}, due to the similar chemical composition of the two alloys. The Co-rich solid solution phase of both alloys was attacked by molten Zn and Al forming diffusion layers and, additionally, they developed a layer of Co-Al-Fe-Zn intermetallic phases on the surface.

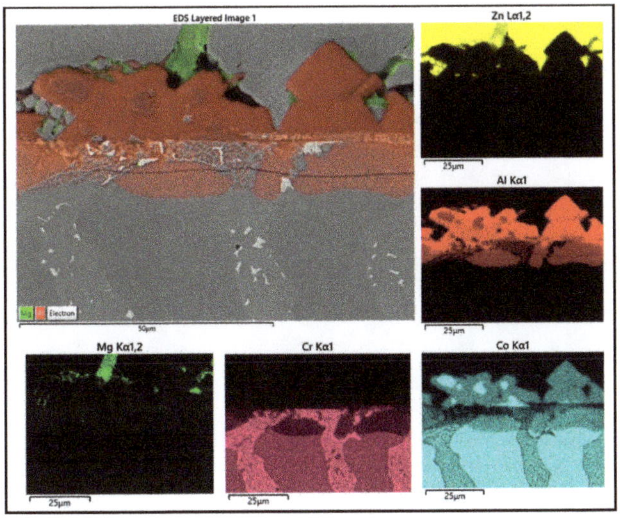

Figure 28. EDS mapping of the elements present in Wallex4™ after testing in Zn-Al-Mg.

3.2.3. SS 316L with HVOF Al_2O_3 and Wallex6™

Visual inspection of the bars tested in Zn-Al and Zn-Al-Mg (Figure 29) after extraction from each molten metal bath evidenced clear signs of damage on the ceramic coatings, leaving areas of the SS 316L specimen unprotected. Examination of the bar with SEM after 48 h of testing in Zn-Al (Figure 30) confirmed that structural damage occurred in the Al_2O_3 coating, as both vertical and horizontal cracks are present. These cracks were not observed in the as-received specimen (Figure 3). The large horizontal crack in the SEM image shows that the coating was forced to separate from the steel bar, in a similar fashion to the mechanism observed during static testing. It was shown in a previous study that cracks and spallation of the coating can be linked to stresses developing due to the large thermal expansion mismatch between Al_2O_3 and SS 316L [17]. The CTEs recorded at ~465 °C for Al_2O_3 and SS 316L were 8.2×10^{-6} and 21.2×10^{-6} K^{-1} respectively and it is believed that this difference resulted in the formation of cracks during dynamic testing. However, despite suffering from significant damage, the ceramic coatings remained completely inert to Zn-Al. No evidence of Zn penetration through cracks and pores was observed during the examination of the cross-sections with SEM and, as a result, reaction products did not accumulate below the coating. Therefore, these observations exclude that the breakdown of the coating occurred due to the build-up of corrosion products on the base metal, suggesting that the coating spalled due to the large thermal expansion mismatch, as discussed. The surface of the coating was free from dross phases, which were observed to build up on WC-Co. This observation agrees with the findings obtained under static conditions in Zn-Al [17].

The performance of Wallex6™ was analogous to that observed in the benchmark configuration. Figure 31 shows that intermetallic particles of Al-Co-Zn-Fe composition are present on the surface of the alloy (1). Furthermore, a subsurface reaction layer developed, and the results of EDS point spectrum analysis (Figure 32) confirmed that Al was contained within this layer (37.2 ± 1.0 wt.%) and at the interface with the Co-solid solution phase of Wallex6™ (23.4 ± 1.3 wt.%), which are, respectively, indicated as location (2) and (3) in Figure 31. This result confirmed that Al diffused from the melt into the alloy. In addition to this, the amount of Co at location (2) is lower (19.7 ± 0.5 wt.%) relative to the as-received sample (47.2 wt.%). On the other hand, at location (3), a Co content closer to that present in the bulk of the material was detected (39.8 ± 3.1 wt.%), suggesting that Co depletion from the diffusion layer occurred.

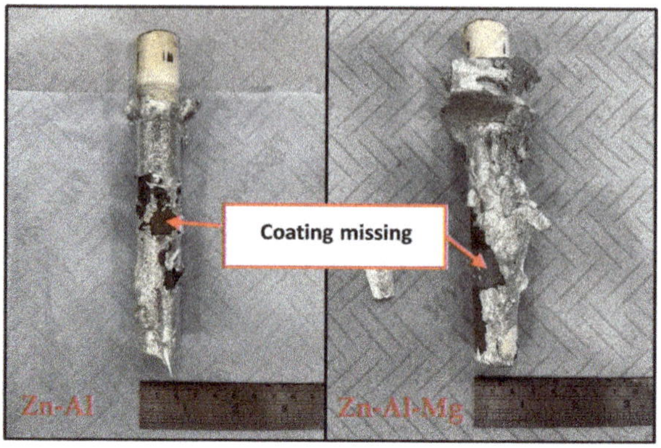

Figure 29. Visual inspection of the bar specimens tested in Zn-Al (**left**) and Zn-Al-Mg (**right**).

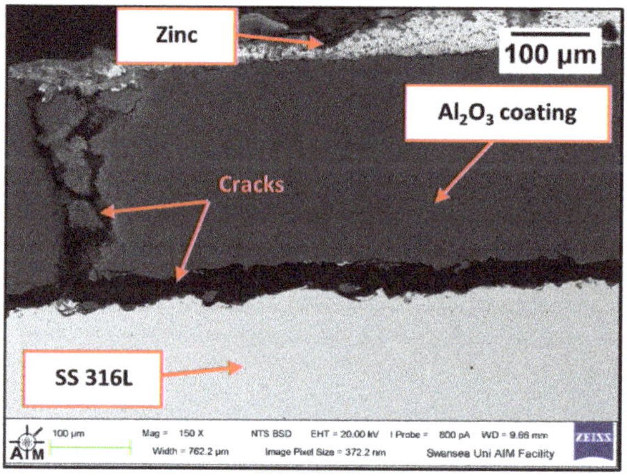

Figure 30. Cross-section of the Al_2O_3/SS 316L bar after dynamic testing in Zn-Al.

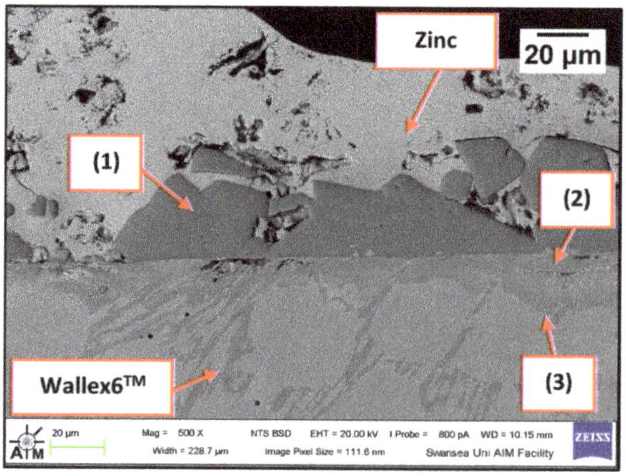

Figure 31. SEM image of the Wallex6^{TM} pad after dynamic testing with SS 316L/Al_2O_3 in Zn-Al: (1) intermetallic particles, (2,3) diffusion layer.

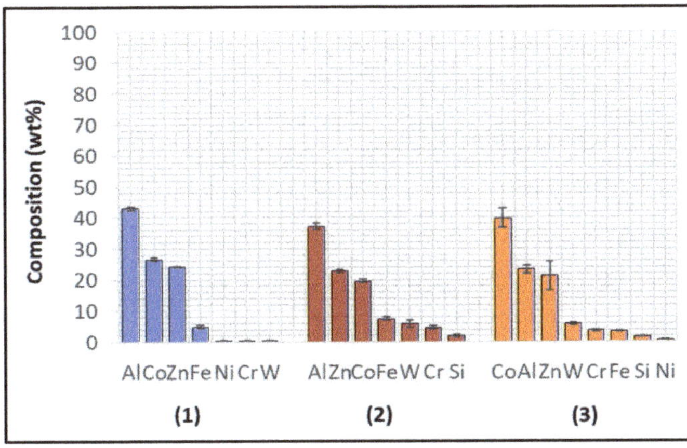

Figure 32. Composition of the phases present in Wallex6^{TM} after dynamic testing with SS 316L/Al_2O_3 in Zn-Al. The numbers refer to the phases shown in Figure 31, namely the (1) intermetallic particles and (2,3) diffusion layer.

The specimens immersed in Zn-Al-Mg showed similar corrosion behaviour. The findings obtained in the previous study, in which the same Al_2O_3 coatings were tested under static conditions for 5 weeks, highlighted a possible reduction of Al_2O_3 by the Mg present in the molten metal bath [17]. However, under dynamic conditions, the SS 316L bar with Al_2O_3 coating remained completely unreactive after exposure to Zn-Al-Mg, as no dross build-up layer was present on the ceramic coating (Figure 33). This behaviour is believed to be linked to the shorter immersion times used in the present work.

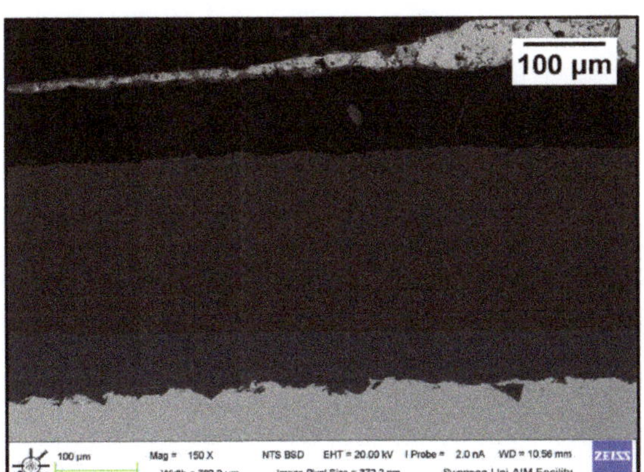

Figure 33. Cross-section of the Al_2O_3/SS 316L bar after dynamic testing in Zn-Al-Mg.

The Wallex6^{TM} pad specimen reacted with the liquid metal bath in a similar fashion to the samples characterised previously. Within the unworn region, dross intermetallic phases are deposited onto the surface. Figure 34 shows two types of reaction products: (1) a CoAl phase containing Zn (16.3 ± 1.1 wt.%) and (2) a different CoAl phase with lower content of Zn (1.6 ± 0.3 wt.%). Co depletion from the diffusion layer occurred, as shown by the results of EDS analysis (Figure 35), which was carried out at the interface with molten metal (3) and the interface between the subsurface reaction layer and the bulk of the material (4).

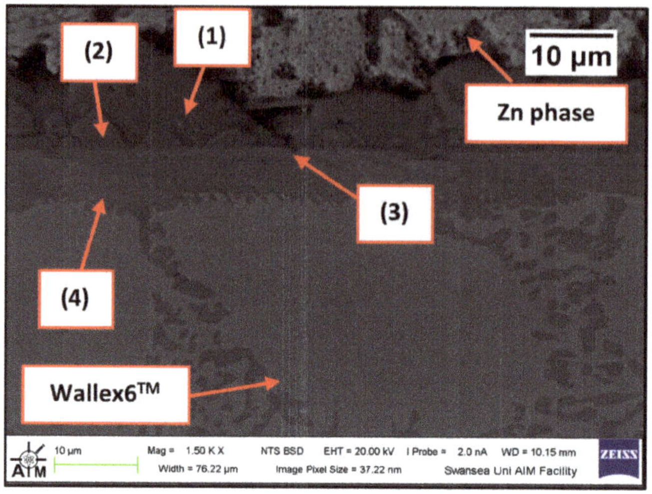

Figure 34. SEM image of the Wallex6™ pad after dynamic testing with SS 316L/Al_2O_3 in Zn-Al-Mg: (1,2) intermetallic particles, (3,4) diffusion layer.

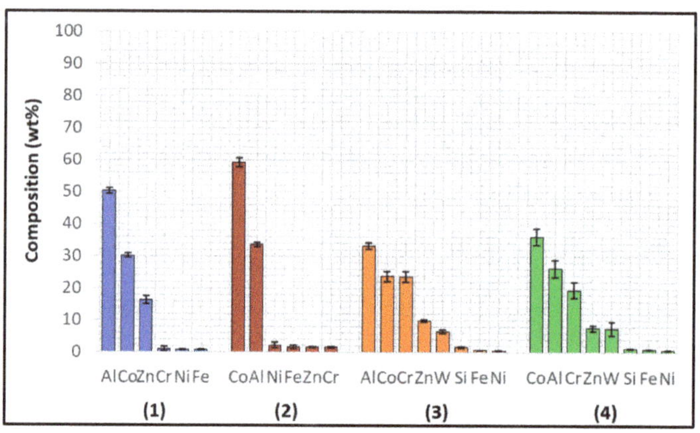

Figure 35. Composition of the phases in Wallex6™ after dynamic testing with SS 316L/Al_2O_3 in Zn-Al-Mg. The numbers refer to the phases shown in Figure 34, which were detected within the (1,2) intermetallic particles and (3,4) diffusion layer.

3.2.4. Effect of Changing Bath Composition

Overall, the results of the dynamic corrosion tests showed that the ceramic coating was the only material that remained inert during the 48 h of dynamic testing. It showed better performance compared to the WC-Co coatings, whose Co matrix reacted with the molten metal bath forming intermetallic compounds at the interface with the molten metal. Although previous studies highlighted that baths with Al content above 0.3 wt.% reduced the corrosion rate of WC-Co [26], the results of the present study showed that the coatings corroded both in Zn-Al and Zn-Al-Mg to a similar extent. The two Co-based alloys, namely Wallex6™ and Wallex4™, were severely corroded after exposure to molten metal. The reactivity of Wallex4™ was found to not significantly differ from that of Wallex6™, due to their similar chemical composition. Therefore, it did not show superior performance at the chosen testing conditions. The introduction of Mg did not affect the corrosion behaviour of the coupling pairs with the surrounding environment.

However, in order to investigate the effect of the higher Al content in Zn-Al-Mg, the amount of Al, Co, and Zn contained within the reaction layers of the pad specimens was tabulated for both Zn-Al (Table 3) and Zn-Al-Mg (Table 4). Figures 36 and 37 show the concentration of these elements and the thickness of the diffusion layers, respectively. It was observed that the content of Al within the reaction layer decreased after changing bath composition from Zn-Al to Zn-Al-Mg. On the other hand, the reaction layers contained more Co, suggesting that less depletion of Co from these layers occurred in the Zn-Al-Mg baths. Furthermore, it was noticed that the Wallex™ pads contained less Zn (~60 %) in the reaction layers after exposure to Zn-Al-Mg compared to the Zn-Al bath, suggesting that the diffusion of Zn into the alloys diminished at higher Al concentration. An analogous behaviour was observed on SS 316L specimens tested in an earlier study, due to the inhibition of the reaction between the SS 316L and the molten Zn in the bath [36]. A decrease in the depth of the diffusion layer was also observed in Wallex6™, as shown in Figure 37, although no significant changes were detected in Wallex4™.

Table 3. The main composition of the reaction layers of the pads after testing in Zn-Al.

Coupling Pair	Pad Material	Al [wt.%]	Co [wt.%]	Zn [wt.%]
Wallex6™/WC-Co vs. Wallex6™	Wallex6™	37.3 ± 1.2	20.6 ± 3.8	23.3 ± 5.1
Wallex6™/WC-Co vs. Wallex4™	Wallex4™	28.6 ± 3.2	36.3 ± 7.0	12.0 ± 3.6
SS 316L/Al_2O_3 vs. Wallex6™	Wallex6™	37.2 ± 1.0	19.7 ± 0.5	22.8 ± 0.5

Table 4. The main composition of the reaction layers of the pads after testing in Zn-Al-Mg.

Coupling Pair	Pad Material	Al [wt.%]	Co [wt.%]	Zn [wt.%]
Wallex6™/WC-Co vs. Wallex6™	Wallex6™	35.2 ± 1.5	30.9 ± 5.3	10.0 ± 6.2
Wallex6™/WC-Co vs. Wallex4™	Wallex4™	8.2 ± 4.5	40.8 ± 1.0	5.0 ± 3.4
SS 316L/Al_2O_3 vs. Wallex6™	Wallex6™	33.2 ± 1.0	23.8 ± 1.6	9.8 ± 0.3

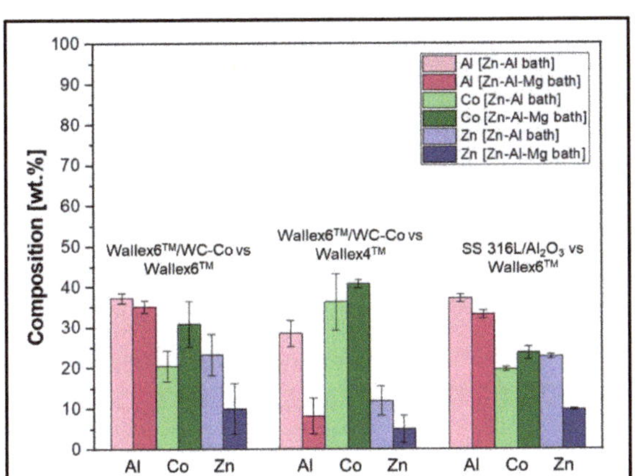

Figure 36. Composition of the diffusion layers developed in the pads for each material pair.

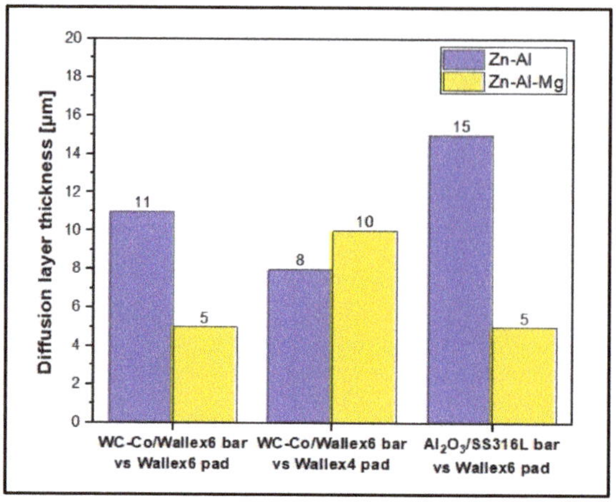

Figure 37. The average thickness of the diffusion layers developed in the pads for each material pair.

3.3. Wear Scar Analysis

The bar displacement and the wear coefficient of each pad specimen after 48 h were calculated according to the procedures previously outlined in Section 2. Materials and Methods. The results are summarised in Table 5 for Zn-Al and Table 6 for Zn-Al-Mg. These values are plotted in Figure 38. Moreover, the percentage change of these parameters after changing the bath from Zn-Al to Zn-Al-Mg was calculated (Table 7).

Table 5. Wear coefficient and displacement after testing in Zn-Al.

Coupling Pair	Pad Material	Wear Coefficient k × 10^{-6} [$mm^3N^{-1}m^{-1}$]	Displacement [mm]
Wallex6™/WC-Co vs. Wallex6™	Wallex6™	26.3	0.80
Wallex6™/WC-Co vs. Wallex4™	Wallex4™	30.9	1.01
SS 316L/Al_2O_3 vs. Wallex6™	Wallex6™	9.7	0.46

Table 6. Wear coefficient and displacement after testing in Zn-Al-Mg.

Coupling Pair	Pad Material	Wear Coefficient k × 10^{-6} [$mm^3N^{-1}m^{-1}$]	Displacement [mm]
Wallex6™/WC-Co vs. Wallex6™	Wallex6™	4.5	0.21
Wallex6™/WC-Co vs. Wallex4™	Wallex4™	1.3	0.10
SS 316L/Al_2O_3 vs. Wallex6™	Wallex6™	3.6	0.21

Table 7. Effect of changing bath composition on the wear coefficient and displacement.

Coupling Pair	Pad Material	Percentage Decrease Wear Coefficient	Percentage Decrease Displacement
Wallex6™/WC-Co vs. Wallex6™	Wallex6™	83%	74%
Wallex6™/WC-Co vs. Wallex4™	Wallex4™	96%	90%
SS 316L/Al_2O_3 vs. Wallex6™	Wallex6™	63%	54%

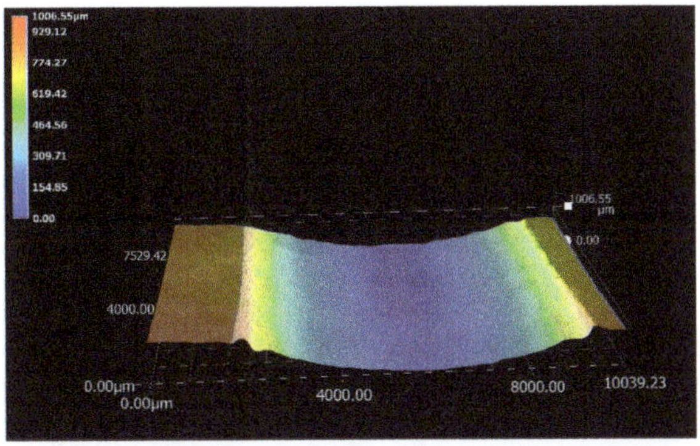

(**a**)

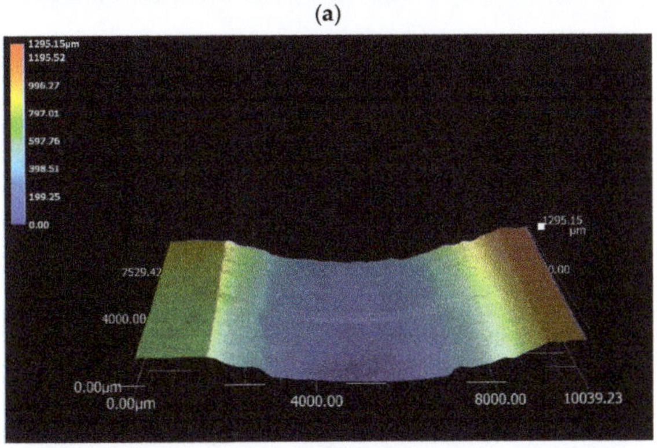

(**b**)

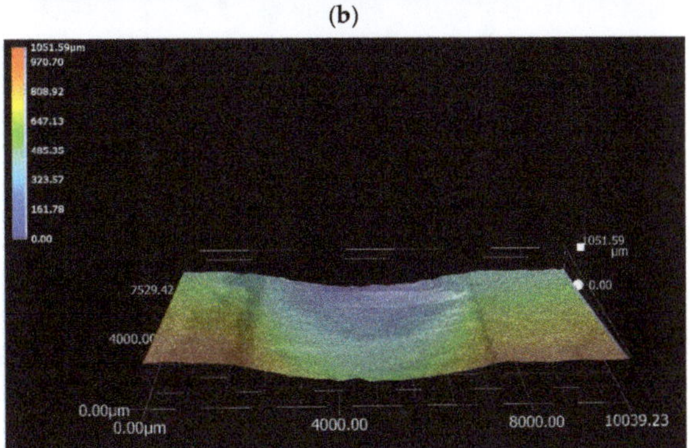

(**c**)

Figure 38. Three-dimensional (3D) imaging of the wear scars produced on the specimens tested in Zn-Al: (**a**) Wallex6^{TM} after sliding with Wallex6^{TM}/WC-Co; (**b**) Wallex4^{TM} after sliding with Wallex6^{TM}/WC-Co; (**c**) Wallex6^{TM} after sliding with SS 316L/Al_2O_3. Deep scars formed after contact with the WC-Co-coated bars, whereas a more superficial scar formed after contact with the Al_2O_3-coated bar specimen.

The results of the wear tests conducted in Zn-Al highlighted that Al_2O_3/SS 316L had the lowest value of displacement (0.46 mm) and inflicted the lowest wear damage on its counterpart (k = 9.7 × 10^{-6} $mm^3N^{-1}m^{-1}$). The hardness of the materials in the current bearing configuration was previously measured by Faulkner [25]. It was found that the hardness of WC was between 2100–2500 HV, which is significantly higher than the hardness of Wallex6^{TM} (400 HV) and Wallex4^{TM} (570 HV). As a result, it is believed that the WC-Co coatings inflicted severe wear damage to their softer counterparts on the contact surface, due to the greater hardness of the WC particles. This observation aligns with the findings of Zhang [37], who reported that WC particles could easily scar the surface of Stellite alloys. The hardness of Al_2O_3 was measured to be approximately 1680 HV [17] and, therefore, it inflicted less wear damage on its counterpart.

However, previous studies demonstrated that the wear process of materials in liquid Zn is complex, due to the chemical reactions with the molten metal bath and the complexity of the intermetallic compounds. For this reason, surface damage or removal of materials during sliding contact can involve multiple mechanisms. It is reported that wear debris can form following the breakup of the materials as well as of the intermetallic compounds. Wear debris was found to react with the elements present in the molten metal, forming particles that could easily plough the contacting surfaces [37,38]. In this work, the CoAl phases observed on the surface of WC-Co were absent on the Al_2O_3 coating, due to the inertness of the ceramic material. The CoAl phase is hard and can groove the bearing surfaces [30]. For this reason, it is theorised that the absence of Co-based aluminides on the Al_2O_3 coating contributed to reducing the wear damage on the Wallex6^{TM} counterpart. Figure 38 illustrates a comparison of the wear scars after testing in Zn-Al with 3D imaging.

The measurements of displacement and wear coefficients obtained for the specimens tested in Zn-Al-Mg revealed that the Wallex4^{TM} pad exhibited the highest wear resistance (0.1 mm displacement, $k = 1.3 \times 10^{-6}$ mm^{3}N^{-1}m^{-1}), although less wear damage was observed in all the pad specimens immersed in Zn-Al-Mg compared to Zn-Al, as shown in Table 7. Similar reaction products formed in the two molten metal baths; however, the composition of the reaction layers formed at high Al concentration was found to vary from that measured on the samples tested in Zn-Al. It was earlier explained that the Wallex6^{TM} and Wallex4^{TM} pads tested in Zn-Al-Mg exhibited less corrosion by molten Zn after 48 h of exposure to the liquid metal. It is reported that during the rotation of submerged pot roll bearings, cracks initiate in the reaction layer and at its interface with the bulk of the material. This process is followed by the detachment of the corrosion products and the bearing materials, resulting in the formation of wear debris, which can inflict further damage to the contact surfaces [37]. Therefore, it is plausible that the specimens submerged in Zn-Al-Mg displayed less wear damage due to their limited corrosion relative to the Zn-Al bath, which reduced the likelihood of breakage of the corrosion products and, consequently, the rate of material removal at the testing conditions chosen for the experiments. The higher wear resistance exhibited in Zn-Al-Mg can be visualised with the 3D images of the contacting surfaces (Figure 39).

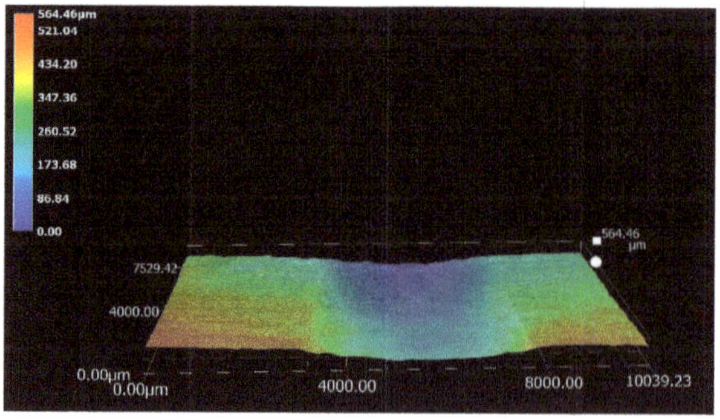

(a)

Figure 39. *Cont.*

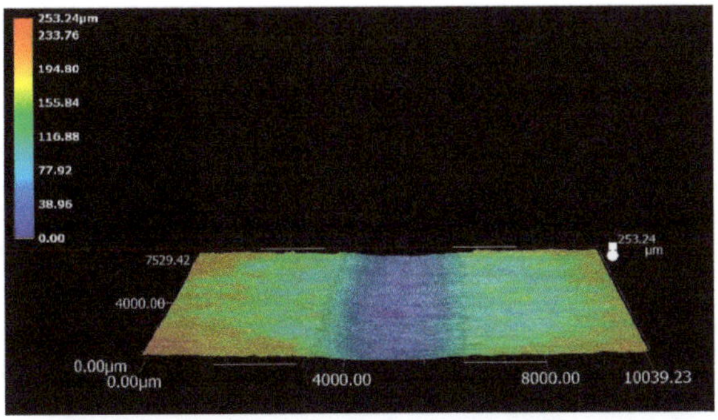

(b)

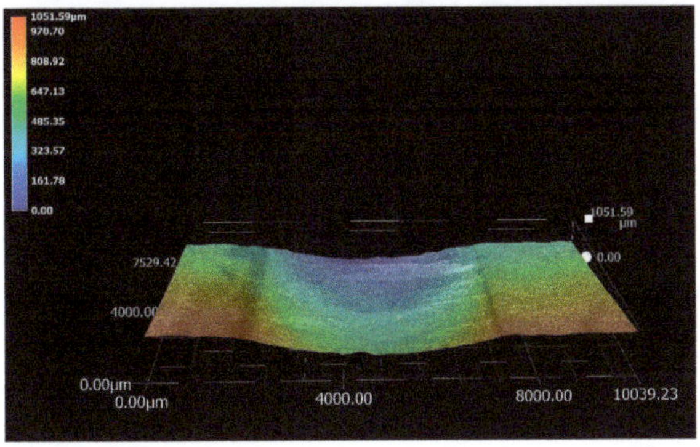

(c)

Figure 39. Three-dimensional (3D) imaging of the wear scars produced on the pad specimens tested in Zn-Al-Mg: (**a**) Wallex6^{TM} after sliding with Wallex6^{TM}/WC-Co; (**b**) Wallex4^{TM} after sliding with Wallex6^{TM}/WC-Co; (**c**) Wallex6^{TM} after sliding with SS 316L/Al_2O_3. The scars are more superficial compared to those formed in the Zn-Al bath.

4. Conclusions

Dynamic wear testing in molten Zn-Al and Zn-Al-Mg was conducted to assess the corrosion behaviour and wear resistance of three material pairs using a bespoke wear testing rig. The following were conducted:

- WC-Co coatings protected the underlying Wallex6^{TM} base metal from the attack of liquid Zn alloy, although they developed intermetallic compounds which mainly contained Al-Co-Zn-Fe.
- Al_2O_3 coatings remained completely inert in liquid Zn alloy. Cracking and spallation occurred due to the large thermal expansion mismatch with the underlying SS 316L base metal.
- Wallex6^{TM} and Wallex4^{TM} were severely corroded in molten metal, showing evidence of Zn and Al diffusion within 5–15 μm from the surface.
- Changing the bath chemistry from Zn-Al to Zn-Al-Mg reduced the diffusion of Zn into Wallex6^{TM} and Wallex4^{TM} by approximately 60%, leading to less wear damage.
- Out of the three material couplings investigated in this study, minimal wear damage in both Zn-Al ($k = 9.7 \times 10^{-6}$ mm^{3}N^{-1}m^{-1}) and Zn-Al-Mg ($k = 3.6 \times 10^{-6}$ mm^{3}N^{-1}m^{-1}) was only obtained by pairing Wallex6^{TM} with Al_2O_3/SS 316L.

The results of dynamic testing confirmed that the introduction of ceramic materials has the potential to improve the performance of galvanising pot roll bearings. However, future research should focus on reducing the thermal expansion mismatch to prevent cracking in the ceramic coatings.

Author Contributions: Conceptualization, G.P.A., D.P., J.S., J.E. and C.M.; methodology, G.P.A., D.P.; validation, G.P.A.; formal analysis, G.P.A.; investigation, G.P.A.; resources, G.P.A., D.P., J.S., J.E. and C.M.; data curation, G.P.A.; writing—original draft preparation, G.P.A.; writing—review and editing, G.P.A., D.P., J.S., J.E. and C.M; visualization, G.P.A.; supervision, D.P., J.S., J.E. and C.M.; project administration, D.P., J.S., J.E. and C.M.; funding acquisition, D.P., J.S., J.E. and C.M. All authors have read and agreed to the published version of the manuscript.

Funding: This project was supported by the Engineering and Physical Sciences Research Council (grant ref: EP/V519601) with Tata Steel.

Institutional Review Board Statement: Not applicable.

Informed Consent Statement: Not applicable.

Data Availability Statement: Data are contained within the article.

Acknowledgments: The authors would like to thank Engineered Performance Coatings (EPC), Cardiff, and The Advanced Imaging Facility (AIM) at Swansea University. Moreover, we thank Clive Challinor, Tata Steel, for advice on the utilisation of bearings in galvanised steel production and David Medway for the technical assistance in building the dynamic wear rig.

Conflicts of Interest: The authors declare no conflicts of interest. The funders had no role in the design of the study; in the collection, analyses, or interpretation of data; in the writing of the manuscript; or in the decision to publish the results.

References

1. Gossuin, T.; Moreas, G. Cleanliness Measurement by Innovative Libs Method. In Proceedings of the 12th International Conference on Zinc and Zinc Alloy Coated Steel Sheet, Virtual, 21–23 June 2021.
2. Beentjes, P.; Bottema, J.; Salgin, B.; Vrenken, J. An Innovative GI with Improved Galling and Surface Properties for Exposed Automotive Applications. In Proceedings of the 12th International Conference on Zinc and Zinc Alloy Coated Steel Sheet, Virtual, 21–23 June 2021.
3. Chakraborty, A.; Ghassemi-Armaki, H. Evaluation of Initiation and Propagation of LME Cracks on the Galvanised 3G-AHSS Using Interrupted Resistance Spot-Welding Method. In Proceedings of the 12th International Conference on Zinc and Zinc Alloy Coated Steel Sheet, Virtual, 21–23 June 2021.
4. Sohn, I.-R.; Kim, T.-C.; Ju, G.-I.; Kim, M.-S.; Kim, J.-S. Anti-Corrosion Performance and Applications of PosMAC® Steel. *Corros. Sci. Technol.* **2021**, *20*, 7–14. [CrossRef]
5. Kim, S.; So, S.; Park, J.; Kim, T.; Han, S.; Park, S.; Kim, H.-y.; Kim, M.; Paik, D. Characteristics of Hot-Dip Znmgal Coatings with Ultra-High Corrosion Resistance. *Corros. Sci. Technol.* **2024**, *23*, 289–295. [CrossRef]
6. Tang, N.Y.; Daniel, L.; Zhang, K. Performance of Submerged Hardware in Continuous Galvanizing. *Corros. Sci. Technol.* **2010**, *9*, 116–121.
7. Zhang, K.; Tang, N.Y.; Goodwin, F. Research and development of pot bearings in continuous galvanizing. In Proceedings of the 6th International Conference on Zinc and Zinc Alloy Coated Steel Sheet, Chicago, IL, USA, 4–7 April 2004.
8. Zhang, K. Effects of test conditions on the tribological behaviour of a journal bearing in molten zinc. *Wear* **2005**, *259*, 1248–1253. [CrossRef]
9. Marder, A.; Goodwin, F. *The Metallurgy of Zinc Coated Steels*; Elsevier Science: Amsterdam, The Netherlands, 2023.
10. Escott, L.J.; Penney, D.J.; Das, A.; Thomas, D. Effects of Heat Treatments on the Morphology and Mechanical Properties of a CoCrW Alloy for Hot Dip Galvanising Applications. In Proceedings of the 12th International Conference on Zinc and Zinc Alloy Coated Steel Sheet, Virtual, 21–23 June 2021.
11. Faulkner, R.; Penney, D.; Bright, M. Offline Simulation of Galvanising Bath Journal Bearings as a Cost Effective Solution to Improve Line Performance and Mitigate Risk. In Proceedings of the 12th International Conference on Zinc & Zinc Alloy Coated Steel Sheet, Virtual, 21–23 June 2021.
12. Nag, A.; Bhadu, M.K.; Bijalwan, P.K.; Pathak, A.S. Investigation of selected HVOF and plasma sprayed coatings for sustained performance in molten zinc. *Corros. Sci.* **2021**, *180*, 109177. [CrossRef]
13. Dong, Y.; Yan, D.; He, J.; Zhang, J.; Li, X. Degradation behaviour of ZrO_2–Ni/Al gradient coatings in molten Zn. *Surf. Coat. Technol.* **2006**, *201*, 2455–2459. [CrossRef]
14. Jarosinski, W.; Quets, J.; Wang, D.; Belov, V.; Klryman, A.S. Thermal Spray Coated Rolls for Molten Metal Baths. US 8,507,105 B2, 13 August 2013.

15. Liu, X.; Zhao, X.; Yang, F. Room-Temperature and High-Temperature Wear Behaviors of As-Sprayed and Annealed Cr_3C_2-25NiCr Coatings Prepared by High Velocity Air-Fuel Spraying. *Coatings* **2020**, *10*, 1090. [CrossRef]
16. Wang, Q.; Zhang, J.W.; Zhang, L.; Liu, H.J.; Zeng, C.L. Stability investigation of electrodeposited zirconium diboride ceramic coatings in molten zinc. *Mater. Corros.* **2019**, *70*, 492–502. [CrossRef]
17. Alparone, G.P.; Penney, D.; Sullivan, J.; Edy, J.; Mills, C. Al_2O_3 Coatings for Protection of Stainless Steel 316L against Corrosion in Zn-Al and Zn-Al-Mg. *Coatings* **2024**, *14*, 606. [CrossRef]
18. Matthews, S.J.; James, B. Review of Thermal Spray Coating Applications in the Steel Industry: Part 2—Zinc Pot Hardware in the Continuous Galvanizing Line. *J. Therm. Spray Technol.* **2010**, *19*, 1277–1286. [CrossRef]
19. Shi, Y.; Tong, J.; Zheng, Q.; Rao, S.; Jin, H.; Rao, S. Analysis About the Failure Mechanism of the Sleeves in Sink Roll System. *J. Fail. Anal. Prev.* **2018**, *18*, 183–188. [CrossRef]
20. Liu, X.; Barbero, E.; Xu, J.; Burris, M.; Chang, K.-M.; Sikka, V. Liquid metal corrosion of 316L, Fe_3Al, and FeCrSi in molten Zn-Al baths. *Metall. Mater. Trans. A* **2005**, *36*, 2049–2058. [CrossRef]
21. Jiang, W.; Wang, W. Sliding Bearing Material Testing Machine under High Temperature Corrosion Conditions. *IOP Conf. Ser. Mater. Sci. Eng.* **2019**, *692*, 012042. [CrossRef]
22. Zhang, K.; Battiston, L. Friction and wear characterization of some cobalt- and iron-based superalloys in zinc alloy baths. *Wear* **2002**, *252*, 332–344. [CrossRef]
23. Yin, L.; Zhao, K.; Ding, Y.; Wang, Y.; He, Z.; Huang, S. Effect of hBN addition on the fabrication, mechanical and tribological properties of Sialon materials. *Ceram. Int.* **2022**, *48*, 7715–7722. [CrossRef]
24. Pulsford, J.; Venturi, F.; Pala, Z.; Kamnis, S.; Hussain, T. Application of HVOF WC-Co-Cr coatings on the internal surface of small cylinders: Effect of internal diameter on the wear resistance. *Wear* **2019**, *432–433*, 202965. [CrossRef]
25. Faulkner, R. Material Selection and Mechanical Design of Submerged Galvanising Bath Journal Bearings. Ph.D. Thesis, Cardiff University, Cardiff, UK, 2020.
26. Seong, B.G.; Hwang, S.Y.; Kim, M.C.; Kim, K.Y. Reaction of WC–Co coating with molten zinc in a zinc pot of a continuous galvanizing line. *Surf. Coat. Technol.* **2001**, *138*, 101–110. [CrossRef]
27. Tani, K.; Tomita, T.; Kobayashi, Y.; Takatani, Y.; Harada, Y. Durability of Sprayed WC/Co Coatings in Al-added Zinc Bath. *ISIJ Int.* **1994**, *34*, 822–828. [CrossRef]
28. Zhang, K.; Tang, N.-Y. Reactions of Co based and Fe based superalloys with a molten Zn–Al alloy. *Mater. Sci. Technol.* **2004**, *20*, 739–746. [CrossRef]
29. Zhang, K. Wear of cobalt-based alloys sliding in molten zinc. *Wear* **2003**, *255*, 545–555. [CrossRef]
30. Zhang, K.; Tang, N.Y. On the wear of a cobalt-based superalloy in zinc baths. *Metall. Mater. Trans. A* **2003**, *34*, 2387–2396. [CrossRef]
31. Tang, N.-Y. Determination of liquid-phase boundaries in Zn-Fe-Mx systems. *J. Phase Equilibria* **2000**, *21*, 70–77. [CrossRef]
32. Ilinca, F.; Ajersch, F.; Baril, C.; Goodwin, F.E. Numerical simulation of the galvanizing process during GA to GI transition. *Int. J. Numer. Methods Fluids* **2007**, *53*, 1629–1646. [CrossRef]
33. McDermid, J.R.; Kaye, M.H.; Thompson, W.T. Fe Solubility in the Zn-Rich Corner of the Zn-Al-Fe System for Use in Continuous Galvanizing and Galvannealing. *Metall. Mater. Trans. B* **2007**, *38*, 215–230. [CrossRef]
34. Kuperus, M. The Delamination Process of the Dross Build-Up Structure on Submerged Hardware in Zn-Al and Zn-Mg-Al Baths: 564 An Empirical Study. Master's Thesis, Delft University of Technology, Delft, The Netherlands, 2018.
35. McDermid, J.; Baril, É.; Goodwin, F. Galvanizing Bath Management During Galvanize to Galvanneal and Galvanneal to Galvanize Product Transitions. In Proceedings of the 6th International Conference on Zinc and Zinc Alloy Coated Steel Sheet, Chicago, IL, USA, 4–7 April 2004.
36. Alparone, G.P.; Penney, D.; Mills, C. Static Immersion of Ceramics in Zn-Al and Zn-Al-Mg for Optimised Galvanising Bath Bearings. In Proceedings of the 13th International Conference on Zinc and Zinc Alloy Coated Steel Sheet, Seoul, Republic of Korea, 15–19 October 2023.
37. Zhang, K. On the Selection of Materials for Improved Performance of Pot Bearings. *AISTech—Iron Steel Technol. Conf. Proc.* **2005**, *2*, 475.
38. Zhang, K.; Battiston, L.; Goodwin, F. Friction and Wear Characteristics of Materials in Molten Zinc. In Proceedings of the 5th International Conference on Zinc and Zinc Alloy Coated Steel Sheet, Brussels, Belgium, 26–28 June 2001.

materials

MDPI

Article

Evaluation of the Wear of Ni 200 Alloy After Long-Term Carbon Capture in Molten Salts Process

Piotr Palimąka [1,*], Stanisław Pietrzyk [1], Maciej Balcerzak [2], Krzysztof Żaba [2], Beata Leszczyńska-Madej [3] and Justyna Jaskowska-Lemańska [4]

1 Department of Physical Chemistry and Metallurgy of Non-Ferrous Metals, Faculty of Non-Ferrous Metals, AGH University of Krakow, al. Adama Mickiewicza 30, 30-059 Cracow, Poland; pietstan@agh.edu.pl
2 Department of Metal Working and Physical Metallurgy of Non-Ferrous Metals, Faculty of Non-Ferrous Metals, AGH University of Krakow, al. Adama Mickiewicza 30, 30-059 Cracow, Poland; balcerzak@agh.edu.pl (M.B.); krzyzaba@agh.edu.pl (K.Ż.)
3 Department of Materials Science and Engineering of Non-Ferrous Metals, Faculty of Non-Ferrous Metals, AGH University of Krakow, al. Adama Mickiewicza 30, 30-059 Cracow, Poland; bleszcz@agh.edu.pl
4 Department of Geomechanics, Civil Engineering and Geotechnics, Faculty of Civil Engineering and Resource Management, AGH University of Krakow, al. Adama Mickiewicza 30, 30-059 Cracow, Poland; lemanska@agh.edu.pl
* Correspondence: palimaka@agh.edu.pl

Abstract: Reducing CO_2 emissions is one of the major challenges facing the modern world. The overall goal is to limit global warming and prevent catastrophic climate change. One of the many methods for reducing carbon dioxide emissions involves capturing, utilizing, and storing it at the source. The Carbon Capture in Molten Salts (CCMS) technique is considered potentially attractive and promising, although it has so far only been tested at the laboratory scale. This study evaluates the wear of the main structural components of a prototype for CO_2 capture in molten salts—a device designed and tested in the laboratories of AGH University of Kraków. The evaluation focused on a gas barbotage lance and a reactor chamber (made from Nickel 200 Alloy), which were in continuous, long-term (800 h) contact with molten salts $CaCl_2$-CaF_2-CaO-$CaCO_3$ at temperatures of 700–940 °C in an atmosphere of N_2-CO_2. The research used light microscopy, SEM, X-ray, computed tomography (CT), and 3D scanning. The results indicate the greatest wear on the part of the lance submerged in the molten salts (3.9 mm/year). The most likely wear mechanism involves grain growth and intergranular corrosion. Nickel reactions with the aggressive salt environment and its components cannot be ruled out. Additionally, the applied research methods enabled the identification of material discontinuities in the reactor chamber (mainly in welded areas), pitting on its surface, and uneven wear in different zones.

Keywords: CO_2 capture; molten salts; CCMS process; CCS/CCU technologies; NDT methods

1. Introduction

Carbon dioxide is one of the greenhouse gases responsible for the increase in Earth's average temperature and the resulting global warming [1–4]. The primary sources of CO_2 emissions into the atmosphere are the energy, transportation, and industrial sectors, which rely on fossil fuel combustion. In the pursuit of climate neutrality, reducing CO_2 emissions across all areas of economic activity, as well as removing it where reduction is challenging, plays a crucial role. By reducing CO_2 emissions to the atmosphere and shifting to zero-emission energy sources, we can prevent the future catastrophic effects of climate warming on our planet [5]. There are four ways to reduce carbon dioxide emissions [6]:

- Limiting the use of fossil fuels (improving energy conversion efficiency, reducing energy demand, and using renewable energy sources);
- Increasing the use of green hydrogen;

- Replacing fossil fuels with gaseous fuels;
- Reducing deforestation, thereby storing more CO_2 in biomass;
- Capturing CO_2 generated from fuel combustion in power plants and other industrial processes and storing it in suitable geological formations (CCS—Carbon Capture and Storage) and/or using it to enhance oil recovery (EOR—Enhanced Oil Recovery).

The Intergovernmental Panel on Climate Change (IPCC) consistently includes CCS technologies as one of the pathways for mitigating CO_2 emissions, in line with the goals set by the Paris Agreement [7]. The main contribution of CCS technologies to achieving climate neutrality lies in their potential to reduce emissions in heavy industries that are considered hard to abate (i.e., where emission reduction is technologically very challenging) and their applicability to existing production facilities.

The primary carbon capture technologies are categorized based on the point at which the capture process occurs relative to combustion. These include pre-combustion capture, post-combustion capture, and oxy-fuel combustion capture [8]. Post-combustion capture is considered the most practical approach to reducing carbon dioxide emissions from the perspective of existing CO_2-emitting installations. Pre-combustion and oxy-fuel combustion capture present challenges such as infrastructure requirements, severe corrosion, and high maintenance and servicing costs [9]. Post-combustion capture methods can be divided into adsorption, absorption, and membrane separation [9]. Currently, the most mature technology for capturing CO_2 from industrial processes is absorption. In this method, CO_2 molecules are dissolved in liquid solvents such as amines (monoethanolamine, diethanolamine, diethylenetriamine), hot potassium and potassium carbonates, chilled ammonia, and ionic liquids. Among these, amine-based systems are the most advanced and widely used [10]. However, according to [11], relying solely on a single technology based on chemical reactions for CO_2 capture is impractical due to the enormous scale of existing emissions. Therefore, every developed and implemented technology contributing to CO_2 emission reductions is essential. Currently, calcium looping is considered a critical technology for industrial decarbonization (though not fully commercialized), particularly in the cement industry [12]. This method is based on the reversible reaction of CaO with CO_2 to form $CaCO_3$, as illustrated in Equation (1) [13–16]:

$$CaO_{(s)} + CO_{2(g)} \rightleftharpoons CaCO_{3(s)} \tag{1}$$

Thermodynamic data [17] for the above reaction indicate that the change in Gibbs free energy equals zero at 886 °C. At this temperature, the pressure of the CO_2 reaches 1 atm. Since the reaction is reversible, its direction can be altered by raising or lowering the temperature, thus enabling the capture or release of carbon dioxide from the solid product. The captured carbon dioxide can then be transported to a storage site, used for Enhanced Oil Recovery, or converted into chemical feedstock. Carbon capture in molten salts (CCMS) is based on the same principles as CaL but with the active substances ($CaO/CaCO_3$) dissolved or partly dissolved in molten salts. This approach allows for faster reaction kinetics and higher CO_2 sorption capacity and helps avoid issues with solid attrition [18]. Moreover, CO_2 absorption reactions are exothermic, which suggests that the capture process could be thermally self-sufficient. There is also the potential for heat recovery during the cooling of molten salts from the desorption process, which are then directed back to absorption. Energy consumption in the process is estimated at 8.8 GJ/tCO_2 [19]. Importantly, the CCMS technology can be applied to existing CO_2-emitting installations without interfering with their structure. The concept of the CCMS process is shown in Figure 1.

The process was experimentally investigated in a single-chamber reactor, involving the passage of an N_2-CO_2 (15 vol. % CO_2) gas mixture through molten salts based on chlorides and fluorides [18,20,21]. Molten salts with a eutectic composition of CaF_2-CaCl_2 (13.8 wt. % CaF_2 in $CaCl_2$) demonstrated the most promising absorption and desorption properties [18]. This composition exhibited a higher carrying capacity value (0.599 g of CO_2/g of CaO at 695 ÷ 705 °C) and complete decarbonation at 927 ÷ 947 °C [18].

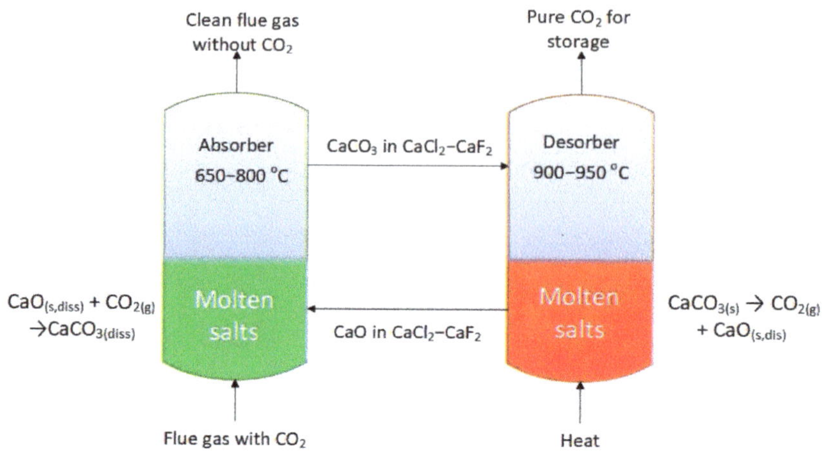

Figure 1. The concept of the carbon capture in molten salts process.

In subsequent studies [22–25] related to the CCMS technology, the following were carried out:

- Evaluation of the impact of other compounds present in the gases on the capture process (SO_2, N_2, H_2O);
- Optimization of the process to achieve maximum efficiency, absorption, etc.;
- Determination of the viscosity of molten salts and its potential impact on the process.

The effect of SO_2 on the CCMS process was investigated in [22]. It was found that similar to the reactions of SO_2 with CaO and $CaCO_3$ in the solid state, sulfation reactions in molten salts are favored over carbonization reactions. Tests in a single-chamber reactor showed that almost all SO_2 is absorbed by molten salts. The authors concluded that CCMS technology could be used to capture both CO_2 and SO_2 from flue gas, potentially avoiding the need for pre-desulfurization. However, it should be noted that the efficiency of the carbonization process might be slightly reduced. The regeneration of SO_2-saturated salts should also be considered.

Olsen et al. [23] studied the effect of H_2O on $CaCl_2$-CaF_2-CaO molten salts in terms of hydrolysis. The results indicated that hydrolysis occurs at elevated temperatures. Both salt components, CaF_2 and $CaCl_2$, form HF and HCl, respectively. However, the presence of CaO significantly reduces the hydrolysis process. It was found that hydrolysis is not a problem when significant amounts of CaO are present (10 wt. % or more) and at temperatures below 850 °C.

The study [24] identified the optimum temperature for the absorption process (approximately 700 °C) and the temperature at which CO_2 release from the carbonate is fastest and complete, occurring at ~930 °C. A sustained CO_2 capture capacity was observed after the initial 4 ÷ 6 carbonation/decarbonation cycles in all multiple tests. After 10 carbonation/decarbonation cycles, the CaO-CaF_2-$CaCl_2$ systems attained carrying capacity values of 0.667 g of CO_2/g of CaO, respectively. A high absorption efficiency (>90%) was also observed in the first 2 h of the process for a CaO content of 15 wt. % of CaO. In the case of desorption, carbonate decomposition occurred with an efficiency of 100%. This shows that the dissolution/dispersion of CaO in molten CaF_2−$CaCl_2$ is a promising pathway for improving the reactivity and stability of CaO in long-term cyclic CO_2 capture operation [24].

For the viscosity of the discussed molten salts, it was stated in [25] that up to 30 wt. % CaO in the eutectic $CaCl_2$-CaF_2 mixture within the temperature range of 700–950 °C, the viscosity should not pose a problem regarding the transport of molten salts.

The latest research was conducted at AGH University of Krakow (Poland) as part of a project in partnership with the Norwegian University of Life Sciences (Norway). This project focused on designing, constructing, and testing a reactor prototype operating in CCMS technology. The main design assumptions for the prototype were based on a structure consisting of two reaction chambers and an additional (auxiliary) tank, with the transport of molten salts forced by gas pressure in the respective tank (pumping). The design solution was patented [26] and described in studies [27,28]. All tanks, barbotage lances, connecting pipes, and valves were made from Nickel Alloy 200. A single reactor chamber was subjected to extended testing to verify the durability of the structural materials used and the feasibility of continuous operation. The purpose of this work was to evaluate the wear of the basic structural components, i.e., the barbotage lance and the reaction tank—components most exposed to prolonged exposure to molten salts $CaCl_2$-CaF_2-CaO-$CaCO_3$, temperatures of 700 and 940 °C, and an N_2-CO_2 gas mixture.

2. Materials and Methods

2.1. Prototype Design

Figure 2 schematically presents the CCMS process utilizing two reaction tanks (absorber and desorber) and an intermediate tank. The diagram also highlights the connection system between the tanks and the flow direction of the molten salts.

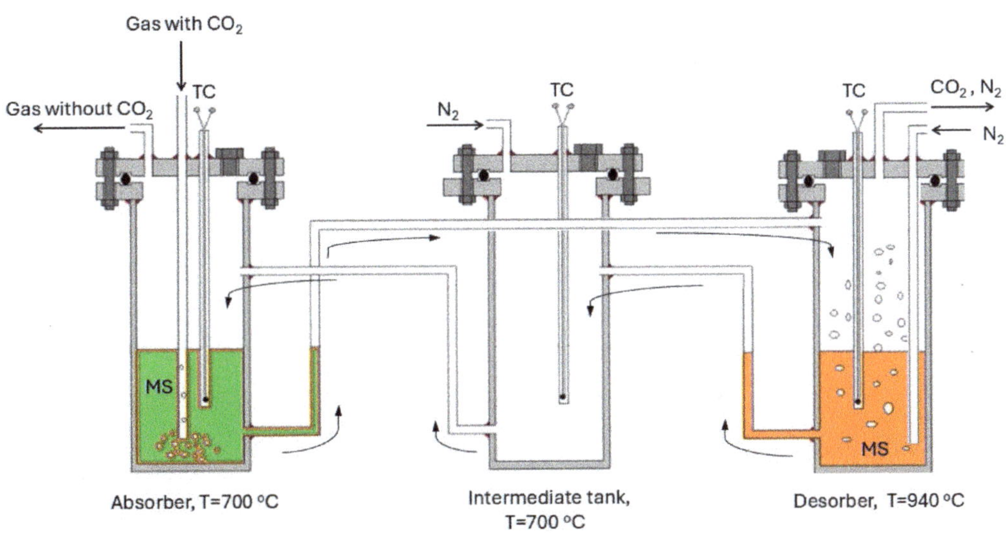

Figure 2. Diagram of the CCMS process using two chambers and an intermediate tank: TC—thermocouple, MS—molten salts $CaCl_2$-CaF_2-CaO-$CaCO_3$.

Figure 3 shows the basic structural elements of a single unit, identical for the absorber, desorber, and intermediate tank. The reactor chambers, along with the transport pipes (horizontal inlet and vertical outlet), were placed within heating modules and closed with tight lids. Gases were introduced and extracted through openings located in the lids. One opening was also left for inserting a thermocouple.

The chambers were connected to each other by horizontal transport pipes, as shown in Figure 4b.

On each transport line between the tanks, there was a high-temperature, gas-tight valve (Figure 4b). This valve allowed the transport line to be shut off, forcing the flow of molten salts to the next tank. To maintain appropriate temperature conditions during salt transport, additional heat sources in the form of removable electric furnaces were installed on the horizontal pipes (Figure 4c). The high-temperature valves were insulated with ceramic mats (Figure 4d). Each reactor chamber heating module, as well as the furnaces installed on the horizontal transport lines, had their own power supply sections (Figure 4a). The construction principle of the prototype is described in detail in studies [27,28].

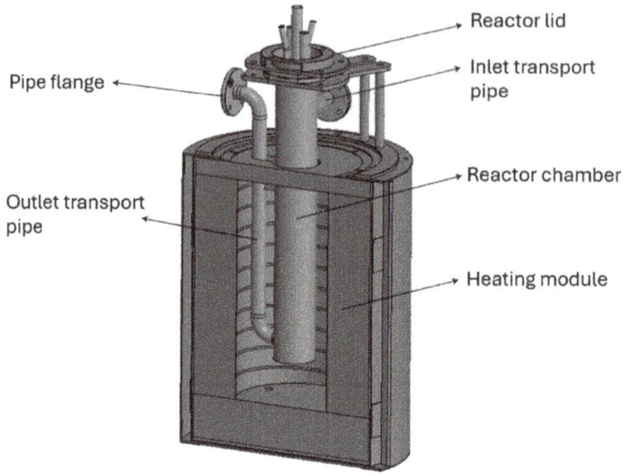

Figure 3. The 3D appearance of a reactor chamber with inlet and outlet transport pipes for molten salts placed within a heating module.

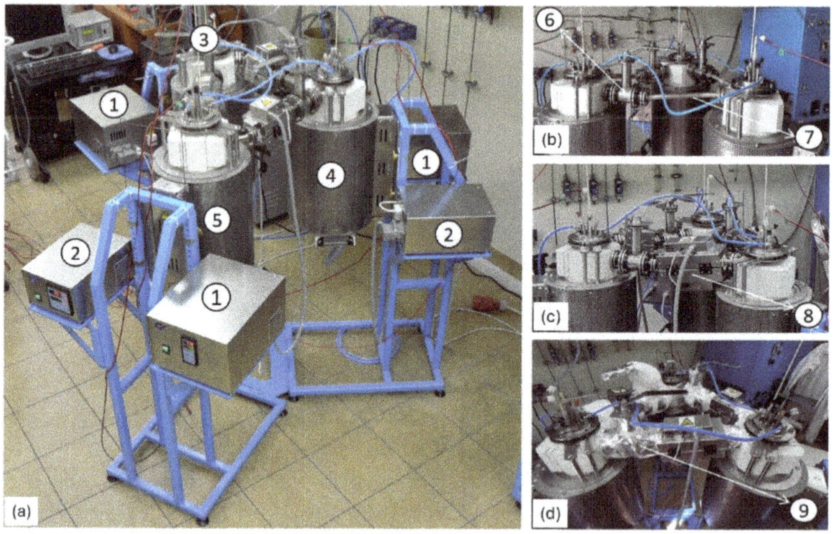

Figure 4. View of the prototype reactor for carbon capture at the AGH laboratory, (**a**) overall view: 1, 2—power supplies; 3—absorber; 4—desorber; 5—intermediate tank, (**b**) view of the system without additional heating modules: 6—high temperature valve; 7—transporting pipe, (**c**) view of the system with additional heating modules: 8—heating module; (**d**) view of the system with valves insulation: 9—ceramic wool insulation.

2.2. Prototype Operation Methodology

The prototype has been designed to operate cyclically [26], with the absorber and desorber working in parallel, aiming for maximum utilization of the active sorbent and complete regeneration of the salts. The operational cycle of the prototype will begin by filling the absorption tank and the intermediate tank with salts of the same composition, containing an eutectic mixture of $CaCl_2$-CaF_2 with a 15 wt. % of CaO. After melting the salts, the absorption chamber will initiate the process of capturing carbon dioxide from an N_2-CO_2 mixture, injected into the salts using a lance. This process will continue until the full utilization of CaO (when the CO_2 concentration in the gases leaving the reactor reaches the level concentration as in the mixture injected into the salts). At this point, the molten salts will be transported to the desorber and preheated to 940 °C. The transport will be carried out by pumping the salts using an inert gas (N_2) injected into the reactor chamber.

After transferring the CO_2-saturated salts to the desorber, the calcination process of calcium carbonate and the release of CO_2 will begin. Simultaneously, the unsaturated salt from the intermediate tank will be transported to the absorber, where the CO_2 capture process from the N_2-CO_2 stream will restart. In this way, the processes of CO_2 capture and release will proceed simultaneously in both tanks (the absorber and the desorber). Upon completion of these processes, the transport of salts will take place first from the desorber to the intermediate tank and then from the absorber to the desorber, initiating the next operational cycle. The reactor chambers will operate under atmospheric pressure during the absorption and desorption processes. Pressure will increase only during the transport of salts between the chambers. The pumpability tests of the salt showed that a pressure of approximately 0.4 atm will be required.

2.3. Testing Conditions for a Single Reactor Chamber

One of the reaction chambers was subjected to high-temperature tests, operating in a cyclic manner with periodic temperature changes. The tests on a single chamber aimed to determine the impact of aggressive operating conditions (rapid temperature changes, the presence of chloride–fluoride salts, continuous gas flow, changes in the chemical composition of salts during the process) on the wear of its key working components—the reaction chamber and the lance. At a temperature of 700 °C, CO_2 capture from the CO_2/N_2 gas mixture (15 vol. % CO_2) occurred for approximately 13 h. Then, the temperature was raised to 940 °C, and with a pure nitrogen flow, the previously absorbed CO_2 in the salts was released. The desorption time was also 13 h. The salts (total mass 800 g) contained a eutectic mixture of $CaCl_2$-CaF_2 (13.8 wt. % CaF_2) and 15 wt. % CaO. During absorption, the N_2-CO_2 gas mixture flowed through the salts at a rate of 24 dm^3/h. Pure nitrogen flowed at the same rate during desorption. The salt level in the reactor was approximately 13 cm. During the capture cycle (700 °C), the CaO concentration decreased as calcium carbonate ($CaCO_3$) formed according to the carbonation reaction (1). During desorption (940 °C), reaction (1) proceeded in the opposite direction, resulting in a decrease in $CaCO_3$ content and an increase in CaO concentration. An example graph of changes in carbon dioxide concentration during a single, complete sorption (carbonation)—desorption (calcination) cycle is shown in Figure 5.

Similar changes in temperature and CO_2 concentration occurred in each cycle during the 800 h of reactor operation. For the first 3–3.5 h, the process operated with very high efficiency, above 96%. This meant that at least 96% of the CO_2 introduced into the molten salts was absorbed as $CaCO_3$. After this time, due to the limited availability of CaO in the salts, the absorption efficiency decreased. The complete depletion of CaO was indicated by the plateau on the curves. On the other hand, the calcination process (release of CO_2) operated at a minimum of 97% efficiency. The CO_2 capture capacity of the prototype was 0.0031 mol/min. The reactor chamber was monitored 24 h a day, particularly the temperature, composition, and flow rate of the N_2-CO_2 gas mixture introduced into the reactor, as well as the composition of the gases leaving the reactor. During the experiments, the reactor chamber operated without issues requiring staff intervention, such as interruption of the experiment, shutdown of the equipment, cooling of the reactor chamber, replacement of salts, etc.

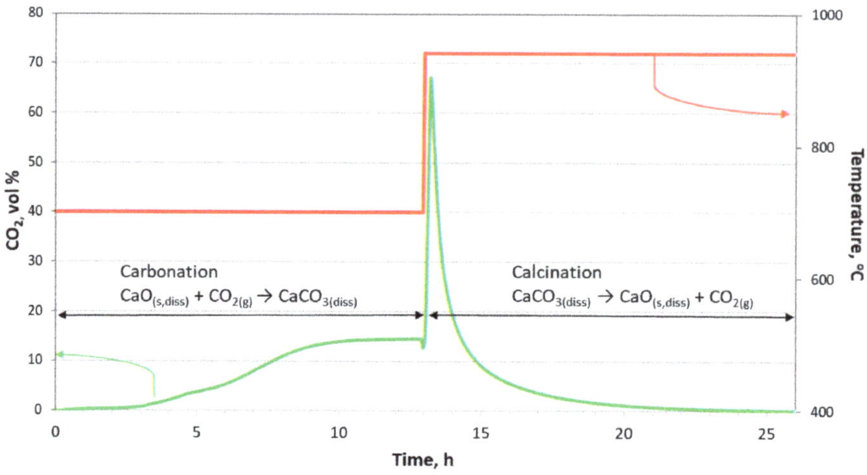

Figure 5. Changes in carbon dioxide concentration during a single, complete operating cycle of the CO_2 capture reactor chamber (green line) and process temperature (red line).

2.4. Materials Subjected to Testing

2.4.1. Characteristics of Ni 200 Alloy

The reactor chambers and the lances introducing the N_2-CO_2 gas mixture, as noted in Section 1, are made of technical nickel (Ni 200 Alloy), whose composition is presented in Table 1.

Table 1. Chemical composition of Ni 200 Alloy, as provided by the vendor.

Elements	Ni	Mn	Cu	Si	C	Fe	S
wt. %	99	0.35	0.25	0.35	0.15	0.40	0.01

It is highly resistant to corrosive environments, especially caustic alkalis, acids, and salt solutions. It also exhibits high thermal and electrical conductivity. The use of technical nickel as the construction material for the main components of the CCMS reactor prototype, whose elements were tested in this study, was mainly motivated by three reasons:

- Nickel 200 alloy has been used as a material for crucibles and lances in all previous experiments at NMBU [19–25], where the CCMS process was tested under various temperature conditions (650–950 °C) and in the presence of molten salts with different compositions (chlorides, fluorides, oxides, carbonates);
- The complexity of the prototype constructed under laboratory conditions required materials suitable for forming individual parts through plastic deformation processes and allowing easy welding to maximize protection during continuous, 24 h high-temperature tests;
- The use of other nickel alloys would have been risky, as alloying elements (mainly Cr) present in nickel/chromium alloys, Inconel, and Hastelloy [29–34] significantly accelerate the rate of corrosion, and grain boundaries are preferred sites for corrosion attack [35,36].

2.4.2. Samples for Testing

After the long-term tests, the molten salts were poured into a separate container, and both the lance and the reactor chamber were cooled. Then, to remove residual crystallized salts, the metal components were submerged in a container filled with distilled water for 24 h, with the water changed three times. After drying, the reactor was subjected to an

X-ray examination. In the next stage, elements of appropriate size were cut from three zones of both the reactor and the lance using a bandsaw and were examined using computed tomography (CT), 3D scanning (reactor), and microscopic analysis (lance). Figure 6 shows the cross-sectional view of the reactor chamber with marked zones with different operating conditions, and samples of lance and chamber reactor subjected to analysis.

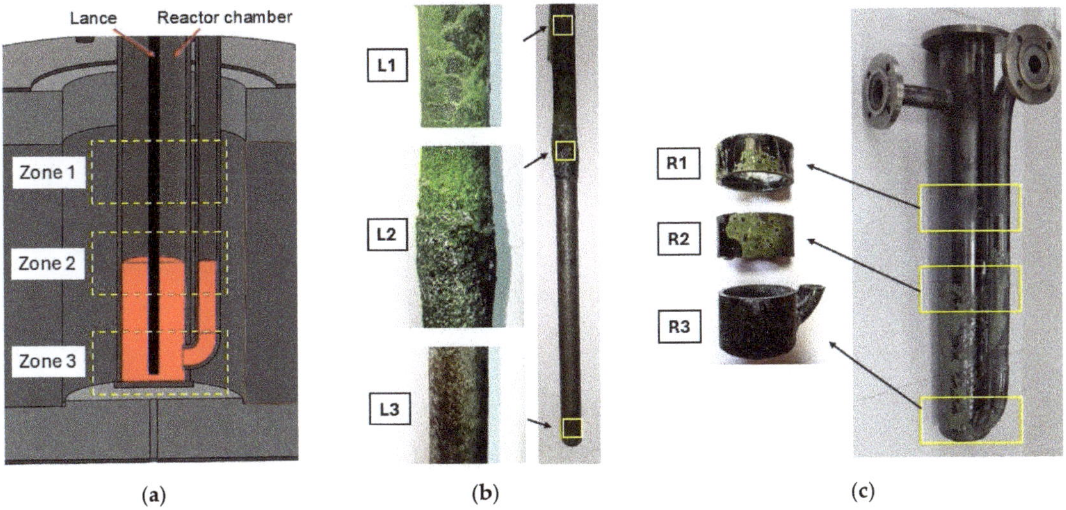

Figure 6. View: (**a**) cross-sectional view of the reactor chamber with marked zones with different operating conditions, (**b**) lance after 800 h tests with marked zones subjected to analysis, (**c**) reactor chamber after 800 h tests with marked areas and cut-out sections subjected to analysis.

The elements marked in Figure 6b,c came from three zones:

- Zone 1: Gas and temperature exposure zone (elements R1, L1). For the reactor tube, this involved continuous exposure to the oxidizing atmosphere on the outer side and the N_2-CO_2 gas mixture on the inner side. During a complete sorption–desorption cycle, the composition of the gas mixture varied according to the curve shown in Figure 5. The outer part of the lance was subject to similar conditions, while the inner side was exposed to the N_2-CO_2 mixture during the low-temperature absorption cycle and pure N_2 during the high-temperature desorption cycle. Contact with splashing molten salts due to sparging could not be ruled out. Due to the temperature gradient at the height of the reactor chamber above the molten salts level, the temperature was slightly lower than in the lower zones of the tank, reaching about 650 °C during sorption and 900 °C during desorption.
- Zone 2: Phase boundary zone between molten salts and N_2-CO_2 gases (elements R2, L2). The contact of both the lance and the salts was similar to that described above; however, it is difficult to precisely determine the exposure time of the elements to salts and gases due to the high intensity of sparging, which caused the molten salt surface to be in constant motion (rising, falling, and turbulent). The temperature in this zone was 700 °C/940 °C.
- Zone 3: Filled with molten salts (elements R3, L3). In the case of the reactor tube, the inner part and the bottom were exposed continuously to the molten salts, while the outer part was exposed to the oxidizing atmosphere (air) in the space between the heating module and the reactor tube. The lance, on the other hand, was exposed to molten salts on its outer side to the N_2-CO_2 gas mixture on the inner side during the low-temperature absorption cycle and to pure N_2 during the high-temperature desorption cycle. The temperature in this zone was 700 °C/940 °C.

The appearance of the reactor chamber before testing and its main initial dimensions are shown in Figure 7.

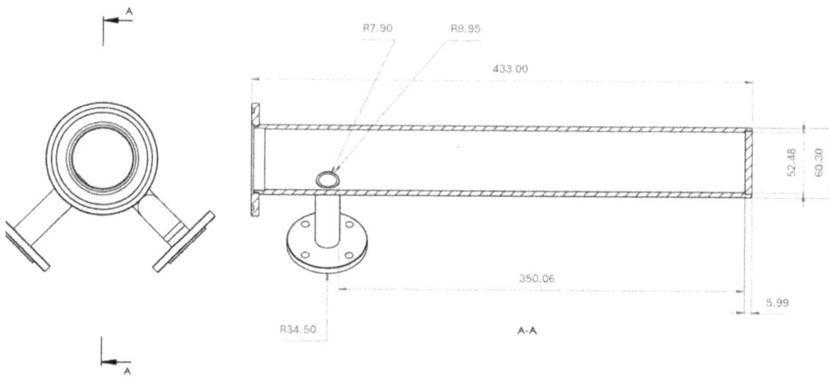

(a)

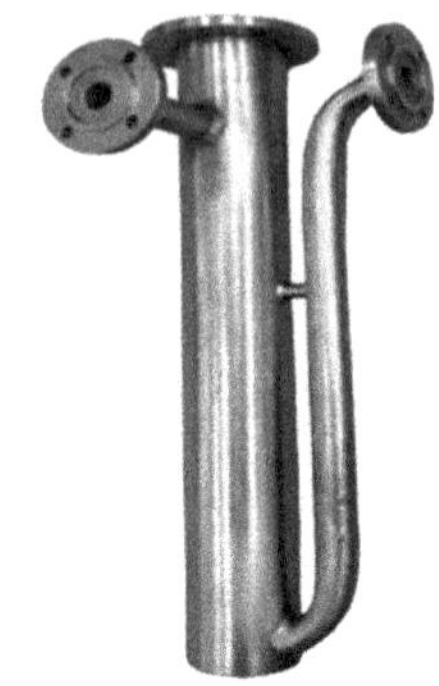

(b)

Figure 7. Reaction chamber: (**a**) main initial dimensions (before experiments), in mm, (**b**) appearance before testing.

In the case of the lance, its internal diameter was 5.6 mm, the outer diameter was 9.8 mm, and the length was 1000 mm.

2.5. Methods of Analysis

In assessing the wear of the reactor chamber, several testing methods were used (X-ray, CT, 3D SCAN) based on several key criteria:

- Direct evaluation of changes in geometric dimensions after testing;
- Identification of internal discontinuities in structural components (defects, voids, pores), quality of welds;
- Detection of wear inhomogeneity, localized pits, and cracks;
- Evaluation of the suitability of individual methods for this type of analysis.

Moreover, the use of X-rays allowed a complete assessment of the reactor without the need to cut it into separate components. X-ray computed tomography, in addition to multi-plane analysis, allowed for the evaluation of wear inhomogeneity across different zones. The 3D scanning was employed as a supplementary technique and made it possible to locate deviations (both positive and negative) from the nominal values of the reactor chamber.

2.5.1. X-Ray

The research utilized the Zeiss BOSELLO 2D X-ray Solutions system (Carl Zeiss IQS Deutschland GmbH, Carl-Zeiss-Straße 22, 73447 Oberkochen, Germany). ZEISS BOSELLO systems are 2D X-ray cabinets equipped with software for Automatic Defects Recognition (ADR) (Software BHT IP Plus). The dedicated software enabled fast image acquisition and fully automated 2D X-ray inspection in accordance with ASTM standards [37].

During the tests, a maximum source voltage of 225 kV and a maximum power of 1800 W was applied, with a detector resolution of 1024 × 1024 and a pixel size of 200 μm. The inspection time was 1 min, and the time for a single image capture was 200 ms.

The entire reactor chamber was subjected to X-ray examination after the tests.

2.5.2. Computed Tomography (CT)

The elements of the reaction chamber were scanned using the GE Phoenix v | tomex | m X-ray tomography system (Waygate Technologies USA, LP 721 Visions Dr, Skaneateles, NY

13152, USA). The object was inspected in a specialized chamber of the device. The element was positioned between the radiation emitter, known as the lamp, and the detector. The device was controlled by setting the position of the object in the space between the lamp and the detector, entering the voltage and current values generating the radiation (power), and specifying the detector's operating characteristics. Proper selection of these parameters allowed for results that enabled further 3D reconstruction of the object.

The next step was the reconstruction of the 3D volume. During a full rotation, the device generated thousands of high grayscale images (e.g., the GE Phoenix v | tomex | m uses 14-bit). It is standard to use complex algorithms in the reconstruction stage that perform image corrections, such as beam hardening correction, automatic geometry calibration, and geometry optimization [38,39]. The reconstructed object can be visually inspected through the analysis of 2D and 3D cross-sectional images and further volumetric analysis.

It should be noted that all measurements were conducted with identical detector settings, voltage, and consistent positioning of the maneuvering table. The results were obtained based on reconstruction with the same algorithm: using the bhc+ filter (coefficient 7.1), auto AGC, and reconstruction optimization. Determining porosity (voids, empty spaces) is widely used when analyzing heterogeneous materials [40,41] or changes in materials due to external factors [42]. The examination of a single reactor tube element consisted of 3000 X-ray projections. The size of a single voxel did not exceed 60 μm. Radiation was generated by a microfocus lamp with a 260 kV and 190 μA characteristic. During reconstruction, a beam hardening correction+ filter with a power of 7.1 and automatic geometry correction were applied.

The reactor components R1, R2, and R3 were subjected to computed tomography.

2.5.3. Scan 3D

A setup consisting of a GOM 3D scanner measuring head (GOM, a Zeiss company, Oberkochen, Germany), a tripod, a movable table, a calibration plate, and a portable Dell computing unit was used for optical 3D scanning.

The ATOS Core 200 scanner head (GOM a Zeiss company, Oberkochen, Germany) used for measurements was equipped with two 5 MP cameras, blue LED lighting, and light with a wavelength range from 450 to 500 nm. The measurement accuracy, verified according to standard [43] and confirmed by an accredited, independent metrology laboratory, is 0.0185 mm. The time required to complete a single scan was less than 2 s. The research was conducted using Atos Scan 2017 software, which enabled the scanning process and the creation of a 3D model, as well as previously described GOM Inspect software for 3D measurement data analysis (2019 Hotfix 6 Rev. 125216). The guidelines for assessing the accuracy and acceptance of optical 3D scanning systems are contained in the standard [44] and the recommendations of the German PTB in accordance with the standard [43].

The reactor component R3 was subjected to 3D scanning.

2.5.4. Microstructural Studies

For microstructure analysis, an Olympus GX51 light microscope (OLYMPUS GX51 microscope, Tokyo, Japan) and a Hitachi SU 70 scanning electron microscope (Hitachi Ltd., Tokyo, Japan) equipped with energy-dispersive X-ray spectroscopy (EDS) detector for chemical composition analysis in micro-areas were used (Thermo Fisher Scientific, Waltham, MA, USA). Images were captured using the secondary electron (SE) and backscattered electron (BSE) detectors. Samples for microscopic examination were embedded in PolyFast conductive resin (Struers, Copenhagen, Denmark), then ground using abrasive papers with the following grit sizes: 220, 500, 800, 1200, 2400, and 4000, and polished in two stages according to the Struers procedure using Dia Duo diamond paste suspension (Struers, Copenhagen, Denmark) with a 3 μm grain size and OPS colloidal silica suspension with a 0.25 μm grain size for final polishing. The samples were not etched.

The lance components L1, L2, and L3 were subjected to microstructural analysis.

3. Results

3.1. Barbotage Lance Tests

The results of the lance wall thickness measurements using light microscopy at various points across the cross-section of the examined samples are presented in Figure 8. Table 2 summarizes the minimum, maximum, and average values, as well as the calculated wall thickness reductions (thinning).

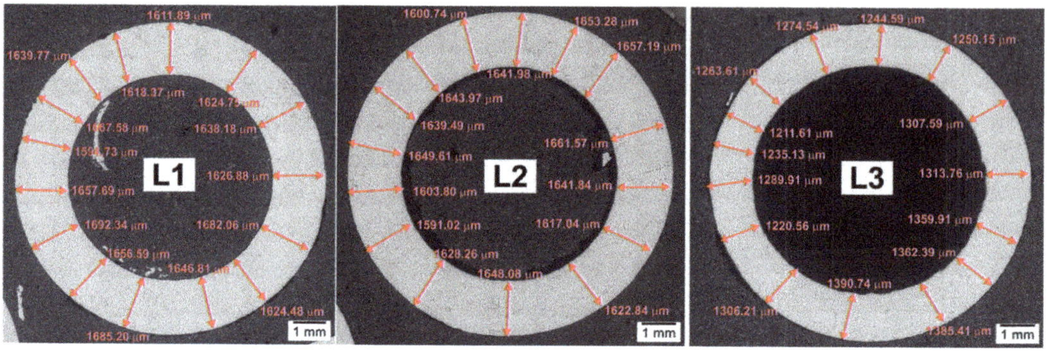

Figure 8. Appearance of samples L1, L2, and L3 and wall thickness measurement points (light microscopy—macro image).

Table 2. Wall thickness measurement results on the cross-section of the examined lance before and after high-temperature tests.

Sample	Wall Thickness, µm	Wall Thinning Relative to the Original Tube, µm
L1	Average: 1645	2 (0.12%)
	Min.: 1595	
	Max.: 1692	
L2	Average: 1637	10 (0.61%)
	Min.: 1591	
	Max.: 1662	
L3	Average: 1294	353 (21.43%)
	Min.: 1212	
	Max.: 1391	
L0 (before tests)	Average: 1647	-
	Min.: 1616	
	Max.: 1678	

The obtained wall thickness results on the cross-section of the examined lance demonstrated thinning relative to the original dimensions. The degree of thinning depends on the lance's operating zone. The average wall thickness on the lance cross-section before experiments was 1647 µm, with wall thickness variations across the cross-section reaching up to 62 µm. The greatest wall thinning after high-temperature prototype tests was observed for the element labeled "L3" (continuous external contact with molten salts $CaCl_2$-CaF_2-$CaCO_3$-CaO, internal contact with N_2-CO_2 gases). In this case, the wall thinned by an average of 353 µm, or 21.43%. Significant variations in wall thickness across the cross-section were also found, reaching up to 180 µm locally.

For samples labeled "L2" and "L1", slight wall thinning of the lance was observed, averaging 10 µm (0.61%) for element L2 and 2 µm (0.12%) for element L1. Wall thickness differences across the cross-section reached up to 70 µm for element L2 and 100 µm for element L1.

When analyzing the L3 section of the lance, it was compared to the material before testing. Images from the light microscope were superimposed, with 50% transparency applied to the sample before testing—L0 (Figure 9).

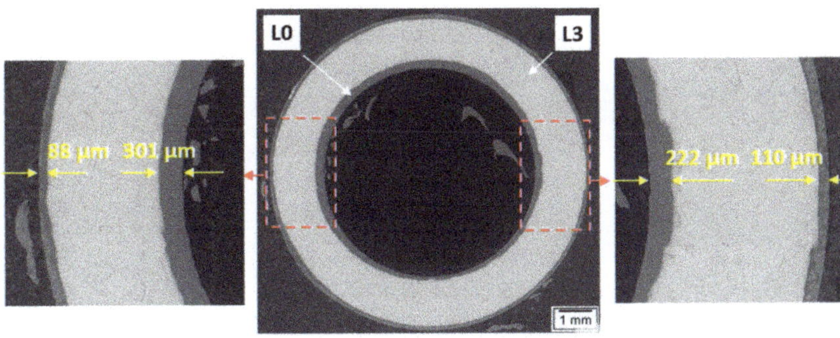

Figure 9. Images of the L3 lance section after testing and before testing (L0). A 50% image transparency was applied for L0.

It was found that the lance experienced significantly greater wear on the inner side. The wear is uneven around the circumference on both the outer and inner sides of the lance.

Figure 10 shows example results of the microstructure analysis of sample L3 performed using a scanning electron microscope, while Figure 11 presents the results of chemical composition analysis in micro-areas using an EDS detector.

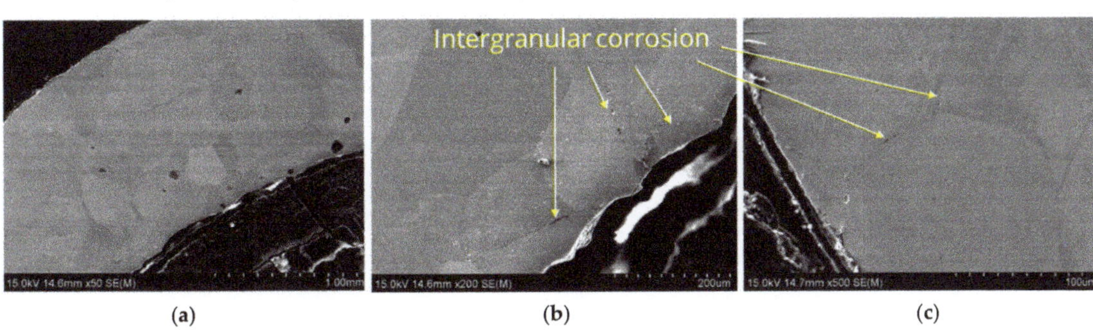

Figure 10. Microstructure images of sample L3, SEM: (**a**) magnification 50×, (**b**) inner section, magnification 200×, (**c**) outer section, magnification 500×.

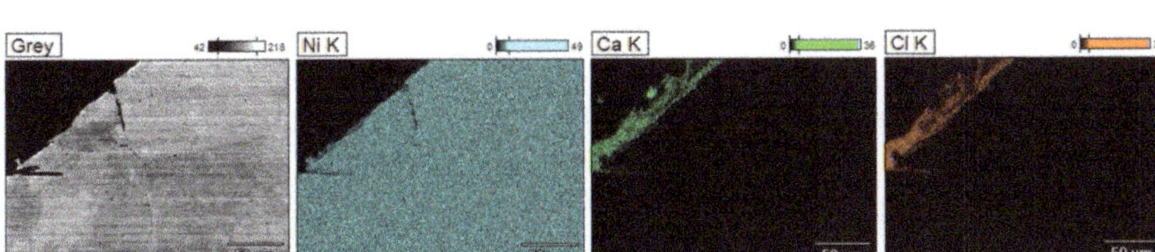

Figure 11. Ni, Ca, Cl distribution maps of sample L3; SEM/EDS.

The microstructure observations on the cross-section of the tube labeled L3 demonstrated the presence of intergranular corrosion on both the inner and outer sides of the tube. The corrosion spreads from the outer and inner surfaces of the lance inward, locally reaching a depth of several hundred micrometers. Corrosion appears as a network of voids spreading along grain boundaries. The SEM/EDS chemical composition analysis results indicate that these areas are nickel-depleted (Figure 11), and some spaces are infiltrated by salt components (Ca, Cl). Salt components were also detected on the analyzed outer surface, which are residues of crystallized salts that did not dissolve in water during lance rinsing.

Similar microscopic observations and analyses were conducted for samples taken from the other two zones, labeled L2 and L1. The results are presented in Figures 12–15.

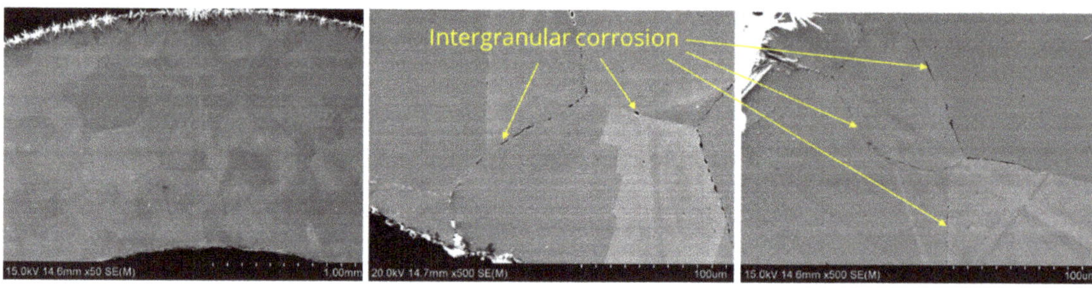

(a) (b) (c)

Figure 12. Microstructure images of sample L2, SEM: (**a**) magnification 50×, (**b**) inner section, magnification 500×, (**c**) outer section, magnification 500×.

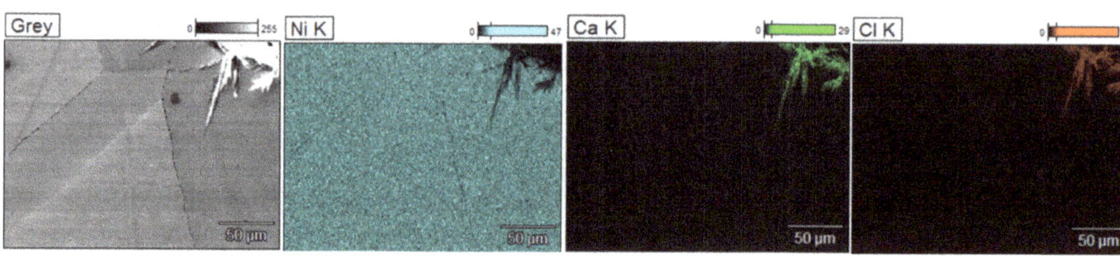

Figure 13. Ni, Ca, Cl distribution maps of sample L2; SEM/EDS.

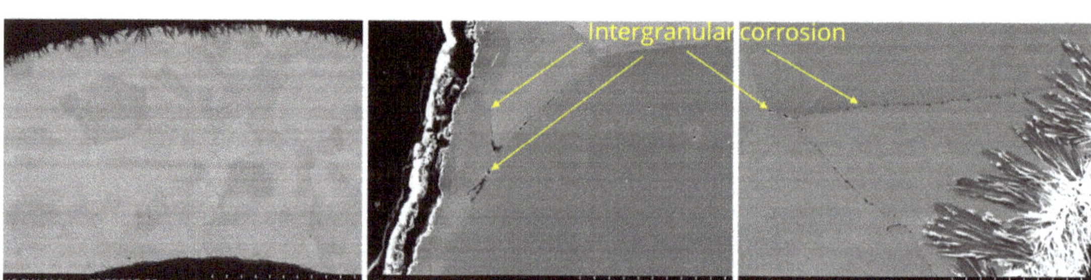

(a) (b) (c)

Figure 14. Microstructure images of sample L1, SEM: (**a**) magnification 50×, (**b**) inner section, magnification 500×, (**c**) outer section, magnification 500×.

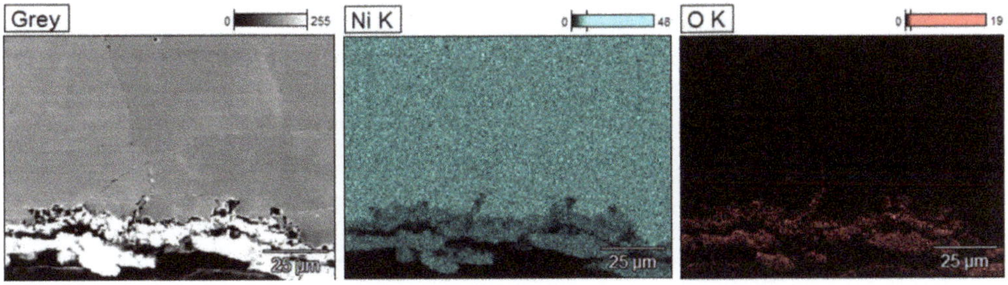

Figure 15. Ni, O distribution maps of sample L1; SEM/EDS.

Analyzing the results presented in Figures 14–17, it was observed that corrosion appears both as a network of fine voids spreading along grain boundaries and as a continuous network at grain boundaries. The SEM/EDS chemical composition analysis results indicate that these areas are nickel-depleted. Additionally, the images show residues of crystallized salts that formed on the sample surfaces and were not dissolved in water after the experiments.

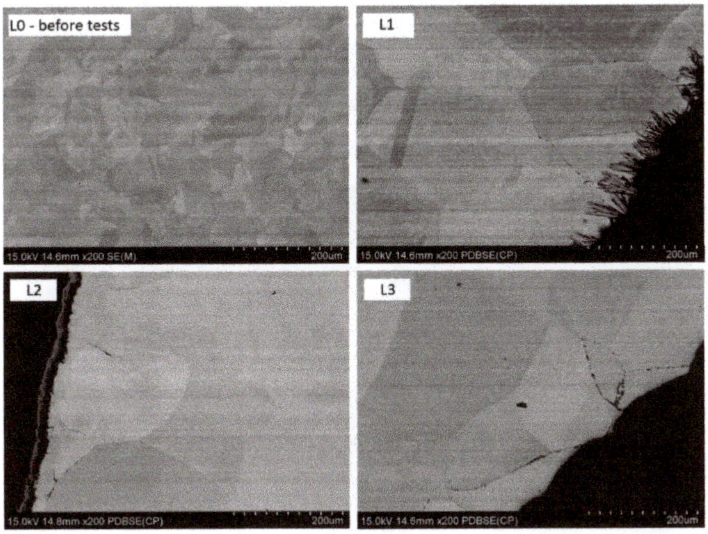

Figure 16. Comparison of grain size in Ni Alloy 200 before testing (L0) and after testing (L1, L2, L3).

Figure 16 presents a comparison between the original Nickel 200 Alloy samples and the samples after testing.

Analyzing Figure 18, it can be observed that the microstructure of the initial tube cross-section, labeled L0, is uniform and consists of grains with regular shapes, averaging about 50–60 µm in size. No precipitates or discontinuities are visible within the examined element. In samples L1–L3, significant grain growth was observed in each tested sample after the tests, regardless of the presence or absence of molten salts or N_2-CO_2 gases. Some grains exceeded 200 µm in size. In each case, the material was exposed to high temperatures; due to the temperature gradient along the reactor, samples L3 and L2 were subjected to temperature fluctuations between 700 and 940 °C, while sample L1 experienced slightly lower temperatures (650/900 °C).

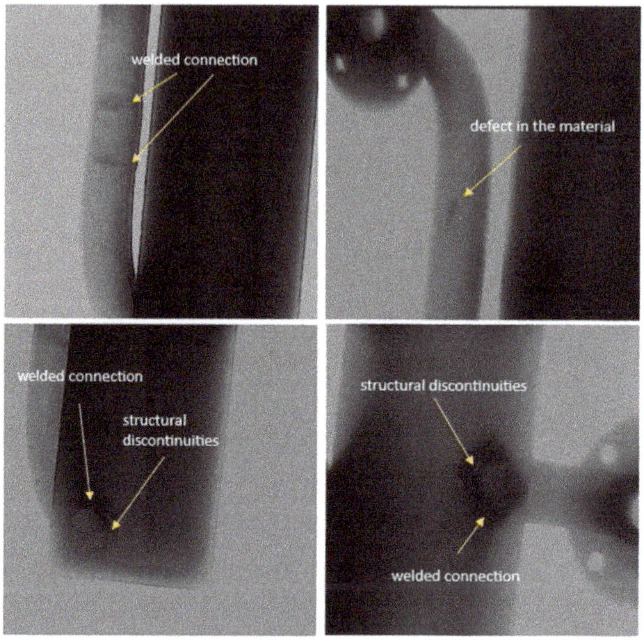

Figure 17. Selected inspection photos of the reactor chamber after testing.

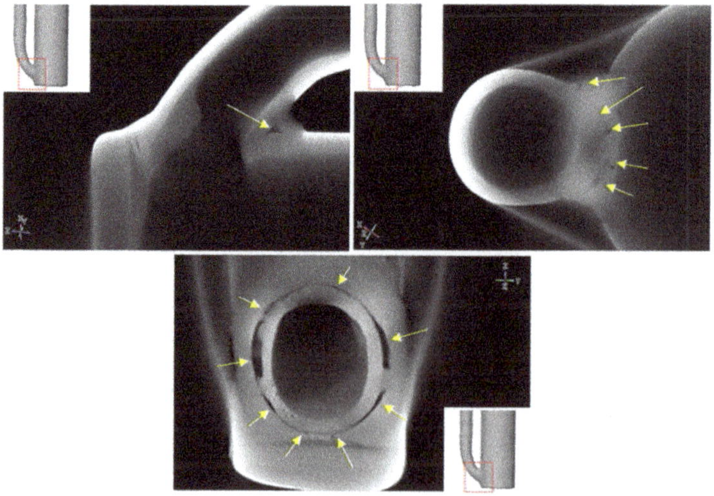

Figure 18. Analysis of the lower part of the reactor chamber after reconstruction to the 3D model. Yellow arrows indicate voids and discontinuities in the material.

3.2. X-Ray Examination of the Reactor Chamber

The X-ray inspection allowed the detection of defects, damage, and discontinuities in the structural material without the need to cut the reactor into smaller components. Figure 17 shows selected inspection images of the reactor chamber after inspection.

After X-rays were taken, reconstruction to a 3D model was performed and the lower part of the reactor chamber was analyzed in detail (Figure 18).

Darker areas marked with yellow arrows in Figures 17 and 18 are local density changes in the analyzed material. Figure 17 shows welded joints and a defect in the material, probably formed during the manufacture of the reactor tube. Most of the identified defects (voids, material discontinuities) are due to the inability to weld structural elements of the chamber from the inside of the reactor. They are not a result of reactor operation or prolonged exposure to the corrosive environment. However, the obtained information is important from the perspective of system sealing and properly securing the chamber against leakage of molten salts at a temperature of 700/940 °C into the heating module, which could lead to a failure of the entire device.

3.3. CT Scan Results

X-ray computed tomography used for the analysis required cutting the reactor components into smaller sections. The study examined reactor fragments operating in three zones, as described in Section 2.1. Figure 19 shows selected cross-sectional images of the lower part of the reactor chamber (R3).

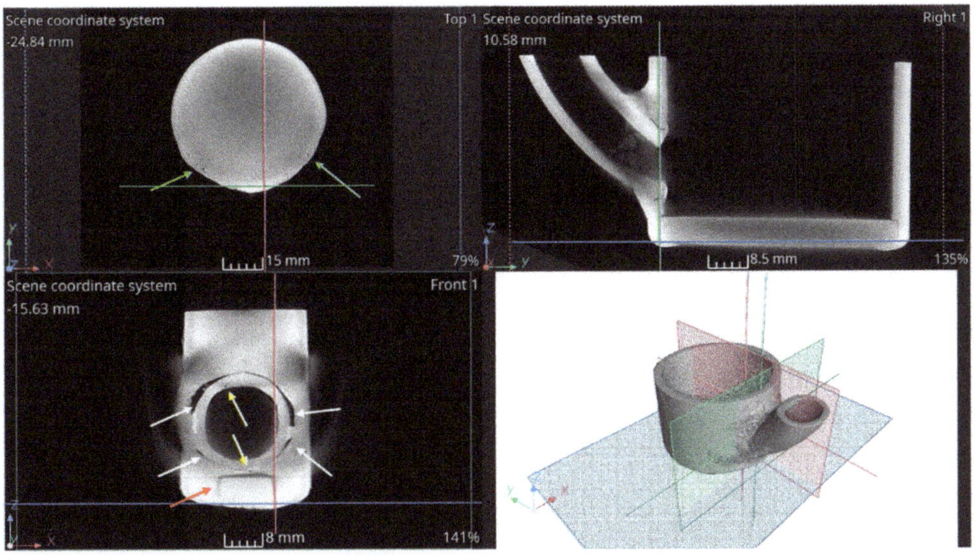

Figure 19. Cross-sectional images of the element R3.

White arrows indicate defects (lack of full remelting) in the joint between two pipe elements. Lack of remelting occurs around the entire circumference at the junction of the smaller-diameter pipe with the reactor pipe (larger diameter). In addition to the lack of full penetration, voids in the weld and numerous pores, marked by yellow arrows, are also visible, some of which form larger clusters. The connection between the reactor pipe and the flat base is also visible. Here, too, there is incomplete remelting (red arrows) and clusters of pores, marked by green arrows. The tricolor lines on the cross-sectional images indicate the positions of the other cross-sections corresponding to the viewing planes shown in the 3D section.

A wall thickness analysis of the R3 element was performed on the reconstructed data. The presented model, created using computed tomography, exhibits noise primarily resulting from the complexity of the shape and geometry of the analyzed component (Figure 20a).

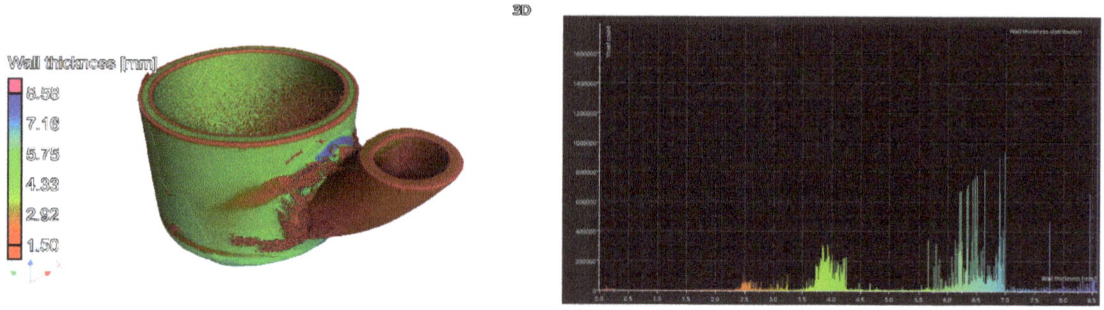

(a) (b)

Figure 20. Element R3: (**a**) 3D view and (**b**) wall thickness distribution.

Artifacts characteristic of this method, such as data-disrupting peaks, hinder the direct interpretation of the results. Nevertheless, the application of advanced filtering procedures has partially mitigated these disturbances, as confirmed by the provided graph. Thickness distribution analysis is further complicated by the presence of noise in the visual representation and the variable thickness of individual components, which becomes apparent when analyzing the data depicted in the graph (Figure 20b). A particular challenge is the attachment of components via welding, which is characterized by uneven thickness on the weld surface, rendering the analysis of wear in specific areas unreliable.

In the case of the R2 element, which has partial contact with molten salts and a gaseous atmosphere, no irregularities, defects, or deep pitting were found in the cross-sectional images presented in Figure 21. The tricolor lines on the cross-sectional images indicate the positions of the other sections, corresponding to the viewing planes shown in the 3D section.

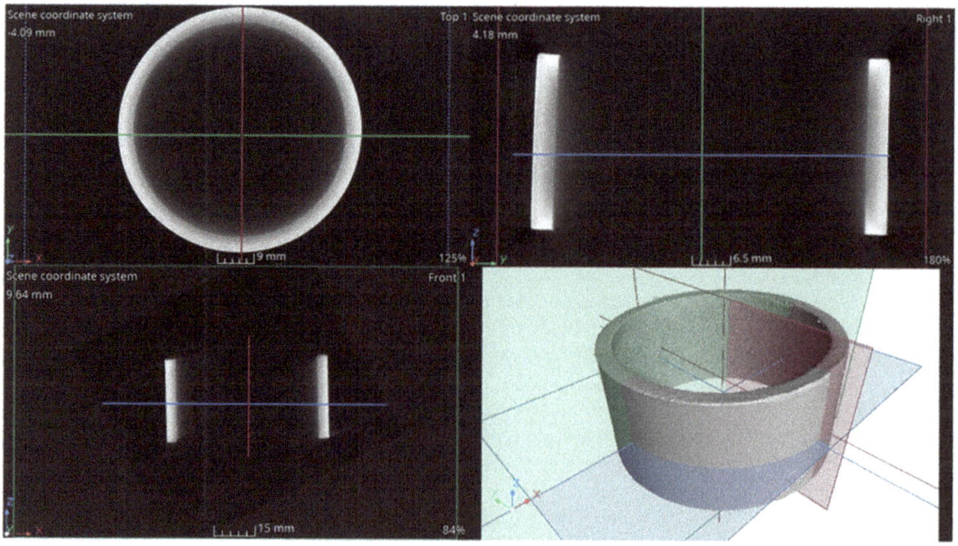

Figure 21. Cross-sectional images of the element R2.

A wall thickness analysis of the R2 element was performed on the reconstructed data. The results of the analysis are shown in Figure 22.

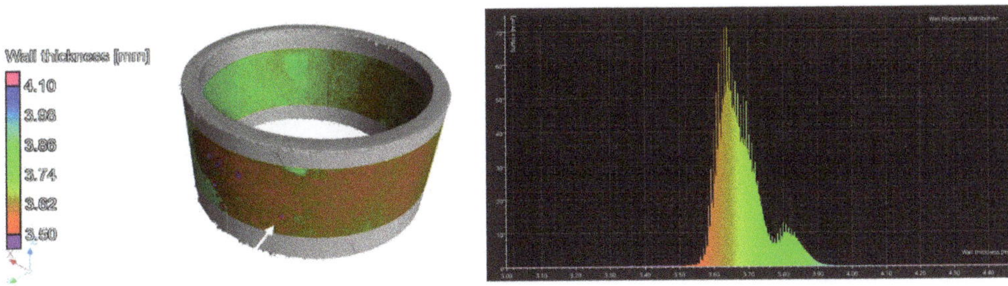

(**a**) (**b**)

Figure 22. Element R2: (**a**) 3D view and (**b**) wall thickness distribution.

The color scale indicates the wall thickness, which in this area is 3.75 mm. Areas of thickening, marked with a white arrow, are clearly visible in the figure, with a thickness of 3.82 mm. The section of the pipe shown in Figure 22a is noticeably thinner. Around the circumference of the element, there are spots colored in shades of red, indicating localized thickness reduction. The thickness change is approximately 0.2 mm. Figure 22b presents a thickness distribution graph of the element. Two peaks are visible—one at a thickness of 3.75 mm and the other at 3.82 mm. These values correspond to the visually identified thicknesses. The wall thickness in 99% of the entire analyzed region falls within the range of 3.51 mm to 3.945 mm.

Cross-sectional images for element R1 are shown in Figure 23. Their analysis indicates the presence of pitting on the inner side of the element, marked by red arrows. Additionally, a deposit is visible on the inner side (green arrows), which may be the result of crystallized and insufficiently flushed salts.

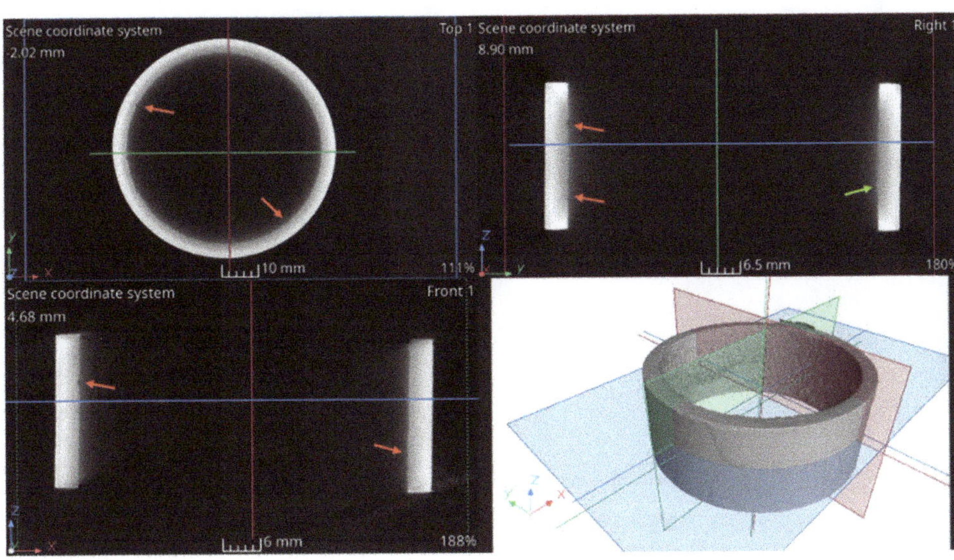

Figure 23. Cross-sectional images of the element R1.

A wall thickness analysis of the pipe element was performed on the reconstructed data. The analysis results are shown in Figure 24. The thickness of element L1 is approximately 3.73 mm. In the 3D images, areas of thickening are clearly visible, marked with a white arrow. Around the entire circumference of the element, there are red-colored spots, indicating localized thickness reductions. The thickness variation is approximately 0.25 mm.

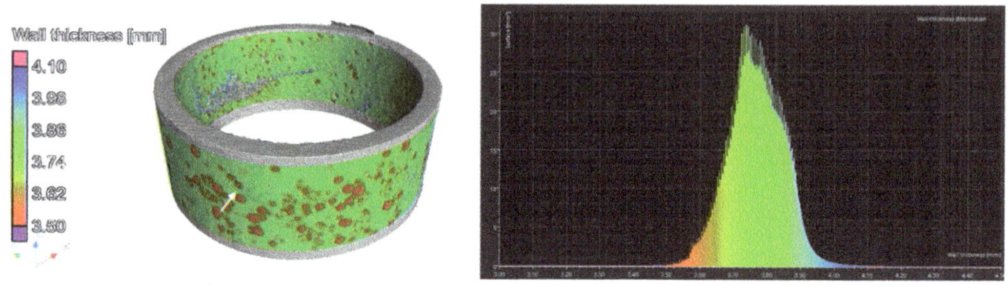

(**a**) (**b**)

Figure 24. Element R1: (**a**) 3D view and (**b**) wall thickness distribution.

3.4. 3D Scanning Method

Due to difficulties in assessing the thickness of element R3 using computed tomography, a thickness analysis was conducted based on 3D scans. The results of this analysis are presented in Figure 25.

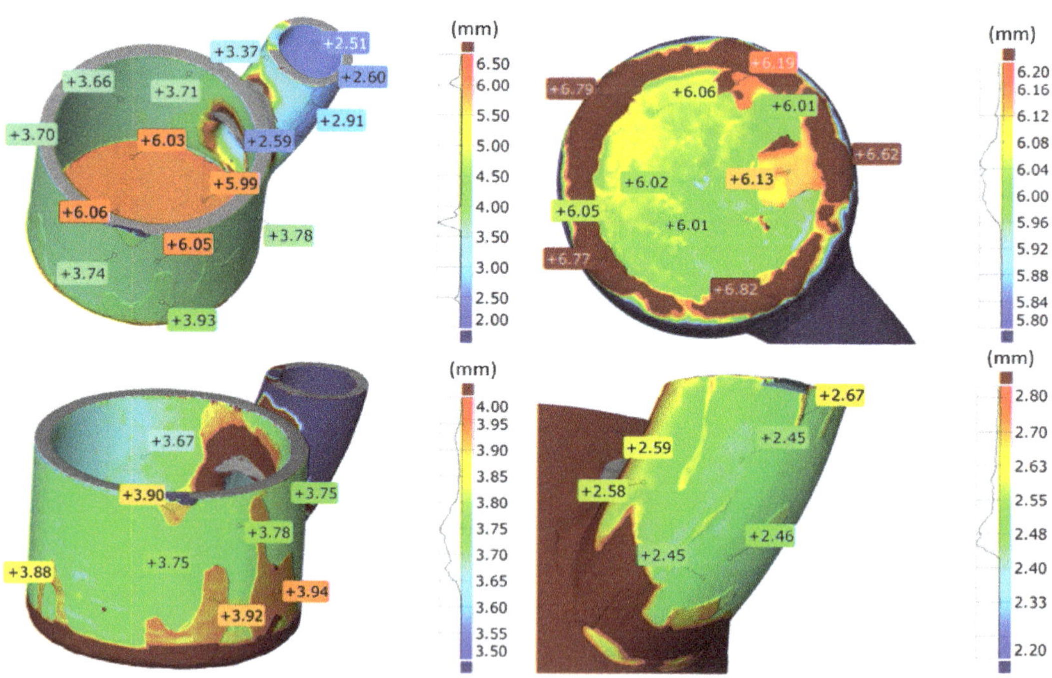

Figure 25. Thickness measurement of element R3 using the 3D scanning method.

For the analysis, it was necessary to adjust the dimension scale to the analyzed fragments. When presenting the results for the entire element, a significant number of thickness values were observed within the ranges of 2.2–2.8 mm, 3.5–4 mm, and 5.8–6.2 mm. These values, visible on the graph presented alongside the measurement scale, correspond to the nominal thicknesses of the bottom, the lateral part of the element, and the welded component of the device. The bottom of the element exhibits a thickness of approximately 6 mm, with significant irregularities resulting from the joining method (welding). The majority of results fall within the range of 5.95 mm to 6.12 mm. Defects on the external surface in the form of uneven corrosion are visible. Additionally, the inner surface shows isolated thickened areas caused by salt residues present in the device during operation. The cylindrical section also exhibits wear due to corrosion, visible on the external surface as large areas with varying thicknesses. The thickness difference averages 0.15 mm. The majority of results fall within the range of 3.65 mm to 3.8 mm. The welded component is characterized by significantly lower thicknesses, ranging from 2.4 mm to 2.6 mm. It contains areas with variable thickness and an elongated shape, which are also the result of corrosion-induced wear. The thickness difference averages 0.14 mm. The weld area was not analyzed due to significant thickness variations resulting from the welding technology itself.

The dimensional comparison of the R3 element after the tests with the CAD model is presented in Figure 26.

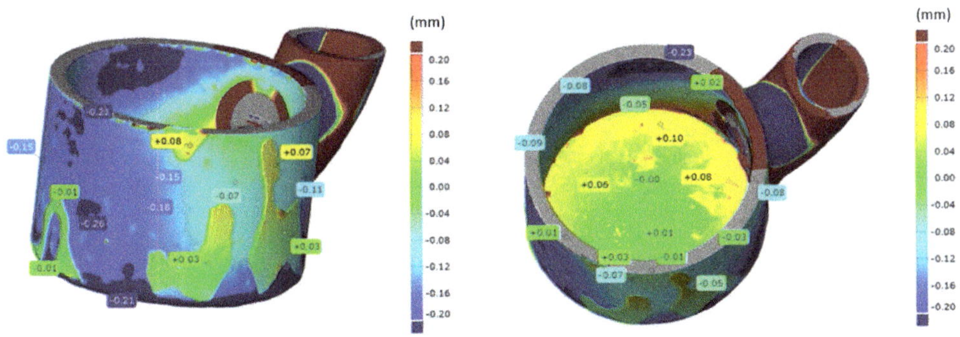

Figure 26. Lower part of reactor chamber (R3) shown from two views—dimensional comparison with CAD model.

The deviations obtained in the working part of the component range from +0.08 mm to −0.26 mm. No significant deformation was observed in the cylindrical area of element R3. A comparison of the scanned element with the CAD model showed a significant shift of the welded element, but this did not affect the device's performance. This shift likely did not result from the prolonged operation of the reactor chamber but rather occurred during the joining (welding) of the elements. The inner surface of the cylindrical area shows minimal wear, with mostly positive values at measurement points, suggesting slight wear and the deposition of molten salts, especially in the lower areas of the reactor (at the bottom). Wear primarily occurred on the outer part of the element, where values reach −0.26 mm. Wear is not uniform, with large areas showing no wear and small localized areas with increased wear, which may suggest a significant influence of corrosion due to operating conditions (mainly high temperature and an oxidizing atmosphere).

4. Discussion

The use of molten salts in the carbon dioxide capture process is a promising and highly efficient technique for CO_2 reduction from diluted industrial gases. One of the greatest challenges of this new technology is finding construction materials with adequate corrosion resistance to chloride–fluoride molten salts and high temperatures (up to 940 °C). The corrosion of materials is caused by the high activity of ions in molten salts, especially at elevated temperatures [45]. This phenomenon is only partially understood, and the mechanisms remain unclear [46–50]. Nickel and its alloys are among the best materials for structural applications in molten salts, as they exhibit very good corrosion resistance and favorable mechanical properties [51–53].

From the perspective of the reactor's long-term operation (with three chambers), the evaluation of the main structural components was the most crucial. In the case of the prototype, these components were the lance and the reactor tank. These elements will be subjected to the highest levels of corrosive environmental stress (molten salts and temperature) during continuous operation. Analysis of the microscopic observations of the lance revealed that wear varied depending on the zone in which the lance material was exposed to specific conditions. Measurements of the lance wall thickness before and after testing allowed for a direct assessment of its wear. The change in wall thickness in the lowest zone (continuous contact with molten salts) was approximately 21.4%. The lance in the intermediate zone (liquid/gas phase boundary) showed a slight thinning, as did the section operating above the molten salts, with thinning in these zones measuring 0.6% and 0.1%, respectively. Based on the results relative to the material's exposure time in molten salts, it can be concluded that in the area most exposed to corrosion, the wear rate of Ni Alloy 200 was 353 μm/800 h, which translates to 3.9 mm/year. For the middle section of the tube, not directly exposed to molten salts, the wear rate was 109 μm/year, while the upper part showed a wear rate of approximately 23 μm/year.

Analysis of the lance thickness changes for section L3 showed uneven wear around the lance circumference and significantly greater wear on the inner side compared to the outer side. It was expected that higher wear would occur on the surface exposed to continuous contact with molten salts rather than in the gaseous atmosphere (N_2-CO_2). It is possible that molten salts were drawn into the end part of the lance, which was analyzed. In the analyzed cross-section of sample L3, characteristic pits were also observed, whose shape suggests the detachment of entire grains due to the potential penetration of molten salts through grain boundaries, as illustrated in the diagram in Figure 27.

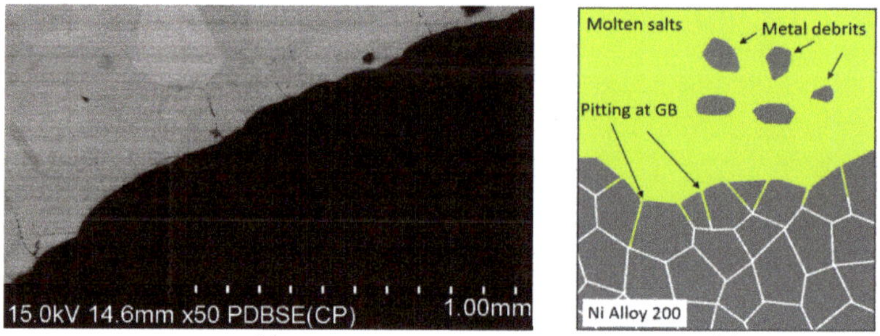

Figure 27. SEM image of the L3 surface (**a**) and schematic illustrating the formation of a pore-salt network for Ni Alloy 200 and molten salts (**b**).

The reaction of Ni with gases that may form in the salt environment, such as HCl or HF produced in reactions (2) and (3), also cannot be ruled out.

$$CaCl_{2(l)} + H_2O_{(g)} \rightarrow CaO_{(s)} + 2HCl_{(g)} \tag{2}$$

$$CaF_{2(l)} + H_2O_{(g)} \rightarrow CaO_{(s)} + 2HF_{(g)} \tag{3}$$

H_2O may be introduced into the system as a contaminant (reagents, construction materials, gases, etc.), especially since $CaCl_2$ is a highly hygroscopic reagent. At process temperatures, the free enthalpy change of reactions (2) and (3) is positive, which would suggest that they should not occur. However, the gaseous products are quickly removed from the reaction zone, resulting in low partial pressures. In such cases, the reaction equilibria will shift to the right, favoring the formation of corrosive and toxic gases. These considerations were experimentally confirmed in the study [20].

The analyzed sections of the lance, L1 and L2, did not show significant wear, as they were only exposed to high temperatures and gases (mainly N_2-CO_2) or potentially to corrosive HCl and HF, as mentioned above. Contact with molten salts was significantly limited and could occur mainly due to gas sparging and salt splashing above the salt level. The green color of the deposit on the outer surface of the lance in sections L1 and L2 may suggest the formation of nickel chloride ($NiCl_2$) as a result of the reaction (4) or nickel oxide (NiO) as a result of a reaction (5):

$$Ni + 2HCl_{(g)} \rightarrow NiCl_2 + H_{2(g)} \tag{4}$$

$$Ni + 1/2O_{2(g)} \rightarrow NiO_{(s)} \tag{5}$$

Oxygen can be introduced into the system as a contaminant, similar to the moisture discussed above. The free enthalpy change for reaction (4) is positive under process conditions (temperature 700–940 °C) [17], whereas for reaction (5), it is negative [17]. Attempts to perform XRD analysis of the green deposit taken from zones L1 and L2 proved very challenging due to the amorphous nature of the material.

The results of the X-ray analysis presented in Figure 17 and reconstructed into a 3D model (Figure 18) revealed minor defects in the structural material (most likely originating during tube manufacturing) and defects arising during the construction of the reactor tank. The most visible of these are in the welded areas, where discontinuities in the metallic structure were detected. These are not a result of prolonged operation of the reactor in the aggressive environment of molten salts. An important finding is that no complete discontinuity occurred in any part of the reactor tank under investigation, which would risk tank leakage and reactor failure.

Using computed tomography, numerous discontinuities were revealed at the welding joints in the lower part of the reactor chamber (R3). In this component, gaps were also found at the junction of the round bottom element with the reactor tube, resulting from insufficient fitting of the components—something not visible in the X-ray images. Proper continuity of the weld was observed on the outer side of the reactor tube. Unfortunately, due to significant noise and interference in the analyzed object, it was not possible to create a three-dimensional thickness map for the R3 component. Such a map is useful, as demonstrated for components R1 and R2, as it allows the identification of wear inhomogeneity.

3D scanning provided additional information on the wear of part R3, which is the most exposed to the corrosive environment and high temperature. In section R3, negative deviations predominate on the outer side. These are mainly caused by the chipping of corrosion products (NiO) from the surface, as this side of the reactor chamber was exposed to oxidizing conditions. It is possible to prevent such strong oxidation of nickel by using an inert gas atmosphere (e.g., argon or nitrogen) injected into the space between the reactor chamber and the heating module, which should significantly limit material oxidation. On the other hand, the inner side, exposed to molten salts, also shows uneven wear. The material is subject to wear due to mechanisms described for the lance, which causes a reduction in wall thickness. However, 3D scanning takes into account the thickness of corrosion products and partially crystallized salts on the walls. Precisely determining the wear of the material itself is problematic and may not be reliable.

5. Conclusions

Studies of the lance operating for 800 h in the reactor chamber revealed wear on both the internal and external surfaces. This wear is primarily dependent on the direct and continuous contact of the lance with molten salts consisting of $CaCl_2$-CaF_2-CaO-$CaCO_3$. Microscopic observations indicated that the upper parts of the lance (above the molten salts and at the salt/gas boundary) experienced minimal wear (109 μm/year and less).

On the external parts of the lance above the salt level, crystallized salts were observed. Their presence during continuous operation may protect the lance, resulting in low wear. For the lance in continuous contact with molten salts, wall thinning amounted to approximately 22%, with an estimated annual wear of 3.9 mm. The significant wear is attributed to the direct contact of metal with the salts and the turbulent motion caused by gas sparging.

The results of microstructure observations on the cross-sections of the lance demonstrated the presence of intergranular corrosion on both the inner and outer sides of the tubes. Corrosion was observed in the form of voids as well as a continuous network along the grain boundaries. The corrosion primarily spreads perpendicularly to the contact surface with the corrosive environment, penetrating to significant depths and locally reaching hundreds of micrometers into the material.

X-ray analysis enabled the assessment of the entire element without the need for its destruction. Discontinuities in the reactor chamber material, caused by incomplete fusion during welding, were identified. The reconstruction of X-ray images into a 3D model allowed for precise localization of these discontinuities. They were only visible on the inner sides of the reactor chamber, where they could not be welded. The external welds were found to be permanent and continuous.

Computed tomography enabled multi-plane analysis of the reactor chamber, revealing additional imperfections from the construction stage (connection of the reactor bottom with the reactor tube). Based on CT results, uneven wear of individual elements on both the inner and outer surfaces was noted. The distribution of the reactor chamber wall thickness indicates thickening in many places (presence of corrosion products) as well as thinning (pitting). Unfortunately, due to the large noise during the analysis, the result of the most complex component, R3, was difficult to interpret.

The 3D scanning served as a complementary analysis and allowed for the assessment of wear on the component most exposed to the corrosive effects of molten salts (R3). This reactor zone exhibited significant material loss on both the external and internal sides. On the external side, this was caused by the oxidizing atmosphere and the formation of nickel oxide, whose layers flaked off the reactor.

The change in the thickness of the inner part of the reactor chamber is caused by material corrosion, with a mechanism analogous to that observed in the lance. This conclusion is reached based on an analysis of part of the results (microstructural studies) and the assumption that both components (reactor and lance) exhibit the same behavior. This is justified when considering that both components were made of the same material and operated at the same temperature and in the presence of the same molten salts.

The outer part of the reactor, exposed to high temperatures and oxidizing conditions, requires inert gas protection to limit nickel oxidation. This is important for the external welds, which are a key place to ensure the reactor chamber is leak-proof.

The condition of the lance used for N_2-CO_2 gas sparging requires monitoring and potential replacement during long-term tests. An alternative is to use a lance with thicker walls. While this would not eliminate wear, it would reduce the frequency of replacements.

The consumption of Ni 200 Alloy found in this study is acceptable for the operation of the lance and reactor chamber. However, it seems that the wear rate found will be too high for the stable operation of more complex industrial plant components (such as molten salt pumps).

Author Contributions: Conceptualization, P.P., S.P. and K.Ż.; methodology, P.P., M.B. and K.Ż.; software, P.P., M.B., K.Ż., B.L.-M. and J.J.-L.; validation, P.P. and K.Ż.; formal analysis, P.P., S.P., M.B., K.Ż. and B.L.-M.; investigation, P.P., M.B., K.Ż., B.L.-M. and J.J.-L.; data curation, P.P., S.P., M.B., K.Ż., B.L.-M. and J.J.-L.; writing—original draft preparation, P.P., M.B. and K.Ż; writing—review and editing, P.P., M.B. and K.Ż.; visualization, P.P., M.B., K.Ż., B.L.-M. and J.J.-L. All authors have read and agreed to the published version of the manuscript.

Funding: The research leading to these results has received funding from the Norway Grants 2014–2021 via the National Centre for Research and Development. The results are part of the Polish–Norwegian project: Carbon Capture in Molten Salts—Prototype, acronym CCMS-P, registration number NOR/POLNORCCS/CCMS-P/0004/2019-00.

Institutional Review Board Statement: Not applicable.

Informed Consent Statement: Not applicable.

Data Availability Statement: The original contributions presented in this study are included in the article. Further inquiries can be directed to the corresponding author.

Conflicts of Interest: The authors declare no conflicts of interest.

References

1. Seneviratne, S.I.; Donat, M.G.; Pitman, A.J.; Knutti, R.; Wilby, R.L. Allowable CO_2 Emissions Based on Regional and Impact-Related Climate Targets. *Nature* **2016**, *529*, 477–483. [CrossRef]
2. VijayaVenkataRaman, S.; Iniyan, S.; Goic, R. A Review of Climate Change, Mitigation and Adaptation. *Renew. Sustain. Energy Rev.* **2012**, *16*, 878–897. [CrossRef]
3. Rahmstorf, S.; Coumou, D. Increase of Extreme Events in a Warming World. *Proc. Natl. Acad. Sci. USA* **2011**, *108*, 17905–17909. [CrossRef]
4. Newell, R.G. *Literature Review of Recent Trends and Future Prospects for Innovation in Climate Change Mitigation*; OECD Environment Working Papers; OECD: Paris, France, 2009; Volume 9.
5. Madejski, P.; Chmiel, K.; Subramanian, N.; Kuś, T. Methods and Techniques for CO_2 Capture: Review of Potential Solutions and Applications in Modern Energy Technologies. *Energies* **2022**, *15*, 887. [CrossRef]
6. Nord, L.O.; Bolland, O. *Carbon Dioxide Emission Management in Power Generation*, 1st ed.; Wiley: Hoboken, NJ, USA, 2020; ISBN 978-3-527-34753-7.
7. Global Status of ccs Report 2023. Available online: https://status23.globalccsinstitute.com/ (accessed on 7 November 2024).
8. Wilberforce, T.; Baroutaji, A.; Soudan, B.; Al-Alami, A.H.; Olabi, A.G. Outlook of carbon capture technology and challenges. *Sci. Total Environ.* **2019**, *657*, 56–72. [CrossRef]
9. Chao, C.; Deng, Y.; Dewil, R.; Baeyens, J.; Fan, X. Post-combustion carbon capture. *Renew. Sustain. Energy Rev.* **2021**, *138*, 110490. [CrossRef]
10. Lian, S.; Song, C.; Liu, Q.; Duan, E.; Ren, H.; Kitamura, Y. Recent advances in ionic liquids-based hybrid processes for CO_2 capture and utilization. *J. Environ. Sci.* **2021**, *99*, 281–295. [CrossRef]
11. Bhown, A.S.; Freeman, B.C. Analysis and status of post-combustion carbon dioxide capture technologies. *Environ. Sci. Technol.* **2011**, *45*, 8624–8632. [CrossRef]
12. Hills, T.; Leeson, D.; Florin, N.; Fennell, P. Carbon capture in the cement industry: Technologies, progress, and retrofitting. *Environ. Sci. Technol.* **2016**, *50*, 368–377. [CrossRef]
13. Shimizu, T.; Hirama, T.; Hosoda, H.; Kitano, K.; Inagaki, M.; Tejima, K. A Twin Fluid-Bed Reactor for Removal of CO_2 from Combustion Processes. *Chem. Eng. Res. Des.* **1999**, *77*, 62–68. [CrossRef]
14. Wang, N.; Feng, Y.; Guo, X. Atomistic Mechanisms Study of the Carbonation Reaction of CaO for High-Temperature CO_2 Capture. *Appl. Surf. Sci.* **2020**, *532*, 147425. [CrossRef]
15. Dean, C.C.; Blamey, J.; Florin, N.H.; Al-Jeboori, M.J.; Fennell, P.S. The Calcium Looping Cycle for CO_2 Capture from Power Generation, Cement Manufacture and Hydrogen Production. *Chem. Eng. Res. Des.* **2011**, *89*, 836–855. [CrossRef]
16. Blamey, J.; Anthony, E.J.; Wang, J.; Fennell, P.S. The Calcium Looping Cycle for Large-Scale CO_2 Capture. *Prog. Energy Combust. Sci.* **2010**, *36*, 260–279. [CrossRef]
17. HSC Chemistry v 7.0, Outotec Research. Available online: https://www.sciencesoftware.cz/userFiles/letaky/outotec/hsc-brochure.pdf (accessed on 7 November 2024).
18. Olsen, E.; Tomkute, V. Carbon Capture in Molten Salts. *Energy Sci. Eng.* **2013**, *1*, 144–150. [CrossRef]
19. Hall, M.; Rigby, S.; Chen, G. Advancing carbon dioxide capture: Investigation into the kinetics and efficiency of absorption in molten calcium oxide–chloride. *RSC Sustain.* 2025. [CrossRef]
20. Tomkute, V.; Solheim, A.; Olsen, E. Investigation of High-Temperature CO_2 Capture by CaO in $CaCl_2$ Molten Salt. *Energy Fuels* **2013**, *27*, 5373–5379. [CrossRef]
21. Tomkute, V.; Solheim, A.; Sakirzanovas, S.; Olsen, E. Reactivity of CaO with CO_2 in Molten CaF_2–NaF: Formation and Decomposition of Carbonates. *Energy Sci. Eng.* **2016**, *4*, 205–216. [CrossRef]
22. Nygård, H.S.; Ruud, N.A.R.; Olsen, E. Investigation of Sulfation of CaO and $CaCO_3$ in Eutectic CaF_2–$CaCl_2$. *Energy Fuels* **2022**, *36*, 6343–6352. [CrossRef]
23. Olsen, E.; Hansen, M.; S. Nygård, H.; Hydrolysis of Molten $CaCl_2$-CaF_2 with Additions of CaO. *AIMS Energy* **2017**, *5*, 873–886. [CrossRef]
24. Tomkute, V.; Solheim, A.; Olsen, E. CO_2 Capture by CaO in Molten CaF_2 –$CaCl_2$: Optimization of the Process and Cyclability of CO_2 Capture. *Energy Fuels* **2014**, *28*, 5345–5353. [CrossRef]
25. Nygård, H.S.; Hansen, M.; Alhaj-Saleh, Y.; Palimąka, P.; Pietrzyk, S.; Olsen, E. Experimental Evaluation of Chemical Systems for CO_2 Capture by CaO in Eutectic CaF_2-$CaCl_2$. *AIMS Energy* **2019**, *7*, 619–633. [CrossRef]
26. Pietrzyk, S. Device for Removing CO2 from Exhaust Gases. Patent P.444465, 21 October 2024.
27. Pietrzyk, S.; Palimaka, P. The Prototype of Reactor for Carbon Capture in Molten Salts. *EPSTEM* **2023**, *24*, 165–170. [CrossRef]
28. Pietrzyk, S.; Palimąka, P. *Carbon Capture in Molten Salts—Prototype of the CCMS Reactor*; Wydawnictwo Naukowe TYGIEL: Lublin, Poland, 2023; pp. 18–39.
29. Raiman, S.S.; Mayes, R.T.; Kurley, J.M.; Parrish, R.; Vogli, E. Amorphous and Partially-Amorphous Metal Coatings for Corrosion Resistance in Molten Chloride Salt. *Sol. Energy Mater. Sol. Cells* **2019**, *201*, 110028. [CrossRef]
30. Phillips, W.; Merwin, A.; Chidambaram, D. On the Corrosion Performance of Monel 400 in Molten LiCl-Li_2O-Li at 923 K. *Met. Mater Trans A* **2018**, *49*, 2384–2392. [CrossRef]
31. Phillips, W.; Karmiol, Z.; Chidambaram, D. Effect of Metallic Li on the Corrosion Behavior of Inconel 625 in Molten LiCl-Li_2 O-Li. *J. Electrochem. Soc.* **2019**, *166*, C162–C168. [CrossRef]

32. Zheng, G.; Sridharan, K. Corrosion of Structural Alloys in High-Temperature Molten Fluoride Salts for Applications in Molten Salt Reactors. *JOM* **2018**, *70*, 1535–1541. [CrossRef]
33. Olson, L.C.; Ambrosek, J.W.; Sridharan, K.; Anderson, M.H.; Allen, T.R. Materials Corrosion in Molten LiF–NaF–KF Salt. *J. Fluor. Chem.* **2009**, *130*, 67–73. [CrossRef]
34. Vignarooban, K.; Pugazhendhi, P.; Tucker, C.; Gervasio, D.; Kannan, A.M. Corrosion Resistance of Hastelloys in Molten Metal-Chloride Heat-Transfer Fluids for Concentrating Solar Power Applications. *Sol. Energy* **2014**, *103*, 62–69. [CrossRef]
35. Ouyang, F.-Y.; Chang, C.-H.; Kai, J.-J. Long-Term Corrosion Behaviors of Hastelloy-N and Hastelloy-B3 in Moisture-Containing Molten FLiNaK Salt Environments. *J. Nucl. Mater.* **2014**, *446*, 81–89. [CrossRef]
36. Ye, X.-X.; Ai, H.; Guo, Z.; Huang, H.; Jiang, L.; Wang, J.; Li, Z.; Zhou, X. The High-Temperature Corrosion of Hastelloy N Alloy (UNS N10003) in Molten Fluoride Salts Analysed by STXM, XAS, XRD, SEM, EPMA, TEM/EDS. *Corros. Sci.* **2016**, *106*, 249–259. [CrossRef]
37. *ASTM E1742*; Standard Practice for Radiographic Examination. ASTM: West Conshohocken, PA, USA, 2024.
38. Brabant, L.; Pauwels, E.; Dierick, M.; Van Loo, D.; Boone, M.A.; Van Hoorebeke, L. A Novel Beam Hardening Correction Method Requiring No Prior Knowledge, Incorporated in an Iterative Reconstruction Algorithm. *NDT E Int.* **2012**, *51*, 68–73. [CrossRef]
39. Nguyen, V.; Sanctorum, J.G.; Van Wassenbergh, S.; Dirckx, J.J.J.; Sijbers, J.; De Beenhouwer, J. Geometry Calibration of a Modular Stereo Cone-Beam X-Ray CT System. *J. Imaging* **2021**, *7*, 54. [CrossRef] [PubMed]
40. Du Plessis, A.; Boshoff, W.P. A Review of X-Ray Computed Tomography of Concrete and Asphalt Construction Materials. *Constr. Build. Mater.* **2019**, *199*, 637–651. [CrossRef]
41. Liu, J.; Ba, M.; Du, Y.; He, Z.; Chen, J. Effects of Chloride Ions on Carbonation Rate of Hardened Cement Paste by X-Ray CT Techniques. *Constr. Build. Mater.* **2016**, *122*, 619–627. [CrossRef]
42. Kaczmarczyk, G.P.; Kinasz, R.; Bilozir, V.; Bidenko, I. Application of X-Ray Computed Tomography to Verify Bond Failures Mechanism of Fiber-Reinforced Fine-Grain Concrete. *Materials* **2022**, *15*, 2193. [CrossRef]
43. *VDI 2634 Part 3*; Optical 3D-Measuring Systems—Multiple View Systems Based on Area Scanning. VDI: Düsseldorf, Germany, 2008.
44. *ISO 10360-8*; Geometrical Product Specifications (GPS)—Acceptance and Reverification Tests for Coordinate Measuring Systems (CMS). ISO: Geneva, Switzerland, 2013.
45. Ozeryanaya, I.N. Corrosion of Metals by Molten Salts in Heat-Treatment Processes. *Met. Sci. Heat Treat* **1985**, *27*, 184–188. [CrossRef]
46. Li, Y.S.; Niu, Y.; Wu, W.T. Accelerated Corrosion of Pure Fe, Ni, Cr and Several Fe-Based Alloys Induced by ZnCl2–KCl at 450 °C in Oxidizing Environment. *Mater. Sci. Eng. A* **2003**, *345*, 64–71. [CrossRef]
47. Ruh, A.; Spiegel, M. Thermodynamic and Kinetic Consideration on the Corrosion of Fe, Ni and Cr beneath a Molten KCl–$ZnCl_2$ Mixture. *Corros. Sci.* **2006**, *48*, 679–695. [CrossRef]
48. Cha, S.C.; Spiegel, M. Local Reactions of KCl Particles with Iron, Nickel and Chromium Surfaces. *Mater. Corros.* **2006**, *57*, 159–164. [CrossRef]
49. Kumar, K.; Gesualdi, J.; Smith, N.D.; Kim, H. Influence of Gaseous Atmosphere on Electrochemical Behavior of Nickel Alloys in LiCl-KCl-Na2SO4 at 700 °C. *Corros. Sci.* **2018**, *142*, 1–11. [CrossRef]
50. Ronne, A.; He, L.; Dolzhnikov, D.; Xie, Y.; Ge, M.; Halstenberg, P.; Wang, Y.; Manard, B.T.; Xiao, X.; Lee, W.-K.; et al. Revealing 3D Morphological and Chemical Evolution Mechanisms of Metals in Molten Salt by Multimodal Microscopy. *ACS Appl. Mater. Interfaces* **2020**, *12*, 17321–17333. [CrossRef] [PubMed]
51. Guo, S.; Zhang, J.; Wu, W.; Zhou, W. Corrosion in the Molten Fluoride and Chloride Salts and Materials Development for Nuclear Applications. *Prog. Mater. Sci.* **2018**, *97*, 448–487. [CrossRef]
52. Sridharan, K.; Allen, T.R. Corrosion in Molten Salts. In *Molten Salts Chemistry*; Elsevier: Amsterdam, The Netherlands, 2013; pp. 241–267, ISBN 978-0-12-398538-5.
53. Sarvghad, M.; Delkasar Maher, S.; Collard, D.; Tassan, M.; Will, G.; Steinberg, T.A. Materials Compatibility for the next Generation of Concentrated Solar Power Plants. *Energy Storage Mater.* **2018**, *14*, 179–198. [CrossRef]

materials

MDPI

Article

Mechanical Characterization and Computational Analysis of TPU 60A: Integrating Experimental Testing and Simulation for Performance Optimization

Luan Lang [1,2], Rodrigo Antunes [1,2], Thiago Assis Dutra [1,2], Martim Lima de Aguiar [1,2], Nuno Pereira [1,2] and Pedro Dinis Gaspar [1,2,*]

[1] Department of Electromechanical Engineering, University of Beira Interior, Rua Marquês de D'Ávila e Bolama, 6201-001 Covilhã, Portugal; luan.lang@ubi.pt (L.L.); rodrigo.antunes@ubi.pt (R.A.); thiago.dutra@ubi.pt (T.A.D.); martim.aguiar@ubi.pt (M.L.d.A.); nuno.pereira@ubi.pt (N.P.)

[2] C-MAST—Centre for Mechanical and Aerospace Science and Technologies, 6201-001 Covilhã, Portugal

* Correspondence: dinis@ubi.pt

Abstract: This study investigates the mechanical properties of thermoplastic polyurethane (TPU) 60A, which is a flexible material that can be used to produce soft robotic grippers using additive manufacturing. Tensile tests were conducted under ISO 37 and ISO 527 standards to assess the effects of different printing orientations (0°, 45°, −45°, 90°, and quasi-isotropic) and test speeds (2 mm/min, 20 mm/min, and 200 mm/min) on the material's performance. While the printing orientations at 0° and quasi-isotropic provided similar performance, the quasi-isotropic orientation demonstrated the most balanced mechanical behavior, establishing it as the optimal choice for robust and predictable performance, particularly for computational simulations. TPU 60A's flexibility further emphasizes its suitability for handling delicate objects in industrial and agricultural applications, where damage prevention is critical. Computational simulations using the finite element method were conducted. To verify the accuracy of the models, a comparison was made between the average stresses of the tensile test and the computational predictions. The relative errors of force and displacement are lower than 5%. So, the constitutive model can accurately represent the material's mechanical behavior, making it suitable for computational simulations with this material. The analysis of strain rates provided valuable insights into optimizing production processes for enhanced mechanical strength. The study highlights the importance of tailored printing parameters to achieve mechanical uniformity, suggesting improvements such as biaxial testing and G-code optimization for variable thickness deposition. Overall, the research study offers comprehensive guidelines for future design and manufacturing techniques in soft robotics.

Keywords: TPU 60A; material characterization; tensile tests; printing orientations; simulation validation; mechanical properties; ISO 37; ISO 527; additive manufacturing

1. Introduction

Due to technological developments in recent decades, the use of robots and the automation of processes has become trivial, from simple tasks to highly complex operations [1]. Robots have become relevant equipment in factories, providing the opportunity to replace 4D (dull, dirty, dangerous, and dear) tasks performed by humans, implementing improvements, and reducing costs in production processes [2].

One of the robot's possible end effectors is a gripper, for grasping and manipulating objects. Grippers, like robots themselves, are fundamental elements in industry, since manipulating objects with various shapes is one of the most complex challenges in robotics [3]. To handle sensitive products or those with complex geometries, soft grippers made from flexible materials have been developed, allowing considerable deformation to adjust to the product without damaging it.

These grippers allow for superior results when manipulating sensitive and irregular objects, compared to rigid grippers [4]. Until now, the prototyping of soft grippers was predominantly carried out through manual processes, such as silicone molding, due to the lack of accessible and developed technologies for production. However, advances in additive manufacturing and flexible materials technologies have significantly enhanced the customization, adaptability, and complexity of soft grippers. While additive manufacturing is not the sole factor making soft grippers viable, it plays a key role in enabling their tailored designs and broader application. These advancements have made soft grippers an increasingly practical and cost-effective solution for complex tasks, such as safe human interaction and handling fragile objects, by offering flexible and innovative design possibilities [5].

The fragility, sensitivity, stickiness, and slipperiness of most food and agricultural products have driven the development of soft robotic grippers that are deformable, flexible, safe, low-cost, and environmentally friendly. Such a model of robotic gripper has the potential to meet these challenges by offering reliable and adaptable solutions for handling agricultural and food products [6], since the main application of the soft gripper is to hold and move objects safely without causing significant deformation or damage to their delicate surfaces [7].

Fused Filament Fabrication (FFF) is a popular additive manufacturing technique that has gained significant relevance [8,9]. According to Sukindar et al. [9], FFF was initially developed to produce and create prototypes. Nowadays, it is adopted in different fields, such as architecture, manufacturing, and medical applications. FFF is a type of additive manufacturing process that uses thermoplastic materials to produce three-dimensional objects. The fundamental principle behind FFF involves the controlled extrusion of a thermoplastic filament through a heated nozzle.

Antunes et al. [10,11] focused on optimizing Fin Ray Effect (FRE)-inspired grippers to enhance their adaptability and grasping strength. This design is optimal for production using additive manufacturing, namely Fused Filament Fabrication (FFF). They employed computational design and simulation to improve the performance of FRE soft robotic fingers, and compared two FRE-based soft robotic fingers, utilizing Ansys software to assess how different designs impact gripping efficacy.

The selection of TPU 60A for this study is justified by its marked advantages over comparable materials, such as TPU 95A. TPU 60A exhibits superior elasticity and softness, characteristics that make it particularly well-suited for applications demanding delicate handling, such as soft robotic grippers designed for interaction with fragile or irregularly shaped objects. In contrast, TPU 95A, while offering greater strength and rigidity, may apply excessive pressure, thereby increasing the risk of damage in such sensitive applications. This comparison highlights the appropriateness of TPU 60A for soft robotics, where flexibility and adaptability are critical, and aligns the material selection with the study's objective of optimizing performance in specialized contexts.

The use of Thermoplastic Polyurethane (TPU) 60A in soft robotics requires an understanding of its mechanical behavior under various conditions. However, the existing literature and supplier-provided data lack comprehensive experimental characterization of TPU 60A across different test speeds and printing orientations. There is an absence of data detailing how varying the test speeds and orientations affect the material's mechanical properties, such as elongation, strength, and stress–strain response. This gap hinders the development of accurate constitutive models for computational simulations. To address this gap, an experimental investigation of TPU 60A was conducted. Tensile tests were performed at three test speeds (2 mm/min, 20 mm/min, and 200 mm/min) and four printing orientations (0°, 45°/−45°, 90°, and quasi-isotropic configurations). By systematically analyzing the effects of these variables, a detailed characterization of the material's mechanical properties is provided, which is necessary for predictive modeling and simulation purposes.

The experimental data reveal the influence of different test speeds and printing orientations on the mechanical behavior of TPU 60A, including the identification of ideal stress–strain curves for each configuration. For constitutive modeling in simulations, the most suitable model to represent TPU 60A's behavior was determined. This model can be used in simulations, using software such as Ansys (version 2023 R1), which will be used to simulate the mechanical characteristics of soft grippers made in TPU 60A. By inputting the experimentally obtained stress–strain data into Ansys, the software's curve-fitting tool was used to generate the required parameters for the selected constitutive model. This ensured that the simulated material behavior closely matched the experimental results. Providing these detailed parameters enables accurate replication of the material's response in computational analyses, easing the adoption of TPU 60A in various engineering applications where simulations closely mirror real-world performance.

The key contribution of this work is the development and validation of a constitutive model for TPU 60A that represents its mechanical behavior when testing different test speeds and orientations. Integrating this model into Ansys software allowed for performing numerical simulations of soft grippers made from TPU 60A, and comparing the simulation results with experimental tensile test data demonstrated that the model predicts material performance, with relative errors between simulated and experimental results less than 5%, confirming the model's validity. By providing empirical data and validated constitutive models, this work explores the design and optimization of devices using TPU 60A. Simulating the material's behavior fastens the development process, reducing the need for extensive physical prototyping and enabling more efficient iterations. In addition, the authors also believe that this research work contributes to the field of soft robotics—particularly to soft grippers—where material properties play a critical role in the functionality and effectiveness of devices.

2. Materials and Methods

Following the recommendations of the datasheet for Filaflex TPU 60A of the manufacturer Recreus. Elda, Spain [12], the 3D printer must be properly configured, since the material has a shore hardness of 60A, which is very elastic with a stretch ability of 950%.

It is important to note that the manufacturer does not provide specific details regarding the exact printing conditions or methods used to achieve this value. The only information available indicates that the tests were conducted following the DIN 53504-S2 standard [13]. This lack of detailed guidance may explain differences in results, regarding factors such as printing parameters (e.g., nozzle temperature, infill density, or print speed), potential imperfections introduced during the printing process (e.g., voids or layer adhesion issues), and deviations in test conditions (e.g., strain rates or environmental factors) that can significantly impact mechanical properties. Furthermore, the use of additive manufacturing introduces anisotropic properties, which could also contribute to the observed variations in elongation at break.

In the present work, the 3D printer Artillery Sidewinder X1 (Artillery 3D Technology Co., Ltd., Shenzhen, China) was used to manufacture the specimens. It has characteristics that, within its market range, facilitate the printing of flexible polymers, such as TPU. Primarily, a direct extrusion system and polytetrafluoroethylene (PTFE)-lined hotend allow for a straightforward printing of flexible thermoplastics [14].

The tensile tests were carried out using a Shimadzu Autograph AGS-X-50 kN high-performance universal testing machine (Shimadzu Group, Kyoto, Japan). It has a maximum force capacity of 50 kN, an effective test width of 425 mm, and a maximum crosshead displacement of 745 mm. It is equipped with TRAPEZIUM LITE X software (version: 1.0) to control tests in real-time and to acquire data [15].

The open-source software UltiMaker Cura (version 5.0) [16], a Computer-Aided Manufacturing (CAM) software, was used to convert the 3D models into the G-code that guides the printing process. Table 1 presents the printing parameters adopted in the present work.

Table 1. Main specifications used in Cura software (version 5.0) for TPU 60A.

Specification	Value	Specification	Value
Nozzle diameter	0.6 mm	Line width	0.58 mm
Layer height	0.25 mm	Printing temperature	230 °C
Infill density	100%	Printing speed	20 mm/s
Infill pattern	Lines	Retraction distance	4.5 mm
Flow	200%	Retraction speed	25 mm/s
Number of layers	16–ISO-527 8–ISO 37	Infill line directions	[0°; 45°, −45°; 90°; Quasi-Isotropic]

The specimens were designed using SolidWorks software according to ISO 37 and ISO 527 standards [17,18], which were also followed when carrying out the tensile tests.

To investigate the time-dependent mechanical response of TPU 60A, the specimens were tested at three different crosshead test speeds (v): 2 mm/min, 20 mm/min, and 200 mm/min. To verify the influence of printing orientations, four sample models were printed, each containing eight specimens, meaning a total of 32 specimens. The samples were printed in four different orientations: 0° (along the Y direction), 45°, and −45°, 90° (along the X direction), and a quasi-isotropic configuration covering the orientations of 0°, 45°, −45°, and 90°. This procedure was based on studies in the literature, such as those by Naveed et al. [19] and Frenkel et al. [20], who investigated different printing orientations to improve the mechanical behavior of specimens.

To comply with ISO 527 and ISO 37 standards, five specimens from each set of eight were considered valid, respecting the requirement to test a minimum of five specimens for each of the established test orientations [17,18], as seen in Figure 1, where it shows the four different orientations in the software Ultimaker Cura. The variation in printing orientation allowed for the analysis of how this variable affects the quality and accuracy of tensile tests, offering a more detailed and representative view of the material's performance. This helps ensure that the material meets the specifications and requirements of technical standards. To facilitate differentiation, the orientations were categorized by color: blue for the longitudinal [0°], green for the transverse [90°] orientations, red for the quasi-isotropic orientation [0°, 45°, −45°, 90°], and yellow for the [45°, −45°] orientation, as shown in Figure 2. These four printing models follow the definitions in Table 1.

Table 2 summarizes the number of valid specimens and respective test speeds tested for each orientation.

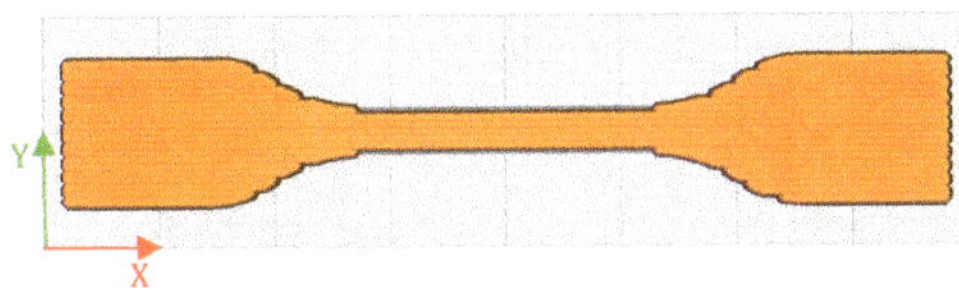

(a) Orientation for the green samples at 0°.

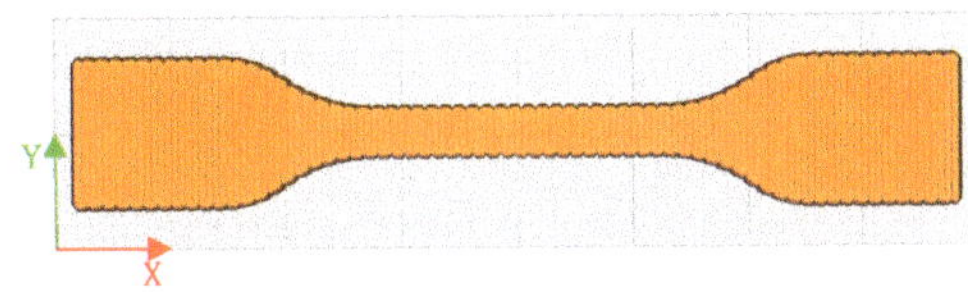

(b) Orientation for the green samples at 90°.

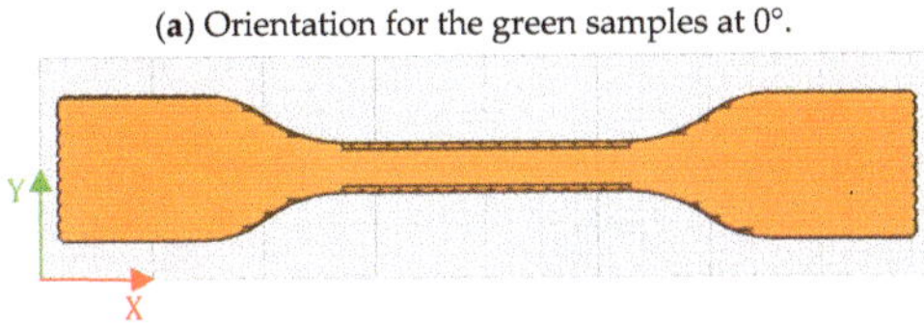

(c) Orientation for the red samples at 0°, 45°, −45°, and 90°.

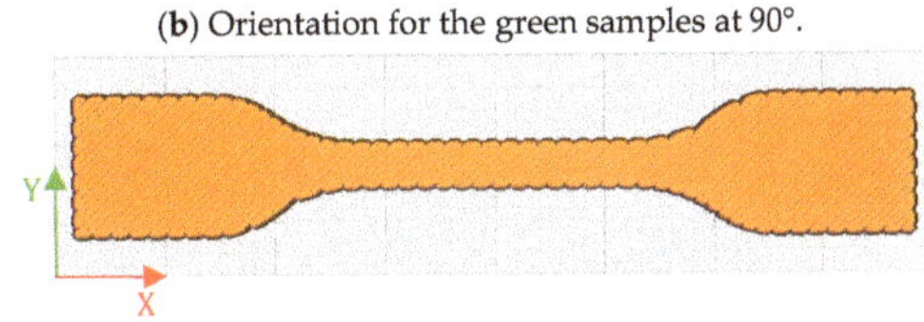

(d) Orientation for the yellow samples at 45° and −45°.

Figure 1. Orientation for the sample ISO 37: (**a**) at 0°; (**b**) at 90°; (**c**) at 0°, 45°, −45°, and 90°; (**d**) at 45° and −45°.

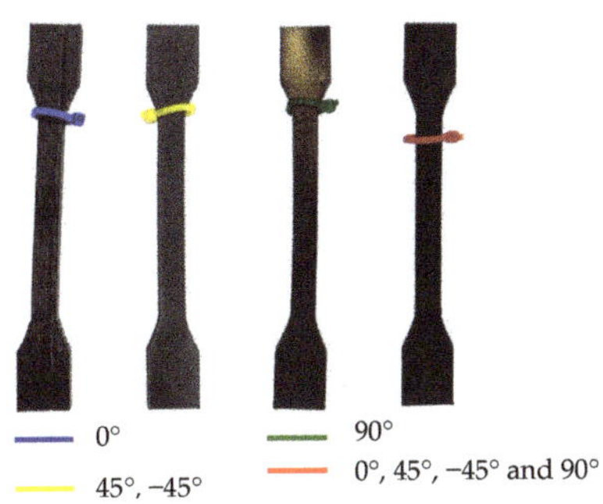

(a) Batch of specimen according to ISO 527, type 1b.

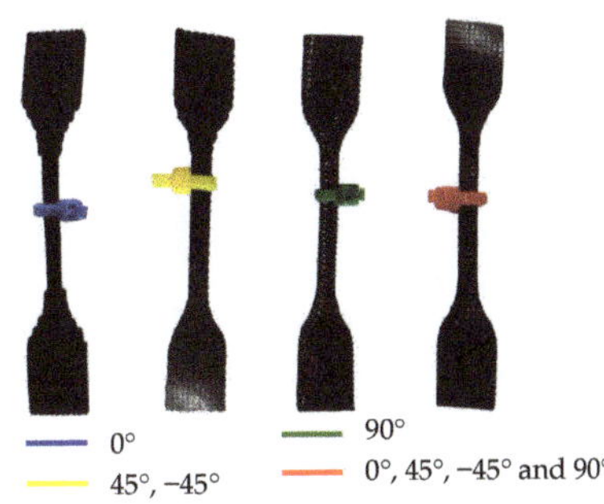

(b) Batch of specimen according to ISO 37, type 2.

Figure 2. Batch of printed samples categorized by color for different printing orientations: (**a**) ISO 527, type 1b.; (**b**) ISO 37, type 2.

Table 2. Summary of the characteristics of the TPU 60A characterization tests.

Tests	Quasi-Isotropic	0°	45° and −45°	90°
Number of valid specimens	13	10	10	10
Nomenclature of the specimens	QI-02-01; QI-02-02; QI-02-03; QI-20-01; QI-20-02; QI-20-03; QI-20-04; QI-20-05; QI-200-01; QI-200-02; QI-200-03; QI-200-04; QI-200-05;	O-20-01; O-20-02; O-20-03; O-20-04; O-20-05; O-200-01; O-200-02; O-200-03; O-200-04; O-200-05;	F-20-01; F-20-02; F-20-03; F-20-04; F-20-05; F-200-01; F-200-02; F-200-03; F-200-04; F-200-05;	N-20-01; N-20-02; N-20-03; N-20-04; N-20-05; N-200-01; N-200-02; N-200-03; N-200-04; N-200-05;
Test speeds (v) tested	2 mm/min; 20 mm/min; 200 mm/min;	20 mm/min; 200 mm/min;	20 mm/min; 200 mm/min;	20 mm/min; 200 mm/min;

The presence of voids in the printed specimens also influences their mechanical properties, leading to premature failure and reduced stiffness. Even though TPU 60A is an elastomeric material, the voids formed during the printing process compromise its resistance. Cross-sectional analysis, as shown in Figure 3, reveals that specimens printed with high-humidity filament exhibit a greater number of voids compared to those printed with low-humidity filament. These voids act as stress concentrators, weakening the material and causing it to fail at lower strains than expected.

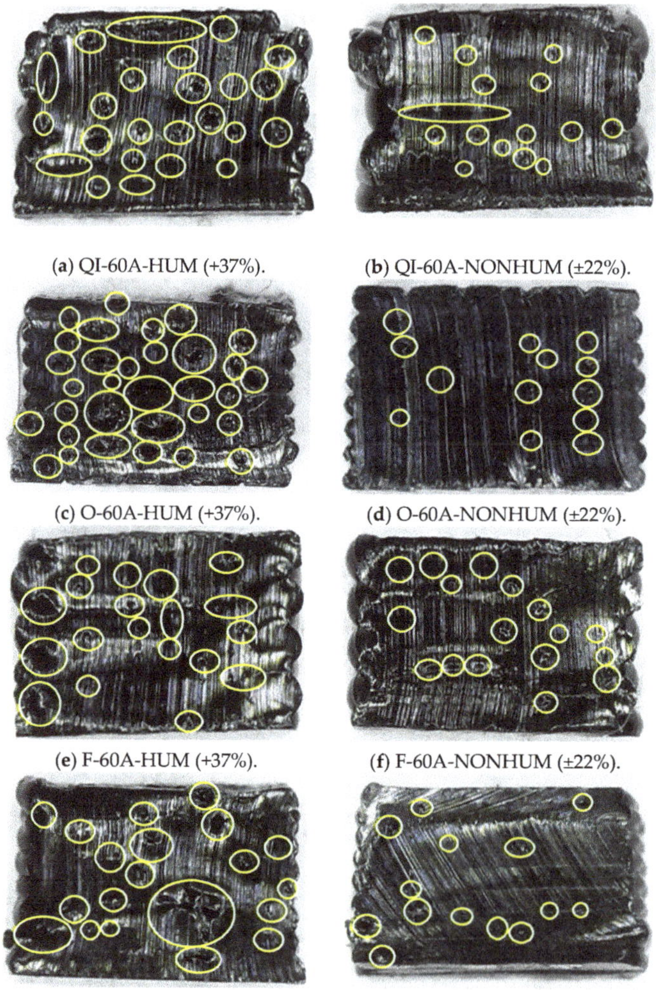

(**a**) QI-60A-HUM (+37%). (**b**) QI-60A-NONHUM (±22%).
(**c**) O-60A-HUM (+37%). (**d**) O-60A-NONHUM (±22%).
(**e**) F-60A-HUM (+37%). (**f**) F-60A-NONHUM (±22%).
(**g**) N-60A-HUM (+37%). (**h**) N-60A-NONHUM (±22%).

Figure 3. Cross-section area of the specimens with high and low humidity levels.

The cross-sectional analysis of the specimens shown in Figure 3 reveals the presence of gaps and voids formed during the printing process. Specimens printed with filament at higher humidity levels (+37%) exhibit a significantly greater number of voids compared to those printed with low humidity filament (±22%). This analysis demonstrates a correlation between filament humidity and void formation in the printed structures. Specifically, samples with higher humidity (Figure 3a,c,e,g) contain markedly more voids and gaps within their microstructure than their dehumidified counterparts (Figure 3b,d,f,h).

The percentage of the void area within the samples was quantified using ImageJ software (version 1.54i) [21], which provides an accurate and efficient method for analyzing particles and void distributions in image data. Table 3 presents a comparison of void percentages for various samples under different humidity conditions, specifically highlighting the significant impact of filament humidity levels on void formation.

Table 3. Comparison between the percentage of voids area inside the samples using ImageJ software.

Samples	Voids Area
QI-60A-HUM (+37%)	12.13%
QI-60A-NONHUM (±22%)	2.80%
O-60A-HUM (+37%)	24.90%
O-60A-NONHUM (±22%)	3.87%
F-60A-HUM (+37%)	9.94%
F-60A- NONHUM (±22%)	3.76%
N-60A-HUM (+37%)	15.12%
N-60A-NONHUM (±22%)	2.98%

Effective control of filament humidity and printing conditions is critical for minimizing void formation and enhancing the mechanical integrity and reliability of printed elastomeric structures. Proper humidity management reduces void density, resulting in improved durability and more consistent mechanical performance. These findings highlight the critical role of filament preparation processes in additive manufacturing, particularly for materials like TPU 60A, and provide practical insights for optimizing production workflows. By ensuring reduced void formation, the material's performance becomes more reliable for applications requiring precise mechanical properties, such as soft robotics and other specialized fields.

The results reveal a consistent trend: samples printed with higher humidity levels (+37%) exhibit significantly greater void percentages compared to their low-humidity counterparts (±22%). For instance, the quasi-isotropic configuration (QI-60A-HUM) exhibits a void percentage of 12.13%, which is more than four times higher than its low-humidity counterpart, QI-60A-NONHUM, at 2.80%. Similarly, the 0° orientation (O-60A-HUM) shows the highest void percentage among all samples (24.90%), compared to only 3.87% for its low-humidity equivalent. Similar patterns are observed for the 45°/−45° (F-60A) and 90° (N-60A) orientations, where high-humidity samples have void percentages of 9.94% and 15.12%, respectively, compared to 3.76% and 2.98% in their low-humidity counterparts.

The increased void density in high-humidity specimens can be attributed to moisture absorption by the filament. During extrusion, the absorbed moisture evaporates, creating bubbles that lead to voids within the material. These voids compromise the overall density and mechanical integrity of the printed samples. Conversely, samples printed with lower humidity levels demonstrate significantly fewer voids, reinforcing the importance of proper filament drying and controlled environmental conditions in reducing void formation.

With this analysis, this test highlights the importance of dehumidifying filament both before and during the 3D printing process. Dehumidification reduces moisture content, thereby minimizing void formation and ensuring more uniform and dense material deposition. By eliminating these defects, the mechanical performance of printed components is significantly enhanced, resulting in increased strength, superior surface finish, and greater reliability. This is particularly important for applications that demand precise and consistent material properties.

Thus, specifically for this work, the TPU 60A filament was dried using a desiccant dryer, and the filament was kept inside the dryer throughout the printing process to maintain the lowest possible humidity levels. During the tests, the ambient humidity level was controlled at approximately 22%.

3. Results

The experimental characterization of tensile mechanical properties was carried out according to ISO 37 and ISO 527 standards [17,18]. During the tests, the data were acquired at a rate of 1 Hz for a crosshead test speed (v) of 2 mm/min, and at a rate of 10 Hz for crosshead displacements of 20 mm/min and 200 mm/min. The data collected during the tests were post-processed using routines developed for this study. For the sake of further reproducibility by third parties, these routines are available on GitHub [22].

3.1. Printing Orientation Quasi-Isotropic

First, three quasi-isotropic specimens, i.e., those with layers printed at [0°/45°/−45°/90°], were tested until failure at a test speed (v) of 2 mm/min. The engineering stress–strain curves obtained throughout the tests are presented in Figure 4, with the peak stresses indicated by the red markers. From Figure 4, it can be verified a good consistency between the mechanical response until the specimens reach strain levels of 300%. From this point, small deviations are noticed, which can be attributed to imperfections in the specimens that were induced by the manufacturing process. After testing the specimens at a test speed (v) of 2 mm/min, considerable stress relaxation effects were observed, which directly affect the mechanical response of the material. For this reason, this test speed was not adopted throughout the other tests.

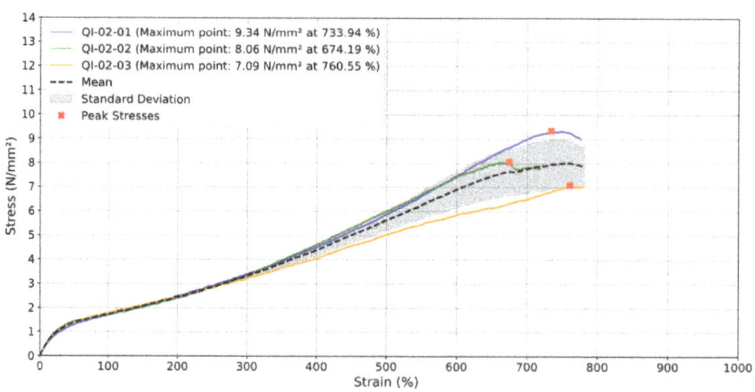

Figure 4. Tensile engineering stress–strain curves for quasi-isotropic samples tested at 2 mm/min.

Additionally, the stress–strain analysis of quasi-isotropic TPU 60A specimens tested at a test speed (v) of 20 mm/min reveals consistent mechanical behavior during the elastic phase, with near-linear stress increases up to around 600% strain and minimal standard deviation, indicating effective load distribution within the material's internal structure. Beyond this point, the material enters the plastic regime, where notable variability emerges among the samples. Maximum stress values range from 7.39 N/mm^2 (QI-20-01) to 8.52 N/mm^2 (QI-20-04), with differences in strain at rupture, suggesting the presence of internal defects like voids or micro-cracks, likely introduced during the 3D printing process. QI-20-04 and QI-20-05, exhibiting the highest stress and strain values, indicate a more robust material structure, while earlier ruptures in QI-20-01 and QI-20-02 point to potential flaws in printing conditions.

The average results for the quasi-isotropic specimens at 20 mm/min are shown in Figure 5, with stress increasing steadily up to approximately 7.81 N/mm^2 around 700% strain. The standard deviation remains low throughout most of the test but increases slightly near maximum stress values, reflecting greater variability as the material reaches its deformation limit. In this analysis, QI-20-01 shows a consistent stress increase up to the point of maximum deformation at around 600% strain. QI-20-02 and QI-20-03 achieve higher maximum stress values, suggesting enhanced material strength before rupture. QI-20-04 and QI-20-05 stand out with the highest stress and strain values, indicating a more robust material structure.

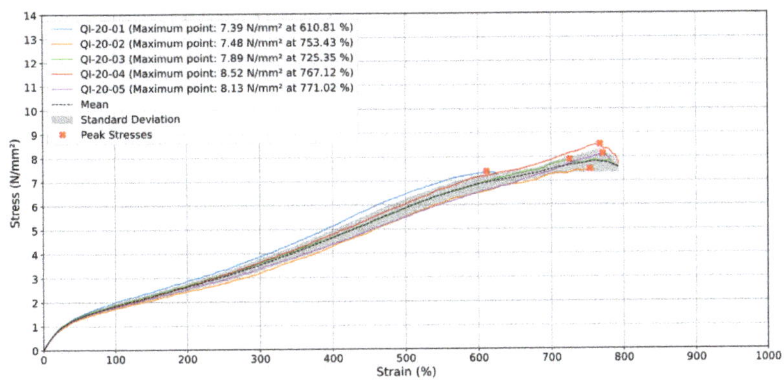

Figure 5. Stress (N/mm^2) vs. Strain (%) for quasi-isotropic samples at 20 mm/min.

The stress–strain analysis for tests conducted at a test speed (v) of 200 mm/min, shown in Figure 6, reveals higher stress levels achieved at lower strain intervals compared to the 20 mm/min tests, indicating the influence of viscoelastic effects at higher testing speeds. The maximum average stress reached approximately 9.78 N/mm^2 at around 727% strain. The standard deviation initially remains low but starts to increase beyond 600% strain, similar to the trend observed in the previous test, indicating greater dispersion of stress values as the deformation nears its upper limit.

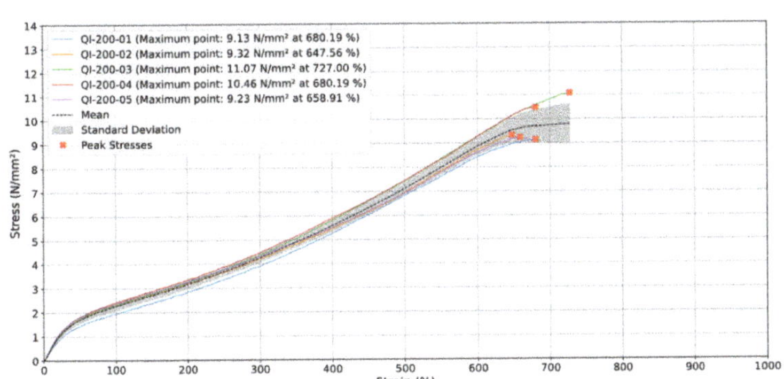

Figure 6. Stress (N/mm^2) vs. Strain (%) for quasi-isotropic samples at 200 mm/min.

The results for this orientation exhibit consistent performance, with most rupture points occurring between 600% and 700% strain, except for one sample breaking between 700% and 800%.

The analysis across all test speeds highlights the significant impact of testing speed on tensile strength which can be seen in Figure 7. The green curve, representing the 200 mm/min rate, shows a higher maximum stress compared to the orange curve (20 mm/min) and the blue curve (2 mm/min), illustrating that lower test speeds result in lower stress. For all test speeds, the material displays a consistent increase in stress up to about 600% strain, followed by a gradual decrease in the stress growth rate as the deformation progresses. Beyond 700–800% strain, the stress values begin to decline, marking the onset of specimen failure.

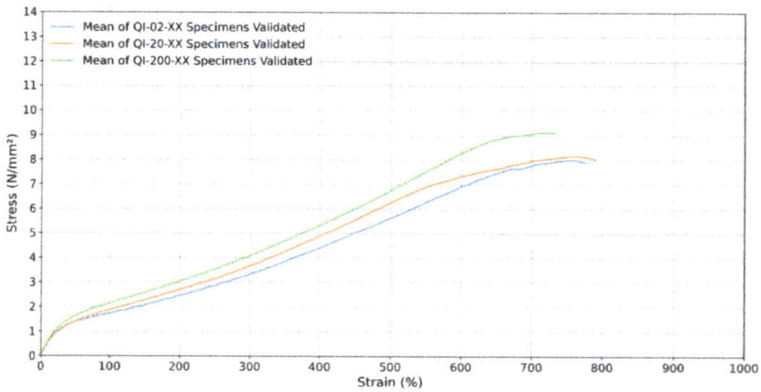

Figure 7. Stress (N/mm^2) vs. Strain (%) for the averages of the quasi-isotropic specimens at 2 mm/min (blue), 20 mm/min (orange), and 200 mm/min (green).

The tests conducted at a rate (v) of 200 mm/min resulted in the highest stresses throughout the test, which suggests that increasing the traction speed can increase the mechanical resistance of TPU 60A. This can be explained by the fact that at higher test speeds, the material has less time to accommodate plastic deformation, resulting in more elastic and resistant behavior. In the case of the (v) of 2 mm/min test speed, on the other hand, it led to a more moderate response, with the material reaching lower stress levels for the same deformation. This may be related to a higher relaxation rate of the material when subjected to a lower loading rate, allowing TPU 60A to deform in a more controlled manner and therefore generate lower stresses. For this reason, the tests in terms of test speed (v) of 20 mm/min show an excellent balance between the two test speeds tested.

The mechanical performance of TPU 60A is strongly influenced by testing speed, as shown in Figure 7, across the three strain rates. At higher strain rates, the material demonstrates increased stiffness, enabling robotic grippers to handle heavier loads and resist sudden impacts, which is essential for tasks requiring strength and durability. Conversely, at lower strain rates, TPU 60A retains sufficient flexibility, making it suitable for the delicate handling of fragile or irregularly shaped objects without causing damage. This balance between stiffness and flexibility is particularly important for optimizing robotic gripper performance across a wide range of applications. The viscoelastic properties of TPU 60A allow designers to adapt gripper designs to specific requirements, such as accommodating varying loading speeds or ensuring precise manipulation of sensitive items. By tailoring gripper configurations to leverage these properties, the material's efficiency and adaptability in industrial, agricultural, and other specialized settings are significantly enhanced. These insights underscore TPU 60A's versatility and its critical role in advancing soft robotics technologies.

3.2. Printing Orientation 0°

For the specimens printed at a 0° orientation and tested at a test speed (v) of 20 mm/min, the average stress curve, seen in Figure 8, follows a relatively constant upward trend until around 700% deformation, where the stress reaches its maximum value close to 12 N/mm^2. This indicates that the material has an elastic and linear behavior during the initial deformation phase. After the maximum stress point, there is a significant decrease in its value, with sharp drops, suggesting a phase of partial rupture and progressive failure of the material, where some strands of deposited material or internal layers may have begun to break down.

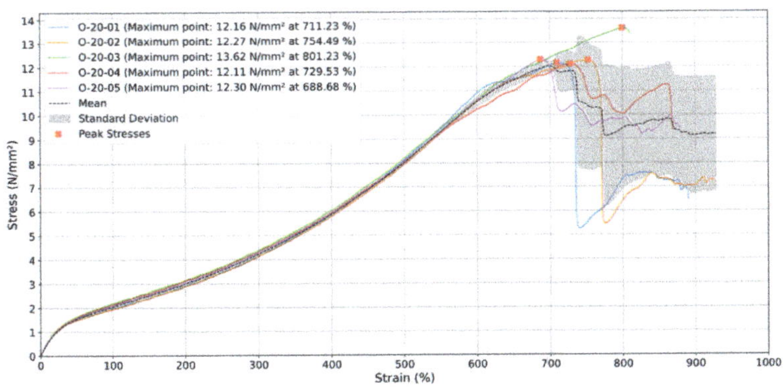

Figure 8. Stress (N/mm^2) vs. Strain (%) for specimens with orientation of 0° at 20 mm/min.

In terms of behavior after the stress peak, i.e., after 700% deformation, the curve shows a series of fluctuations, with abrupt drops and subsequent increases, which may indicate the rupture of different layers of material. During this phase, the material still retains some of its structural integrity, but the fluctuations indicate intermittent failures, probably due to the orientation of the layers in relation to traction. In addition, the standard deviation, represented by the shaded area, increases significantly after the point of maximum tension, suggesting greater variability between the specimens tested. This phenomenon can be attributed to the increased fragility of the material as deformation progresses, causing the internal layers to fail unevenly. The presence of voids within the specimen collaborates with this non-uniform failure, further compromising the material's integrity, such as what happened with the tests O-20-01 and O-20-02.

Along the 0° orientation the printed layers are aligned to the direction of the load, and it is then expected to make the material stronger to a certain extent, but also more susceptible to failure due to delamination between layers. Regarding the final deformation phase occurring after 700%, the sharp drops indicate a widespread material failure, where multiple rupture points seem to occur in sequence. The stress curve tends to drop sharply, and the standard deviation expands further, indicating the unpredictability of the mechanical behavior at this extreme deformation stage. The post-peak behavior reveals the gradual rupture of the deposited material strands, with the material continuing to give way under the load until complete failure.

Figure 9 shows that the 0° specimens subjected to a test speed (v) of 200 mm/min exhibit predictable elastic behavior up to around 700% strain, with stress close to 10 N/mm^2.

A very similar range of results obtained with the 0° specimens tested at a test speed (v) of 20 mm/min.

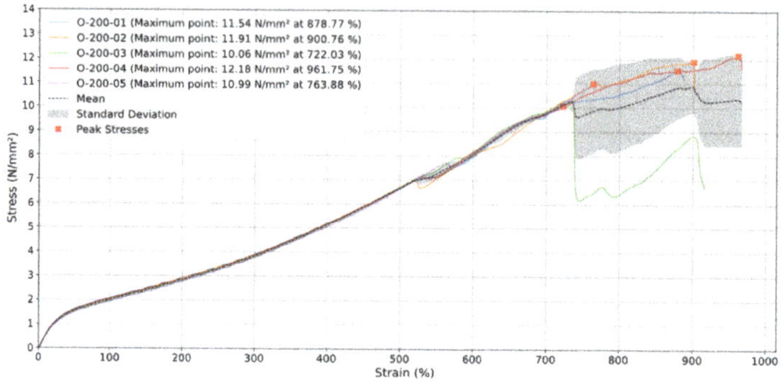

Figure 9. Stress (N/mm^2) vs. Strain (%) for specimens with orientation of 0° at 200 mm/min.

After this point, the material enters a progressive failure phase, with fluctuations in stress and a considerable increase in standard deviation, reaching a maximum stress of 10.89 N/mm^2. This suggests variability between the specimens regarding failure behavior. For example, in specimen O-200-03, a minor failure occurs between 700% and 800% deformation, after which the test continues until complete failure is reached at 900% deformation. This happens because the deviation observed in sample O-200-03 likely stems from early, partial failures within its deposited material strands. The higher test speed seems to contribute to a slightly better tensile performance of the material, although the final failure phase continues to show significant variations in the mechanical response of the specimens.

In this case, the localized, partial failures in the specimen's deposited material strands can dominate the expected viscoelastic response. Typically, lower speeds allow for more time-dependent deformation and relaxation, which would reduce the overall stress. However, when certain strands fail prematurely, the load redistributes irregularly, forcing the remaining intact strands to carry more stress and exhibit more "elastic-like" behavior. This disruption masks the usual viscoelastic effects and results in unexpectedly higher stress values at lower testing speeds.

It is also worth noting that, in these cases, despite the partial rupture of the deposited material strands of the specimens indicated in the graphs by the drops in stress values, the test continues until the final rupture of all the deposited material strands.

3.3. *Printing Orientation 45° and −45°*

In tests involving specimens oriented at 45° and −45° with a test speed (v) of 20 mm/min (Figure 10), failure occurred in two different zones: specimens F-20-01, F-20-02, and F-20-05 broke between 500% and 600% strain. F-20-03 and F-20-04 ruptured between 600% and 700%, and once after exceeding 700% strain. Throughout these tests, stress values remained relatively constant, ranging from 5.7 to 6.17 N/mm^2. The average stress curve exhibited a linear increase up to approximately 550% strain, reaching a peak of around 5.72 N/mm^2, followed by a stabilization phase where stress levels stayed between 5 and 6 N/mm^2 up to 700% strain. The low standard deviation indicates high consistency among the specimens. Unlike 0° orientations, the 45° and −45° specimens did not display a sharp stress drop after reaching maximum stress. Instead, the inclined orientation of the material strands allowed for a more gradual load distribution and controlled deformation, resulting in sustained strength after the stress peak.

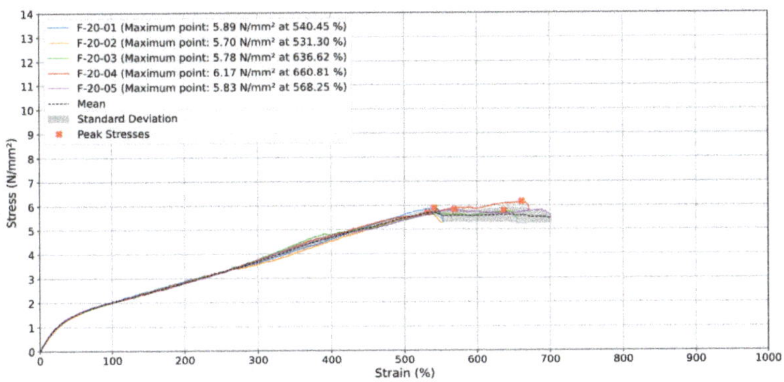

Figure 10. Stress (N/mm^2) vs. Strain (%) for specimens with orientation of 45° and −45° at 20 mm/min.

When tested at a higher test speed (v) of 200 mm/min (Figure 11), three specimens (F-200-02, F-200-03 (around 500%), and F-200-04) ruptured between 500% and 600% strain with stress values between 6.6 N/mm^2 and 7.59 N/mm^2. The remaining two specimens broke after 700% strain, exhibiting a significant decrease in maximum stress to 5.83 N/mm^2 and 6.17 N/mm^2, respectively. This behavior highlights the influence of strain rate on the failure characteristics of 45° and −45°-oriented specimens, showing their consistent performance at lower rates and variable outcomes at higher rates.

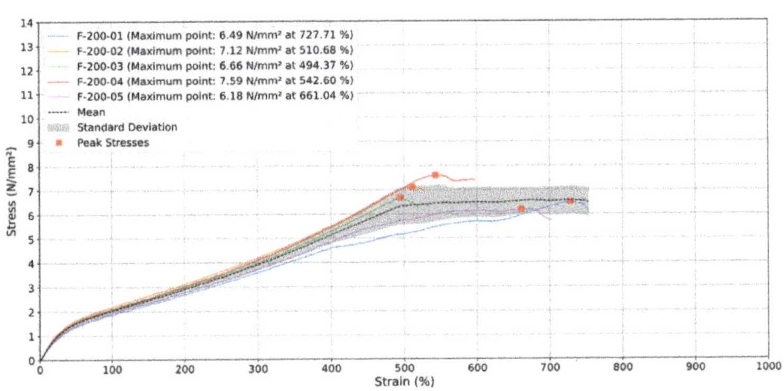

Figure 11. Stress (N/mm^2) vs. Strain (%) for specimens with orientation of 45° and −45° at 200 mm/min.

Additionally, the average stress curve increased linearly up to approximately 500% strain, reaching a peak of around 7 N/mm^2. Beyond this point, the stress stabilized, fluctuating slightly between 6 and 7 N/mm^2 until the test concluded at around 750% strain. Similar to tests conducted at a lower test speed of 20 mm/min, the 45° and −45° specimens did not exhibit an abrupt stress drop after reaching maximum stress. Instead, the material maintained its tensile resistance, indicating controlled deformation. This sustained strength suggests that the layer orientation facilitates even load distribution during the final deformation phase, ensuring stability and preventing sudden failure.

The specimens showed high consistency with minimal variability up to around 400% strain. The 45° and −45° orientations were less resistant than 0° due to misalignment with the tensile direction, leading to earlier failure. Testing at a higher test speed of 200 mm/min slightly increased maximum stress and resulted in a smoother stress stabilization phase, indicating improved load management and reduced risk of abrupt failure compared to a lower rate of 20 mm/min.

3.4. Printing Orientation 90°

The specimens printed at 90° tested at a strain rate of 20 mm/min shown in Figure 12 ruptured at lower strain percentages due to the perpendicular orientation of their material strands relative to the applied load. This alignment reduces their tensile resistance, leading to earlier failure compared to specimens with other printing orientations. The delamination between layers, caused by this perpendicular alignment, significantly limits the specimens' ability to withstand higher stress and strain levels. Specifically, specimens N-20-01, N-20-02, and N-20-05 failed between 200% and 300% strain, while N-20-03 and N-20-04 ruptured between 300% and 400% strain. This result indicates that rupture occurs more quickly under these conditions. In addition to reaching lower maximum stresses, ranging between 2.77 N/mm^2 and 3.47 N/mm^2, this happens precisely because the deposited material strands of the specimens are oriented in the opposite direction to the test, resulting in less resistance and leading these specimens to break earlier than the others.

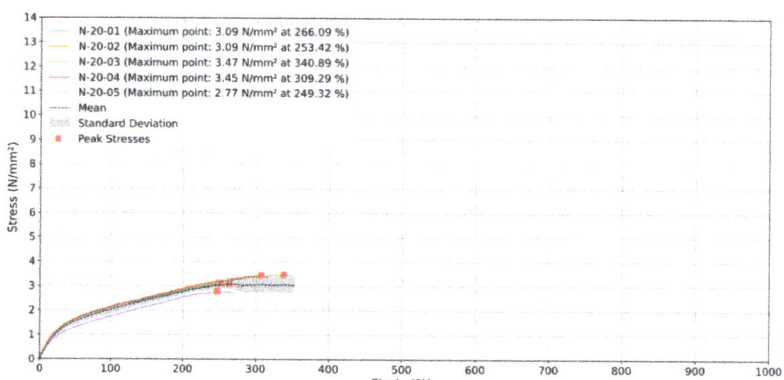

Figure 12. Stress (N/mm^2) vs. Strain (%) for specimens with orientation of 90° at 20 mm/min.

When analyzing the average results for specimens printed at a 90° orientation, the stress–strain curve exhibits a linear increase up to approximately 250% strain, reaching a maximum stress of about 3.09 N/mm^2. Beyond this point, the stress stabilizes at around 3 N/mm^2, indicating a zone where the material cannot sustain further strain increases with corresponding stress growth. The standard deviation remains low and consistent throughout most of the test, demonstrating uniform behavior among specimens, with only slight variability in the later stages.

Compared to other orientations (0°, 45°, and −45°), the 90° specimens exhibit significantly lower strain capacity. This reduced performance is attributed to the perpendicular alignment of the printed layers relative to the tensile direction, which promotes delamination and layer separation, thereby limiting the material's ability to withstand stress and strain.

For specimens printed at 90° and tested at a higher test speed (v) of 200 mm/min (Figure 13), two specimens (N-200-01 and N-200-04) failed between 500% and 600% strain with stress values ranging from 5.99 N/mm^2 to 6.34 N/mm^2, while the remaining specimens N-200-02, N-200-03, and N-200-05 broke between 300% and 500% strain. The average stress curve maintained a linear trend up to approximately 500% strain, reaching nearly 5 N/mm^2, indicating consistent elastic behavior. Unlike tests conducted at 20 mm/min, the higher tensile speed enhanced the material's tensile strength, allowing strains up to over 550%. However, the standard deviation increased significantly after 400% strain, reflecting greater variability due to layer separation. Despite the higher test speed improving stress resistance, the 90° specimens still demonstrated limited strain capacity and increased vulnerability to failure compared to other orientations.

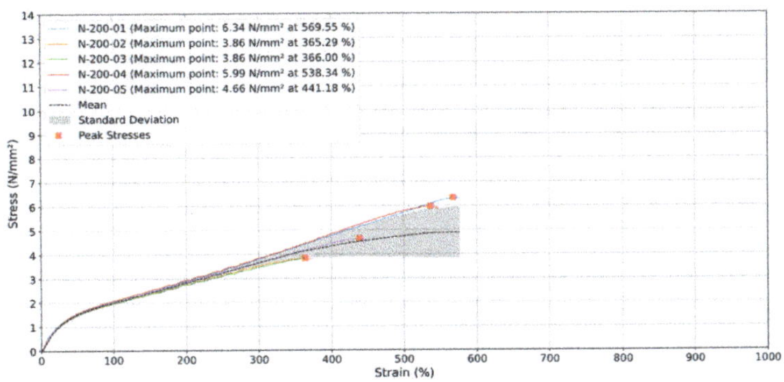

Figure 13. Stress (N/mm^2) vs. Strain (%) for specimens with orientation of 90° at 200 mm/min.

Testing at a higher test speed of 200 mm/min increased maximum stress and strain capacity slightly but also introduced greater variability in results. Overall, the 90° orientation compromises material strength and deformation capacity, especially under higher strain rates.

Based on the graphs presented in Figures 12 and 13, it can be observed that the 90°-oriented specimens face an additional challenge, as the print orientation is perpendicular to the direction of the applied load during the tensile tests. This results in lower structural resistance since the printed layers are arranged perpendicular to the axis of force, making the specimens more susceptible to failure due to delamination or rupture between layers, compared to different orientations. The anti-parallel layer arrangement directly affects the ability to withstand higher stress, as evidenced by the test results.

4. Discussion

4.1. Selection of Printing Orientation

When analyzing the graphs of all orientations at v = 20 mm/min, the orientation that withstands the greatest stress × strain is the orientation when the material was printed at 0°. However, this is not a uniform relationship, as throughout the process, the deposited material strands were breaking, as can be seen in the fluctuations that occur in this sample. Additionally, since the orientation aligns with the machine's process (vertically), the specimen withstood the stress and continued to perform until the total rupture of its deposited material strands. It is worth noting that the quasi-isotropic orientation, after the 0° orientation, shows the best results at v = 20 mm/min, and it is also the best orientation at v = 200 mm/min (Figures 14 and 15), in addition to being a more uniform curve and thus providing greater reliability in its results. This orientation also showed a smoother and more predictable stress–strain response, indicating a state of greater "equilibrium". It better represents the 3D printing process, as it incorporates multiple printing directions, effectively distributing stress more uniformly across the material. These qualities make quasi-isotropic orientation more suitable for practical applications and for use in computational simulations where consistency and predictability are critical.

Therefore, it was selected to be used in the computational simulations performed in the Ansys software.

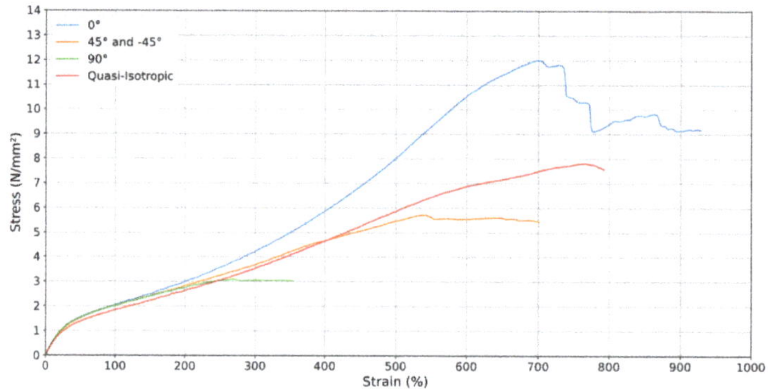

Figure 14. Stress–Strain of printing orientations at (v) of 20 mm/min.

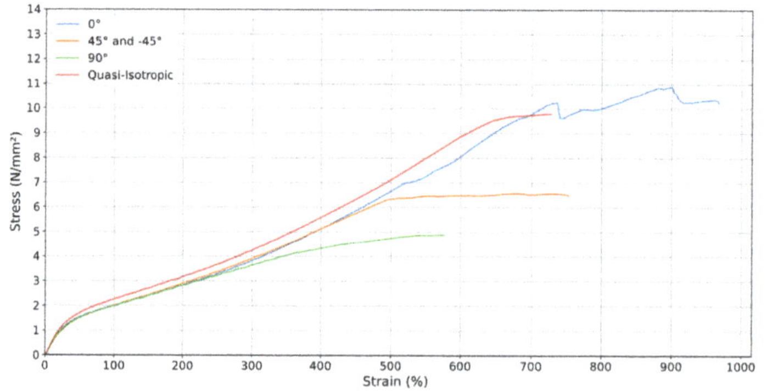

Figure 15. Stress–Strain of printing orientations at (v) of 200 mm/min.

Analyzing Figures 14 and 15, it is evident that the printing orientations significantly influence the material's mechanical behavior. To enhance uniformity, the top and bottom layers of the specimens, which were originally aligned parallel to the printing bed in a 0° orientation, were removed during the printing process. These layers often have a distinct infill pattern and density compared to the rest of the specimen, potentially leading to variations in mechanical properties. Removing them ensures a more consistent structure and allows for a more accurate assessment of the material's performance.

By removing these layers, the uniformity of the material's structure is enhanced, as the remaining infill and layers maintain consistent orientation and density throughout the specimen. This approach ensures that the mechanical behavior observed during testing is not influenced by the additional stiffness or strength contributed by the top and bottom layers, allowing for a more accurate evaluation of the material's anisotropic properties and its response to tensile loading.

There may also have been manufacturing process failures, such as:

1. Material accumulation or shortage in certain areas during sample fabrication, especially in regions with higher concentrations of localized stresses.
2. Variations in the stiffness of the specimens.
3. Batches of specimens printed on different days.
4. The printers, being frequently used, may have experienced undesirable configuration changes, resulting in slack that causes variations and affects the fabrication of the specimens.

The orientation for the quasi-isotropic TPU 60A specimens was chosen with the objective of obtaining more predictable and consistent material behavior up to rupture. When analyzing the stress vs. strain graphs, the 0° orientation showed greater resistance up to around 700% strain, reaching the highest maximum stress. However, the rupture that occurred after this point was not uniform, with sharp drops and visible fluctuations in the curve, which indicates gradual and less controlled failures in the material.

On the other hand, the quasi-isotropic orientation showed more uniform behavior throughout the test. Despite withstanding slightly lower stresses than the 0° orientation, the material showed a smoother and more continuous curve, with a clean and less abrupt rupture. This suggests that the material distributes the stress better in different directions, resulting in a more predictable and controlled failure. The consistency of the quasi-isotropic curve also indicates greater material reliability in this orientation, which is important for applications where failure prediction and material integrity are essential.

Therefore, the quasi-isotropic orientation was chosen for the simulations as it offers a more stable and reliable representation of the behavior of TPU 60A up to rupture, without the fluctuations observed in the 0° orientation, which can compromise the accuracy of the results and performance analysis in critical applications.

The quasi-isotropic orientation, identified as the most balanced and predictable configuration in this study, offers valuable insights for real-world applications in soft robotics. Its ability to uniformly distribute stress across multiple directions enhances the reliability and adaptability of grippers in handling tasks, particularly those involving irregularly shaped or fragile objects. This characteristic is crucial for applications where mechanical failure could result in damage to sensitive items or inefficiencies in automated systems. Additionally, the uniform mechanical behavior associated with the quasi-isotropic orientation minimizes stress concentrations, which could optimize material usage by reducing wear and extending the operational lifespan of soft robotic components.

In industrial and agricultural settings, the benefits of the quasi-isotropic orientation are particularly pronounced. For example, soft robotic grippers designed with this configuration can provide the strength required for tasks like automated harvesting of delicate produce, such as fruits or vegetables, while maintaining the flexibility needed to adapt to varying shapes and textures. Similarly, in assembly lines, grippers configured with quasi-isotropic orientations can handle sensitive items without compromising precision or causing damage, thus ensuring both efficiency and product integrity.

Furthermore, the findings from this study highlight the potential for TPU 60A, when paired with quasi-isotropic printing, to drive innovations in soft robotics. By offering a reliable and versatile material configuration, these results can guide the development of more robust and adaptable gripper designs tailored to specific applications. This integration of experimental data and practical applications highlights the critical role of material and orientation optimization in advancing the field of soft robotics.

Despite TPU 60A being an elastomeric material, practical tensile tests have confirmed that the printing orientation greatly affects its mechanical performance. The experimental results showed significant variations in tensile strength and maximum elongation based on the layer deposition direction during additive manufacturing. This indicates the material's inherent anisotropy when fabricated via FFF, requiring consideration of printing orientation in the design and practical use of TPU 60A components.

4.2. Computational Simulation Validation

For computational simulations, the finite element method was used. In the context of the numerical simulations performed in the commercial software Ansys [23], the equilibrium equations are applied to each finite element that makes up the mesh of the model. The software solves these equations at each node of the mesh, ensuring that the system is in static or dynamic equilibrium during the analysis. In the case of these simulations, the internal forces generated by material stresses are counterbalanced by the applied external forces, which are represented as displacements.

In addition, the Ansys software, when carrying out simulations of non-linear structural behavior, uses the Newton–Raphson method to solve the system of non-linear equations associated with the analysis. This method is used in non-linear problems due to its efficiency in guaranteeing convergence [24]. The procedure involves an iterative approximation of the solution, where at each iteration the system of differential equations is linearized around an estimated equilibrium point, and the internal and external forces are adjusted. During each step of the simulation, Ansys recalculates the deviation between the applied forces and the internal reactions, making successive corrections until the difference is small enough to reach convergence. The exact solution to the problem occurs when equilibrium between the forces is reached [25].

After reaching convergence using the Newton–Raphson method, the system is solved using numerical methods for solving systems of linear equations, such as matrix decomposition or iterative methods, depending on the complexity of the model and the number of variables involved. This process ensures that the responses generated, such as displacements and stresses, are calculated accurately and conform to the conditions imposed in the model, providing reliable results that are consistent with the simulated physical reality.

The cube and the specimen simulations serve distinct but complementary purposes in this study. The simulation of the cube is primarily aimed at determining the most suitable constitutive relationship for the material, given its simplicity and efficiency in simulation. This approach allows for an effective comparison of different material models and the calibration of parameters to closely match experimental data. On the other hand, the simulation of the specimen is performed to validate the robustness and reliability of the selected constitutive model. By demonstrating that the adopted model accurately represents the material's behavior under realistic conditions, it ensures that the simulation framework can effectively predict the performance of the material in practical applications. This dual approach—simplifying with the cube and verifying with the specimen—provides a comprehensive understanding of the material's mechanical response and strengthens the confidence in its use for design and application in soft grippers or similar devices.

In Ansys, for both the unit cube and the specimen, a high-density mesh with automatic resolution and level 3 refinement (maximum) was applied. This configuration ensures that the mesh quality is balanced, providing sufficient resolution to accurately capture the material's behavior in the regions of interest without compromising computational efficiency. The element size was standardized and controlled by the program, with adaptive behavior and quick transitions between mesh regions to ensure precision and numerical convergence.

For the unit cube, quadratic hexahedral elements (Hex20) were automatically selected by the program. This type of element is used for regular geometries and performs well in terms of accuracy and computational efficiency. The cube's mesh consists of 8281 nodes (2197 corner nodes and 6084 intermediate nodes) and 1728 elements, with 3 degrees of freedom per node. The unit cube schematic representation of boundary conditions is shown in Figure 16a. Due to the geometry of the adopted specimen, second-order tetrahedral elements (Tet10), with 3 degrees of freedom per node, were used in the discretization. In fact, these types of elements are known to be particularly suitable for meshes with complex geometries. In terms of mesh properties, the specimens were discretized into 6217 elements with 12,602 nodes in total (2185 corner nodes and 10,417 intermediate nodes). The region of greatest interest is the gauge length section, and an example of the schematic representation of the adopted boundary conditions is shown in Figure 16b.

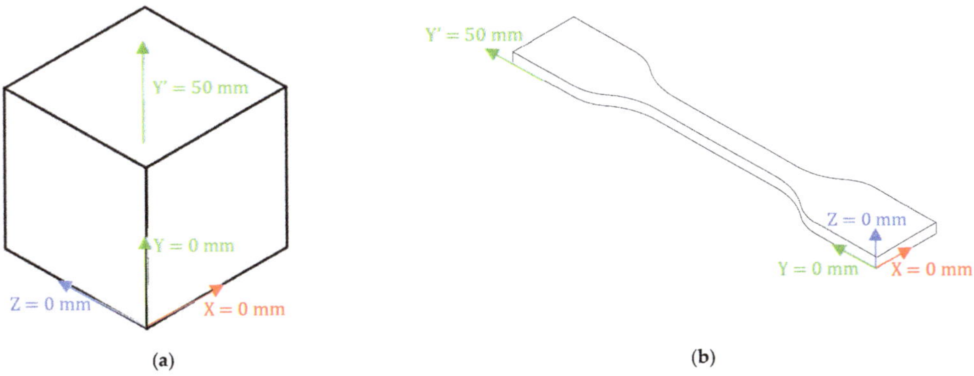

Figure 16. Schematic representation of Boundary conditions for: (**a**) Unit Cube; (**b**) Specimen.

For the unit cube model, four constitutive models were selected for testing: Neo-Hookean, 2nd Order Polynomial, 3rd Order Yeoh, and 3rd Order Ogden. These models were identified for their ability to accurately represent the non-linear elastic behavior of flexible gripper materials. The stress–strain curve of the four constitutive models considered for selection is shown in Figure 17.

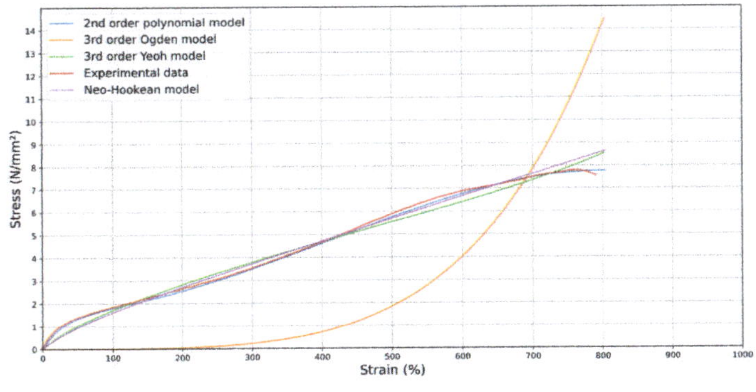

Figure 17. Constitutive models and average experimental test data for TPU 60A with quasi-isotropic orientation at v = 20 mm/min.

By analyzing Figure 17, it becomes evident that the 2nd Order Polynomial model is the most promising among the evaluated options. Additionally, the coefficient of determination

(R^2) values for all the tested mathematical models, as presented in Table 4, further support this conclusion.

Table 4. Coefficient of determination (R^2) values for different material models.

Models	Values of Coefficient of Determination (R^2)
2nd Order Polynomial model	0.9993
3rd Order Yeoh model	0.9843
3rd Order Ogden model	−0.8049
Neo-Hookean model	0.9857

Thus, the 2nd Order Polynomial model has the highest R^2 value, demonstrating the best fit to the experimental data. This strong correlation supports our selection of this model for accurately representing the mechanical behavior of TPU 60A in the simulations. And, to validate its accuracy, a coefficient of determination (R^2) value of 0.9986 was achieved between the experimental data and the model, demonstrating that it captures nearly all the variability in the data and provides a highly reliable representation of the stress–strain behavior. Additionally, as highlighted by Jaiswal et al. [26], recent studies have shown that this model effectively represents the mechanical response of similar materials, accurately capturing viscoelastic deformations under varying loading conditions. This has been consistently validated through multiple computational simulations and comparisons with experimental results.

Figure 18 shows the comparison between the stress observed in the experimental study and the simulations of the unit cube and the specimen. The close correspondence between the curves confirms that the constitutive model is appropriate for predicting the material's mechanical behavior, ensuring the model's reliability.

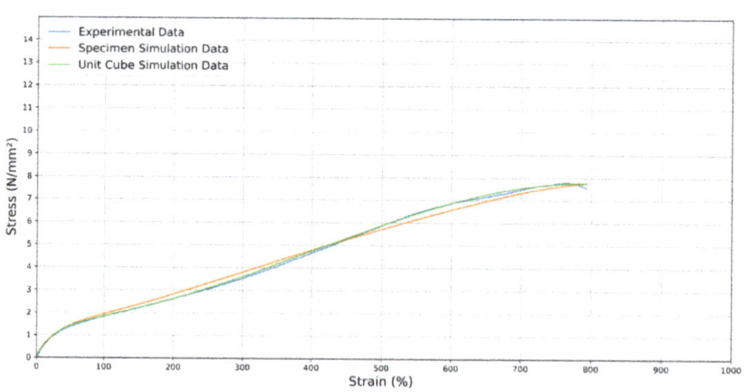

Figure 18. Comparison between the average stresses of the experimental data, the simulation data of the specimens, and for the unit cube.

As a result, the graph in Figure 18 was generated, showing three curves, one for the actual average stress (in blue), other for the interpolated stress that occurred in the computational simulation for the specimen (in orange), and the computational simulation of the unit cube data (in green).

According to Ansys [27], the equation for the 2nd Order Polynomial model is as follows:

$$\psi_P = \sum_{m+n=1}^{N} C_{mn} \times \left(\overline{I_1} - 3\right)^m \times \left(\overline{I_2} - 3\right)^n + \sum_{k=1}^{N} \frac{1}{d_k} \times (J-1)^{2k}, \; N = 2 \quad (1)$$

where ψ_P = strain-energy potential; C_{mn} = Material constants; $\overline{I_1}$ is the first deviated principal invariant; $\overline{I_2}$ is the second invariant of the isochoric left or right Cauchy–Green deformation tensor; d_k = material incompressibility parameters; J = determinant of the elastic deformation gradient F; N = material constant. The initial shear modulus is defined as Equation (2):

$$\mu_P = 2 \times (C_{10} + C_{01}) \tag{2}$$

and the initial bulk modulus for this case is defined as Equation (3):

$$K_P = \frac{2}{d_k} \tag{3}$$

Material constants are used to define the parameters of the 2nd Order Polynomial constitutive model, which describes the material's mechanical behavior under stress. Together, these constants calibrate the 2nd Order Polynomial model to match the experimental data, ensuring the simulation accurately reflects the mechanical behavior of the material under various loading scenarios. These values are listed in Table 5.

Table 5. Material constants (C_{mn}) and incompressibility parameters (d_1, d_2) of the 2nd Order Polynomial constitutive model, as related to Equation (1).

Constants	Value
Material constant 10 [Pa]	2.6643×10^5
Material constant 01 [Pa]	6.6007×10^5
Material constant 20 [Pa]	−7852.1
Material constant 11 [Pa]	80,523
Material constant 02 [Pa]	-1.6973×10^5
Incompressibility parameter d_1 [Pa^{-1}]	0
Incompressibility parameter d_2 [Pa^{-1}]	0

It is important to note that the coefficients used in the constitutive models are generated automatically by Ansys in the materials tool, based on the experimental data entered. The software accurately adjusts the data to calculate the necessary coefficients. It is then necessary to select the curve that best represents the material's constitutive behavior, ensuring equivalence with the tensile tests carried out.

Thus, the constitutive model that best represents the curve is defined using Equation (4).

$$y = -7 \times 10^{-6}x^2 + 0.018x + 0.7173 \tag{4}$$

Equation (4) describes the relationship between stress and strain in the material as predicted by the model. This equation allows for interpolation and extrapolation of data, enabling the prediction of material behavior under different strain conditions. Moreover, it provides a mathematical representation that can be directly implemented in computational simulations, ensuring consistency and accuracy when comparing experimental and simulated results. The high R^2 value associated with this equation further demonstrates its reliability in capturing the material's mechanical response.

It is important to note that, although the software automatically selected different types of elements for the cube and the specimen, the simulation results remain valid. In the case of the cube, Hex20 elements were used, while the specimen was modeled with Tet10 elements. This is because both types of elements, when properly configured and refined, can capture the mechanical behavior of the material in the regions of interest. Figure 19 displays the distribution of normal stress values during the numerical test at 0.5 s, as visualized in the Ansys software.

The stress–strain diagrams presented in Figure 18 show three distinct datasets: (1) the results from the elongation of the unit cube in numerical simulations, (2) the results from the elongation of the ISO specimen in numerical simulations, and (3) the tensile test results of ISO specimens obtained experimentally.

The unit cube simulation was conducted to identify the most suitable constitutive relationship for the material, leveraging its simplicity and simulation efficiency. This approach facilitated an effective comparison of different material models and enabled the calibration of parameters to closely match experimental data. By using the stress–strain data obtained experimentally, the numerical simulations of the unit cube established a solid foundation for selecting the most appropriate constitutive model. Once calibrated, this model was applied to simulate the tensile behavior of the ISO specimens, enabling a direct comparison with the experimentally obtained tensile test data. This process successfully validated the constitutive model's ability to accurately predict the material's mechanical response under practical conditions.

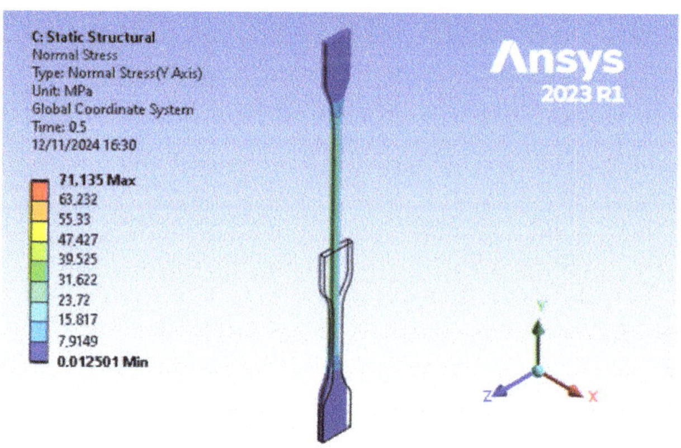

Figure 19. Specimen during a simulation at 0.5 s with scale of 1:2.

On the other hand, the numerical simulation with the specimen is conducted to validate the robustness and reliability of the selected constitutive model. By demonstrating that the adopted model accurately represents the material's behavior under realistic conditions, it ensures that the simulation framework can effectively predict the material's performance in practical applications. This dual approach—simplifying with the unit cube and verifying with the specimen—provides a comprehensive understanding of the material's mechanical response and reinforces confidence in its use for different designs, such as the application of soft grippers or similar devices.

In Figure 18, the stress–strain curves compare the results from the numerical simulation of the ISO specimen, unit cube, and the experimental tensile test data. It is important to note that the stress values in this figure correspond to the averaged global response of the material and are not directly comparable to the localized stress values shown in Figure 19, which shows the distribution of normal stress during the numerical tensile test of the ISO specimen at 0.5 s, as simulated in Ansys software. The legend in this figure displays the range of stress values, with a maximum value of 71.135 MPa at the specified moment in the simulation. This value represents the localized stress at specific nodes and is derived from the computational model.

The R^2 value for the comparison between the cube data and the specimen data is 0.9927. Also, the R^2 value was calculated for the comparison between the cube data and the experimental data, which is 0.9714, and the R^2 value for the comparison between the specimen data and the experimental data, which is 0.9892. These results indicate a good fit and high agreement between the two datasets. Having high R^2 value confirms that the simulation model is well-calibrated and effectively captures the material's mechanical behavior. Furthermore, it validates that the material properties defined for the simulation align closely with the experimental observations, ensuring the reliability and accuracy of the simulation in predicting the material's stress–strain response.

Then, in this simulation, the maximum force reached was 62.132 N. All the data was divided by the initial cross-section area (A_0 = 8.00 mm^2), reaching a maximum stress of 7.81 N/mm^2 for the experimental values and a maximum stress value of 7.77 N/mm^2 for the simulation results. This means that although the specimen breaks at approximately 372.50 mm of displacement in practice, the maximum force reached occurs at approximately 360.50 mm, approximately 13.00 mm after that predicted in the simulation. The simulation is experimentally validated, with only an absolute stress error of 0.04 N/mm^2, corresponding to a relative error of approximately 0.5%.

To validate the model, the relative errors of both force and displacement were calculated. The Relative Error of the Maximum Stress (E_{rSm}) is given by Equation (5):

$$E_{rSm}(\%) = \left(\frac{7.81 - 7.77}{7.77}\right) \times 100$$
$$E_{rSm}(\%) \cong 0.51\% \tag{5}$$

The Relative Error of the Displacement where the Stress is Maximum (E_{rDm}) is given by Equation (6):

$$E_{rDm}(\%) = \left(\frac{360.54 - 346.00}{360.54}\right) \times 100$$
$$E_{rDm}(\%) \cong 4.03\% \tag{6}$$

In addition, the force error was calculated at the exact point of 346 mm, and it was first necessary to carry out a linear interpolation of the average of the realized force, which is given by Equation (7):

$$y = y_1 + \frac{(x - x_1)}{(x_2 - x_1)} \times (y_2 - y_1) \tag{7}$$

where, according to the data taken from the average stress × strain, the following values are approximated to 15 decimal places: x_1 = 345.61 mm; $y_1 = \frac{61.57}{8.00}$ N/mm^2; x_2 = 346.35 mm; $y_2 = \frac{61.61}{8.00}$ N/mm^2; x = 346.00 mm. This is given by Equation (8):

$$y = \frac{61.57}{8.00} + \frac{(346.00 - 345.61)}{(346.35 - 345.61)} \times \left(\frac{61.61 - 61.57}{8.00}\right)$$
$$y \cong 7.70\ \text{N/mm}^2 \tag{8}$$

Then, the relative error where the displacement of 346 mm occurs is given by Equation (9):

$$E_{rd}(\%) = \left|\frac{61.59 - 62.13}{61.59}\right| \times 100$$
$$E_{rd}(\%) \cong 0.88\% \tag{9}$$

In the Ansys software, the TPU 60A material was created in the Engineering Data section and the 2nd Order Polynomial model was added. In the properties table, it was noted that the material has a density (ρ) of 1070 kg/m^3 and that the tests were carried out at room temperature ($T = 22$ °C). The average displacement (mm/mm) and stress (N/mm^2) values for 500 points of the quasi-isotropic specimens at $v = 20$ mm/min were added to the uniaxial test section. The curve was adjusted and the results copied to the creation of the material.

Thus, it was noticed that exploring different constitutive models helps to accurately capture the complex mechanical behavior of materials under various loading conditions. Models such as the 2nd order polynomial, 3rd order Ogden, Neo-Hookean, and 3rd order Yeoh provide distinct approaches to describing the stress–strain relationship, each suited to specific material characteristics. Comparing these models against experimental data helps identify the most accurate representation, optimize simulation performance, and understand the limitations of each model. This ensures reliable predictions and better alignment with the material's real-world behavior.

4.3. Real-World Applications and Challenges of TPU 60A

TPU 60A has proven its utility in various real-world applications, particularly in industries requiring flexibility and adaptability. In soft robotics, it serves as a material of choice for grippers due to its capacity to handle delicate objects without causing damage, as observed by Gaafar et al. [28], who introduced the design for a JD-printed soft pneumatic bending actuator, material characterization, and 3D printing parameter optimization of flexible material with 60A shore hardness. This makes this application valuable in agriculture for harvesting fruits and vegetables and in the food industry for manipulating fragile items during packaging processes. Similarly, TPU 60A is used in the manufacturing of wearable medical devices and assisted technologies, since its elasticity ensures comfort and functionality. Despite these advantages, practical applications face some challenges, such as material fatigue under cyclical loads, sensitivity to environmental factors like humidity, and limitations in large-scale manufacturing. Addressing these challenges requires further material optimization, particularly in terms of humidity control during processing and post-treatment methods to enhance durability [28,29]. While this study offers significant insights into the mechanical properties of TPU 60A, some limitations should be acknowledged. The research primarily focused on uniaxial tensile tests and specific printing orientations, which may not fully capture the material's behavior under real-world stress conditions. Future studies could explore biaxial or multiaxial stress conditions to better understand its performance in practical applications. Additionally, the long-term behavior of TPU 60A under cyclic loading remains unexplored, which is critical for applications involving repetitive mechanical stress. To address these gaps, future research should incorporate fatigue testing to develop advanced constitutive models [29]. Further, the development of tailored G-code algorithms for graded material deposition could minimize stress concentrations and enhance structural integrity. Exploring hybrid materials or composites involving TPU 60A also presents promising research paths for achieving superior performance in demanding applications. Recent studies, for example, demonstrate improvements in mechanical properties and environmental resistance through the addition of nanofillers like carbon nanotubes or graphene further underscore the potential of such advancements [30].

5. Conclusions

It is evident in the characterization of TPU 60A material, carried out through tensile tests for the orientations of 0°, 45° and −45°, 90°, and quasi-isotropic, that both the printing orientation and test speed have a significant impact on the material's performance. These factors directly influence key mechanical properties, including force, stress, strain, and the overall deformation behavior. While some orientations exhibit greater resistance, others prove to be less effective for these purposes. Among the tested orientations, the quasi-isotropic configuration was found to be the most suitable for representing the material's behavior in a balanced manner.

Despite the TPU 60A filament having well-defined properties provided by the manufacturer, such as an elongation at break of 950% [12], factors related to the printing process can significantly alter these characteristics. Variables such as printing orientation, environmental humidity, and the geometric design of the printed structure significantly affect the final mechanical properties of the material, including its strength, ductility, and strain capacity. These modifications can lead to deviations from the nominal values, highlighting the importance of carefully considering these factors in the context of application. This ensures that the desired performance is achieved and aligns with the requirements of the specific use case. Also, in the present study, the material preparation and printing process were calibrated based on the manufacturer's recommendations, provided in the datasheet for Filaflex TPU 60A. Key parameters such as nozzle temperature (230 °C), infill density (100%), and layer height (0.25 mm) were set in accordance with these guidelines to ensure consistency with standardized practices, but future work could explore the interaction of these parameters to build upon the findings presented here.

From the average data of the five valid tensile tests, the uniaxial behavior required to create the TPU 60A material library in the Ansys software was obtained. Using the second-degree polynomial method, these data were used to calibrate and validate the material model through two approaches: first, a numerical simulation of a unit cube to evaluate and refine the constitutive model parameters, and second, a simulation of the tensile test using ISO specimens to validate the model against experimental results. It is important to note that the unit cube test was not conducted experimentally; it served as a numerical tool for calibration, while the experimental tensile tests provided the primary validation of the material model.

Reducing the presence of voids in printed specimens is important for enhancing mechanical properties and reliability, and controlling filament humidity significantly decreases void formation by optimizing the printing parameters such as extrusion settings, nozzle diameter, print paths, and infill patterns. Fine-tuning these factors leads to stronger, more consistent specimens, resulting in improved performance of the TPU 60A. In future work, additional experimental data focusing on precise measurements of void density under varying humidity levels and their direct impact on mechanical properties such as tensile strength and elongation at break should be analyzed.

The findings presented in this study provide a strong foundation for the use of TPU 60A in soft robotics applications, particularly in the design and optimization of soft robotic grippers. To further establish the practical relevance of these results, it is recommended for future work to conduct practical tests using a robotic arm equipped with TPU 60A grippers. Such tests would allow for evaluating the grippers' performance in real-world scenarios, including their adaptability to delicate or irregularly shaped objects. Additionally, laboratory experiments designed to test the grasping ability and effectiveness of the grippers under various conditions would provide deeper insights into their practical capabilities. These tests would demonstrate how the material's mechanical properties translate into functional advantages, ensuring a clearer alignment between the material characteristics and their application in soft robotics.

This work offers significant contributions to the development of elastomer grippers, as well as providing valuable insights for future projects. Among the suggestions for future work, one key recommendation is the manual development of a G-code that allows for varying the material deposition thickness during the printing process, resulting in a variable thickness. This approach reduces the formation of defects in the material, promoting stress relief and significantly reducing stress concentrations. Consequently, it contributes to the production of a more uniform part, yielding more consistent and reliable results.

The inclusion of biaxial tests for a more comprehensive and accurate characterization of elastomeric materials, such as TPU, is also suggested. By applying stresses in two directions simultaneously, these tests allow for a more reliable capture of the materials' real behavior, especially in applications involving complex deformations, as is the case with soft grippers. This would enable the acquisition of richer experimental data, which can be used to refine computational models, improve the simulation of grippers under multiple stress conditions, and optimize performance in real-world scenarios.

Author Contributions: Conceptualization, L.L., M.L.d.A. and P.D.G.; methodology, L.L., M.L.d.A., R.A., N.P., T.A.D. and P.D.G.; validation, L.L., M.L.d.A., T.A.D. and P.D.G.; formal analysis, L.L., M.L.d.A., R.A., N.P., T.A.D. and P.D.G.; investigation, L.L.; resources, P.D.G.; data curation, L.L. and M.L.d.A.; writing—original draft preparation, L.L., M.L.d.A., T.A.D. and P.D.G.; writing—review and editing, M.L.d.A., R.A., N.P., T.A.D. and P.D.G.; supervision, M.L.d.A., T.A.D. and P.D.G.; project administration, P.D.G.; funding acquisition, P.D.G. All authors have read and agreed to the published version of the manuscript.

Funding: These results are within the research activities of project "ROBOTA-SUDOE—Robotics, Automation, and Digitalization as Drivers of Competitiveness and Growth for SMEs" (S1/1.1/P0125), which is co-funded by the European Union through the European Regional Development Fund (ERDF) and national funds, under the territorial cooperation Interreg Europe Programme 2021–2027 (eSUDOE 2021–2027). This research was also funded by Fundação para a Ciência e Tecnologia (FCT), C-MAST (Centre for Mechanical and Aerospace Science and Technologies), under the project UIDB/00151/2020 (https://doi.org/10.54499/UIDB/00151/2020; https://doi.org/10.54499/UIDP/00151/2020, accessed on 25 October 2024).

Institutional Review Board Statement: Not applicable.

Informed Consent Statement: Not applicable.

Data Availability Statement: The original contributions presented in the study are included in the article, further inquiries can be directed to the corresponding author.

Conflicts of Interest: The authors declare no conflicts of interest.

References

1. Campilho, R.D.S.G.; Silva, F.J.G. Industrial Process Improvement by Automation and Robotics. *Machines* **2023**, *11*, 1011. [CrossRef]
2. Javaida, M.; Haleema, A.; Singhb, R.P.; Suman, R. Substantial capabilities of robotics in enhancing industry 4.0 implementation. *Cogn. Robot.* **2021**, *1*, 58–75. [CrossRef]
3. Youn, J.-H.; Jeong, S.M.; Hwang, G.; Kim, H.; Hyeon, K.; Park, J.; Kyung, K.-U. Dielectric Elastomer Actuator for Soft Robotics Applications and Challenges. *Appl. Sci.* **2020**, *10*, 640. [CrossRef]
4. Dinakaran, V.P.; Balasubramaniyan, M.P.; Muthusamy, S.; Panchal, H. Performa of SCARA based intelligent 3 axis robotic soft gripper for enhanced material handling. *Adv. Eng. Softw.* **2023**, *176*, 103366. [CrossRef]
5. Zhai, Y.; Yan, J.; Shih, B.; Faber, M.; Speros, J.; Gupta, R.; Tolley, M.T. Desktop fabrication of monolithic soft robotic devices with embedded fluidic control circuits. *Sci. Robot.* **2023**, *8*, eadg3792. [CrossRef]
6. Liu, Y.; Hou, J.; Li, C.; Wang, X. Intelligent Soft Robotic Grippers for Agricultural and Food Product Handling: A Brief Review with a Focus on Design and Control. *Adv. Intell. Syst.* **2023**, *5*, 2300233. [CrossRef]
7. Kim, D.; Baek, S.R.; Kim, M.S.; Park, C.Y.; Lee, I.H. Design and Manufacturing Process of Pneumatic Soft Gripper for Additive Manufacturing. *Sens. Actuators A Phys.* **2024**, *370*, 115218. [CrossRef]
8. Ahmad, M.; Javaid, M.; Haleem, A. A study on fused deposition modeling (FDM) and laser-based additive manufacturing (LBAM) in the medical field. *Intell. Pharm.* **2024**, *2*, 381–391. [CrossRef]

9. Sukindar, N.A.; Azhar, M.A.M.; Shaharuddin, S.I.S.; Kamaruddin, S.; Azhar, A.Z.A.; Choong, Y.C.; Adesta, E.Y.T. A Review Study on the Effect of Printing Parameters of Fused Deposition Modelling (FDM) Metal Polymer Composite Parts on Mechanical Properties and Surfaces Roughness. *Malays. J. Microsc.* **2022**, *18*, 281–297.
10. Antunes, R.; Lang, L.M.C.; de Aguiar, M.L.; Dutra, T.A.; Gaspar, P.D. Enhancing the Performance of Fin Ray Effect Soft Robotic Finger via Computational Design and Simulation. In Proceedings of the 20th ASME/IEEE International Conference on Mechatronic and Embedded Systems and Applications (MESA 2024), Genoa, Italy, 4–6 August 2024.
11. Antunes, R.; Lang, L.; de Aguiar, M.L.; Dutra, T.A.; Gaspar, P.D. Design of Fin Ray Effect Soft Robotic Gripper for Improved Mechanical Performance and Adaptability: Numerical Simulations and Experimental Validation. In Proceedings of the IEEE/ASME MESA 2024, Genoa, Italy, 4–6 August 2024.
12. Recreus. *FILAFLEX 60A 'PRO'*; Recreus: Elda, Spain, 2022.
13. *DIN 53504-S2*; Testing of rubber - Determination of tensile strength at break, tensile stress at yield, elongation at break and stress values in a tensile test. German Institute for Standardisation (Deutsches Institut für Normung): Berlin, Germany, 2017.
14. Addify3D. Artillery Sidewinder X1 3D Printer—Ultra Quiet & Fast. Available online: https://www.addify3d.com/product/artillery-sidewinder-x1-3d-printer/ (accessed on 21 October 2024).
15. Shimadzu. Autograph AGS-X Series—Specs. Available online: https://www.shimadzu.com/an/products/materials-testing/uni-ttm/autograph-ags-x-series/spec.html (accessed on 21 October 2024).
16. Ultimaker. Ultimaker Cura. Available online: https://ultimaker.com/software/ultimaker-cura/ (accessed on 22 October 2024).
17. *ISO 37*; Rubber, Vulcanized or Thermoplastic—Determination of Tensile Stress-Strain Properties. International Organization for Standardization (ISO): Geneva, Switzerland, 2017.
18. *ISO 527-2*; Plastics—Determination of Tensile Properties—Part 2:Test Conditions for Moulding Andextrusion Plastics. International Organization for Standardization (ISO): Geneva, Switzerland, 2012. Available online: https://cdn.standards.iteh.ai/samples/56046/c5c3ec20cc0741289bf6def861e0a40b/ISO-527-2-2012.pdf (accessed on 21 October 2024).
19. Naveed, N. Investigate the effects of process parameters on material properties and microstructural changes of 3D-printed specimens using fused deposition modelling (FDM). *Mater. Technol.* **2020**, *36*, 317–330. [CrossRef]
20. Frenkel, D.; Ginsbury, E.; Sharabi, M. The Mechanics of Bioinspired Stiff-to-Compliant Multi-Material 3D-Printed Interfaces. *Biomimetics* **2022**, *7*, 170. [CrossRef]
21. ImageJ. ImageJ: Image Processing and Analysis in Java. Available online: https://imagej.net/ij/index.html (accessed on 13 December 2024).
22. Lang, L.M.C. Plotting-Test-Specimen-Graphs. Available online: https://github.com/LuanLang98/Plotting-Test-Specimen-Graphs (accessed on 21 October 2024).
23. Ansys. Available online: https://www.ansys.com/ (accessed on 11 November 2024).
24. Feng, H.; Zhou, J.; Gao, S.; Jiang, L. Finite element simulation of the viscoelastic behavior of elastomers under finite deformation with consideration of nonlinear material viscosity. *Acta Mech.* **2021**, *232*, 3935–3952. [CrossRef]
25. Chapter 202411: Material Models Used in Explicit Dynamics Analysis. 11.6. Hyperelasticity. Available online: https://ansyshelp.ansys.com/public/account/secured?returnurl=////Views/Secured/corp/v242/en/exd_ag/ds_non_linear.html (accessed on 26 November 2024).
26. Jaiswal, M.; Jadhav, V. Numerical Study of Iterative Methods for Solving the Non-Linear Equation. *Int. J. Creat. Res. Thoughts* **2021**, *9*, 659–670.
27. Analysis Tools—Newton-Raphson Procedure. Available online: https://ansyshelp.ansys.com/public/account/secured?returnurl=/////////Views/Secured/corp/v242/en/ans_thry/thy_tool10.html (accessed on 11 November 2024).
28. Gaafar, A.H.; Zagloul, A.S.; El-Shae, Y. Design, Modeling, and Characterization of 3D Printed Soft Pneumatic Actuator using TPU 60A. In Proceedings of the 6th Novel Intelligent and Leading Emerging Sciences Conference (NILES2024), Cairo, Egypt, 19–21 October 2024.
29. Viccica, M.; Giordano, M.; Galati, M. Additive Manufacturing of Flexible Thermoplastic Polyurethane (TPU): Enhancing the Material Elongation through Process Optimisation. *Prog. Addit. Manuf.* 2024, *in press*. [CrossRef]
30. Mirzapour, M.; Cousin, P.; Robert, M.; Benmokrane, B. Dispersion Characteristics, the Mechanical, Thermal Stability, and Durability Properties of Epoxy Nanocomposites Reinforced with Carbon Nanotubes, Graphene, or Graphene Oxide. *Polymers* **2024**, *16*, 1836. [CrossRef] [PubMed]

materials

Article

Microstructure, Hardness, and Wear Behavior of Layers Obtained by Electric Arc Hardfacing Processes

Sebastian Balos [1], Danka Labus Zlatanović [2], Petar Janjatović [1], Milan Pećanac [1], Olivera Erić Cekić [3,4], Milena Rosić [5,*] and Srećko Stopić [6,*]

1 Department of Production Engineering, Faculty of Technical Sciences, University of Novi Sad, Trg Dositeja Obradovica 6, 21000 Novi Sad, Serbia; sebab@uns.ac.rs (S.B.); janjatovic@uns.ac.rs (P.J.); pecanac.milan@uns.ac.rs (M.P.)
2 Department of Production Technology, Technische Universität Ilmenau, 98693 Ilmenau, Germany; danka.labus-zlatanovic@tu-ilmenau.de
3 Innovation Center, Faculty of Mechanical Engineering, University of Belgrade, Kraljice Marije 16, 11120 Belgrade, Serbia; oeric@mas.bg.ac.rs or eric.o@mfkv.kg.ac.rs
4 Faculty of Mechanical and Civil Engineering in Kraljevo, University of Kragujevac, Dositejeva 19, 36000 Kraljevo, Serbia
5 "Vinča" Institute of Nuclear Sciences, National Institute of the Republic of Serbia, University of Belgrade, Mike Petrovića Alasa 12-14, 11351 Belgrade, Serbia
6 IME Process Metallurgy and Metal Recycling, RWTH Aachen University, Intzestraße 3, 52056 Aachen, Germany
* Correspondence: mrosic@vin.bg.ac.rs (M.R.); sstopic@ime-aachen.de (S.S.)

Abstract: Hardfacing is a welding-related technique aimed at depositing a harder and tougher layer onto a softer, less wear-resistant substrate or base metal. This process enhances the abrasion resistance of the component, increasing its durability under working conditions. A key feature of hardfacing is dilution, which refers to the mixing of the hardfacing layer and the base metal. In this study, shielded metal arc welding (SMAW) was employed to hardface structural steel using chromium carbide vanadium consumables, with results compared to AISI D2 cold-work tool steel. Four different SMAW parameters were tested, and the abrasive test was conducted against SiC discs. Wear rate, represented by the wear loss rate, was correlated to microstructure, scanning electron microscopy, energy-dispersive X-ray spectroscopy, hardness, microhardness, and surface roughness. The results showed that key SMAW parameters, such as welding speed and current, significantly influence wear resistance. Specifically, slower welding speeds and higher currents, which result in greater heat input, led to the increased wear resistance of the deposited layer through the mechanism of the inoculation of larger and harder carbides.

Keywords: hardfacing; microstructure; microhardness; abrasive wear resistance; shielded metal arc welding

1. Introduction

Hardfacing is a deposition technique that can be applied to various base materials, most commonly relatively affordable structural materials, such as steel. The aim of hardfacing is twofold: to restore dimensions, i.e., re-establish the shape and size of the worn or corroded part, or to impart entirely new properties to a newly produced component [1,2].

This technological process can significantly enhance resistance to wear caused by abrasion, impact, adhesion, corrosion, heat, or a combined effect of any of the mentioned properties [3,4]. Thus, hardfacing can improve the service life of fabricated parts and prevent fractures and breakdowns in a cost-effective manner for both maintenance and manufacturing. Maintenance costs are reduced, since hardfaced parts have a longer life and are less prone to failure. Therefore, instead of replacing these failed parts, they can be repaired or regularly maintained through on-site hardfacing, thereby reducing both planned and unplanned downtime. Furthermore, production costs can be lower, since only the surface layer is made of expensive material, while base materials are typically made of low-cost materials that are easy to fabricate and join by welding if needed [5–7]. Hardfacing is generally a cost-effective technology and can be performed using existing welding equipment.

The most common welding techniques used for hardfacing include arc welding, gas-shielded arc welding, and powder-based welding. Among arc welding processes, the most common are shielded metal arc welding (SMAW) [8,9] and submerged arc welding [10,11]. Gas-shielded metal arc processes include gas tungsten arc welding (GTAW) [5,6], gas metal arc welding (GMAW), as well as flux cored arc welding (FCAW), and metal cored arc welding (MCAW) [12–14]. Finally, for powder-based processes, plasma transfer arc welding (PTAW) [15–17] and laser cladding/welding (LC/LW) are the most common, with the latter becoming increasingly attractive due to a drop in cost [15,18].

A wide range of hardfacing techniques corresponds to the diverse welding processes that encompass the majority of industrial joining technologies. Furthermore, this diversity is reflected in the variety of materials that can be deposited, which in turn, influences applications in various industries, such as agriculture, mining, nuclear, construction, general manufacturing, and railways. Even the defense industry has benefited from hardfacing, as several studies have explored its potential to enhance ballistic protection against kinetic energy projectiles [19,20].

Hardfacing consumables can generally be categorized into four main groups, closely corresponding to their base alloy: cobalt-based, nickel-based, tungsten-based, and iron-based [9,21]. Cobalt-based alloys also contain Cr, W, Ni, Mo, and C, providing good wear resistance combined with corrosion and heat resistance, offering excellent versatility, but at a relatively high cost. Furthermore, Co is considered a critical raw material in the European Union and is often targeted to be fully or partially replaced by Ni. Compared to Ni-based alloys, Co-based hardfacing alloys offer better weldability [22–24]. Ni-based alloy deposits (Ni-) combine abrasion resistance (due to additions of C, B, and Nb) with corrosion resistance (from Cr and Al additions), particularly at high temperatures (with W and Mo) [25,26]. Tungsten alloys, often referred to as cemented carbides, consist of tungsten carbide (WC) particles with a binder, mainly Co. These deposits have extreme abrasion resistance due to their very high hardness. However, unlike other high-hardness materials, such as sintered ceramics, these alloys have a Co binder that offsets the brittleness of WC and provides resistance to impact loading [25,27]. On the other hand, there is ongoing interest in partially replacing WC and Co as critical raw materials in the EU [24].

Iron-based alloys are the most cost-effective and widely used hardfacing material. These alloys primarily contain C, Cr, Mo, and other elements, similar to those found in ferrous alloys. Fe-Cr-C hardfacing alloys are predominantly used for parts subject to grinding abrasion combined with light impact, such as digging teeth, conveyor screws, mixer wings, scraper blades, rendering plant equipment, and sand pumps (around 3% C and 30% Cr) [UTP Voestapline WEARstick XD60 brochure, accessed on 19 December 2024; UTP Voestapline WEARstick XD61 brochure, accessed on 19 December 2024]. Alloys containing around 4.5% C and 20–25% Cr are suitable for extreme sliding abrasion, including at temperatures up to 500 °C, due to their extremely high abrasion resistance from the very high content of special carbides based on Mo, V, W, and Nb. Their main applications are in earth-moving equipment, the cement and brick industry, and steel mills for radial breakers and revolving bar screens in sintering plants [UTP Voestapline WEARstick XD65 brochure, accessed on 19 December 2024]. Alloys with around 6.5% C and 25% Cr are suitable for hardfacing parts subjected to heavy abrasion with moderate impact. Typical applications include parts for cement presses, brick presses, refractory press screws, conveyor screws, and mixer blades, with service temperatures up to 450 °C [UTP Voestapline WEARstick XD63 brochure, accessed on 19 December 2024]. Their properties are closely related to their microstructure, where Cr-C or Cr-C-B carbides are mixed with the matrix containing one or more common microstructures in ferrous alloys, such as austenite, ferrite, and martensite [1,25,26]. Besides basic alloying elements, a small amount of Nb can be added. Correa et al. [11] reported positive results regarding the formation of primary NbC carbides randomly distributed in the matrix. These findings, despite the marginally lower hardness of the bulk material, resulted in the increased wear resistance of the SAW deposit. This confirmed an earlier study by Berns and Fischer [1,26]. Singh and Pandey [2] investigated the influence of several passes and SAW current, finding that an increased number of passes and a lower deposition current improved wear resistance. This was attributed to a higher hardness achieved in the third pass compared to the first. Additionally, a lower current resulted in higher cooling rates, leading to a transformation from a pearlitic to a martensitic matrix surrounding the carbides.

The primary aim of this research was to vary the microstructure and evaluate and compare the intensive abrasion resistance of four Fe-Cr-C deposits produced using manual SMAW processes with varying parameters. The performances of these deposits were benchmarked against AISI D2 cold-work tool steel in the heat-treated condition. The obtained wear rate results (weight loss rates) were correlated to the hardness and microhardness of the deposits, microstructures, particularly carbides, and the local chemical composition.

2. Materials and Methods

Hardfacing was performed on S235JR structural steel, with dimensions of 100 mm in length, 50 mm in width, and 30 mm in thickness. The 100 × 50 mm area was hardfaced. The chemical composition of the base metal is shown in Table 1. The consumable used was a rutile-coated electrode (Shielded Metal Arc Welding—SMAW process) from the WEARstick XD 60/UTP Ledurit 60 series (Voestalpine, Linz, Austria), belonging to the Fe-Cr-C type (E Fe14), with the nominal composition presented in Table 2. Hardfacing parameters for Specimens 1 to 4 are shown in Table 3, along with the deposit thickness range. The arc length was manually maintained at approximately 3 mm, with no preheating. Prior to the hardfacing process, the electrodes were dried at 300 °C for 2 h. The SMAW process was conducted on an Iskra E10 device (Iskra, Ljubljana, Slovenia). The hardfacing direction was longitudinal, with a push/forward orientation of the electrode, and the width was covered using stringer beads.

The reference material, which served the base material for the deposited material, was a rolled homogeneous plate made of a cold-work tool steel of the AISI D2 type in a heat-treated, hardened condition. This was selected as an alternative to FeCrC hardfacing on structural steel for application in rendering mills. In this application, a combination of wear and impact is expected, which is why the cold-work FeCrC alloy was chosen as the hardfacing material. This specimen was designated as Specimen D2. The chemical composition is presented in Table 4. Heat treatment involved austenitization at 980 °C, quenching in oil, and tempering at 450 °C to achieve a uniform hardness of 58 HRC. Hardness was measured using an HP-250 Rockwell hardness testing device (WPM-Werkstoffpruefmachinen, Leipzig, Germany).

The wear test was conducted on a customized metallographic laboratory polishing machine, the DP-U2 (Struers, Bellerup, Denmark). The customization involved replacing the polishing wheel with SiC grinding paper, which had a grit size of P100. According to the FEPA (Federation of European Producers of Abrasives), the corresponding SiC abrasive paper particle size is 162 μm, with a loading of 1500 g. Basic surface roughness parameters of the counterbody, based on five measurements, were as follows: Ra (arithmetic mean roughness value) of 3.261 ± 0.237 μm and Rz (maximum height of the profile) of 35.825 ± 3.415 μm. The specimens were mounted in a brass holder and placed in a Struers PdM-Force (Struers, Bellerup, Denmark) specimen mover–rotator. Prior to each wear test, each of the five tested specimens was ground using a full set of SiC abrasive papers: P100, P150, P240, P360, P500, P600, P800, P1000, P1500, and P2000.

Table 1. Chemical composition of S235JR (wt.%).

C	Si	Mn	S	Cr	P	Cu	Ni	Fe
0.15	0.18	0.74	0.035	0.02	0.02	0.02	0.05	Balance

Table 2. Nominal chemical composition of E Fe14 electrode (wt.%).

C	Si	Cr	Fe
3.0	1.3	29.0	Balance

Table 3. Hardfacing parameters for Specimens 1 to 4 (electrode E Fe14).

	Specimen 1	Specimen 2	Specimen 3	Specimen 4
Electrode diameter × length	Ø3.2 × 320 mm			
Arc voltage [V]	25	25	24	24
Welding current [A]	95	95	115	115
Current type	DC+			
Welding speed [cm/min]	14	12	14	12
Preheating	No			
Electrode inclination towards horizontal [°]	80			
Deposit thickness range [mm]	5.5–6	6.5–7	6.5–7	7–7.2
Heat affected zone [mm]	1–1.2	1.4–1.8	1.2–1.6	1.6–1.8

Table 4. Chemical composition of the AISI D2 cold-work tool steel (wt.%).

C	Si	Mn	S	Cr	P	Ni	Mo	V	Co	Fe
1.56	0.46	0.34	0.004	12.81	0.024	0.17	0.97	0.90	0.021	Balance

The abrasive paper base in the wear testing spindle was set to a speed of 250 rpm, with a wear time of 60 s. The specimen axis was 70 mm away from the centerline of the spindle, giving a sliding distance of 109.9 m. The wear test was performed in a water environment at 15 °C, with a constant flow rate maintained at 100 mL/min. Specimen mass was measured using a Type 2615 analytic balance (Tehtnica, Zelezniki, Slovenia) with an accuracy of 0.1 mg, before and after the wear test. The results were calculated as the average weight loss rate of the five specimens. After wear testing, surface roughness was measured using the SJ-301 (Mitutoyo, Kawasaki, Japan) device, with the average for all five tested specimens reported. The measurement parameters were set to λc = 0.8 × 5 mm and λs = 2.5 mm, and the Gaussian filter method was applied with a tip radius of 2 μm and a tip angle of 60°. Two roughness parameters were measured, Ra and Rz.

The metallographic examination was conducted using a light microscope (LM) Orthoplan (Leitz, Wetzlar, Germany), equipped with an ocular micrometer to measure the heat-affected zone (HAZ) width. Additionally, metallographic examination and chemical composition analysis of phases were carried out using a JSM-6460LV scanning electron microscope (SEM) equipped with an Inca Microanalysis system (Oxford Instruments, Abingdon, UK) for energy-dispersive X-ray spectroscopy (EDS), to evaluate the chemical composition of different phases. Prior to SEM analysis, the specimens were coated with gold using the Leica SCD-005 apparatus (Ball-Tech, Leica—Leitz, Wetzlar, Germany). Metallographic preparation before microscopic examination included grinding with SiC abrasive papers (P150, P240, P360, P500, P600, P800, P1000, P1500, and P2000), polishing with 6, 1, and $\frac{1}{4}$ µm diamond suspensions, and etching the hardfaced layer with Aqua Regia (17% HNO_3, 50% HCl in glycerol). Base metal in the hardfaced specimens and the reference cold-work D2 tool steel were etched with Nital (3% HNO_3 in ethanol). SEM was also used to examine the surface topography of worn surfaces.

To evaluate the hardness of the specimens, the Vickers method was applied. The HPO-250 (WPM, Leipzig, Germany). Vickers hardness measurement device was used. The microhardness of specific phases was tested using the Wilson Tukon-1101 (Buehler, Uzwil, Switzerland), with a minimal load of 10 g. Both methods used a dwell time of 15 s. For each hardness test, five measurements were taken, and the average value was reported.

3. Results and Discussion

The microstructure of the base metal, obtained through light microscopy, is shown in Figure 1. It reveals a structure composed of ferrite and pearlite, which is characteristic of the base metal (structural steel S235JR) to which the hardfacing layer was deposited.

Figure 1. Microstructure of base metal in the form of S235JR structural steel using light microscopy (LM).

This section is divided into subheadings and provides a concise and precise description of the experimental results, their interpretation, and the experimental conclusions that can be drawn.

The microstructures of the subsurface finishing layer, which is critical for wear resistance, are presented in Figure 2. The light microscope analysis reveals a similarly shaped white phase, with an approximately hexagonal cross-section and elongated morphology. The growth of these phases, believed to be M_7C_3 carbides, occurs in various directions, forming star-shaped formations. Hardfacing alloys of the Fe-Cr-C type rely on large carbides to provide wear resistance. There are no significant differences in the morphology of the white phases observed in Specimens 1 to 4. However, in Specimen 4, these white phases appear larger compared to Specimens 1 to 3. This could be the result of a higher hardfacing current and a lower welding speed, which influence a higher heat input. In turn, higher heat input increases the cooling time, allowing for the more pronounced diffusion of elements like Cr and C, which combine with Fe to form larger complex M_7C_3 carbides ($(Cr,Fe)_7C_3$). This can be correlated with the deposit thickness and heat-affected zone, as shown in Table 3, where the deposited layer and heat-affected zone are thickest in Specimen 4. When analyzing Figure 2a–d (Specimens 1 to 4), the minimal dimension of the cross-section should be compared between specimens, as the elongated bright phases are merely a result of the cutting direction during metallographic preparation. To support this, the minimal dimensions of the carbide cross-sections were also measured and reported for Specimens 1 and 4. In Specimen 1, the average carbide dimension was 15 μm, while in Specimen 4, it was 10 μm. On the other hand, in AISI D2 cold-work tool steel (Figure 2e), the microstructure consists of M3C carbides ($(Fe,Cr)_3C$), appearing as white strips in the tempered martensitic matrix, which is typical for heat-treated D2 steel, with the heat treatment parameters as explained in the experimental section.

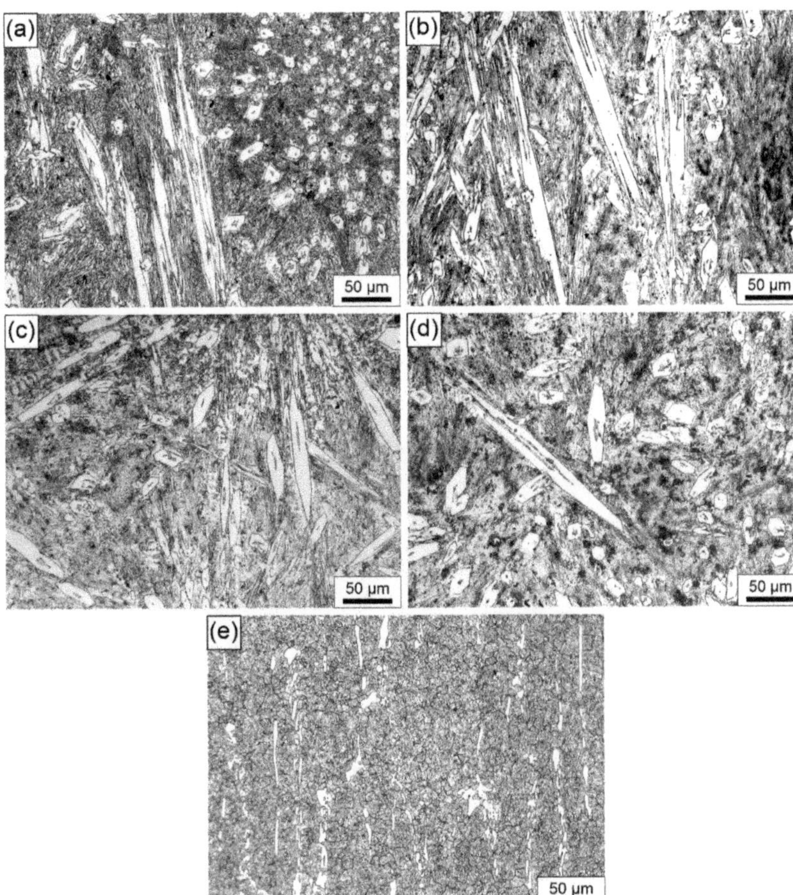

Figure 2. LM images of microstructure of the final layer: (**a**) Specimen 1; (**b**) Specimen 2; (**c**) Specimen 3; (**d**) Specimen 4; (**e**) Specimen D2.

In addition to light microscopy, scanning electron microscopy (SEM), specifically energy-dispersive X-ray spectroscopy (EDS), was applied to further describe the microstructures of the hardfaced layers. The results of the EDS analysis are shown in Figures 3 and 4. These reveal a similar composition of primary hexagonal carbides in Specimens 1 and 4 (Spectra 1 and 2 in Figure 3 and Spectrum 2 in Specimen 4). In both specimens, there is a hollow structure within the carbide, as described by Ito et al. [28]. It is proposed that hollow primary carbides form when rapid cooling occurs due to the solidification dynamics during the hardfacing deposition process, which is the main difference between the two specimens analyzed in this work. The key difference is the lower current and welding speed in Specimen 1 compared to Specimen 4. This is evidenced by the core morphology, which is similar to the eutectic matrix consisting of austenite and fine M_7C_3 carbides in Specimen 1, and a composition with less Cr in the core compared to the rest of the carbide in Specimen 4 (Spectrum 1). On the other hand, Spectra 3 in Figures 3 and 4 reveal the composition of the eutectic matrix, which contains significantly less Cr and more Fe, resembling the central section of the carbide. This is in agreement with the results obtained by Buytoz [29].

To further explore the properties of the hardfaced layers, the hardness and microhardness of certain phases were measured. Vickers hardness of the bulk material is shown in Table 5. It can be seen that the hardness of the hardfaced layers increases from Specimen 1 to Specimen 4, with Specimen 4 being the hardest. The bulk material hardness of the reference Specimen D2 is the lowest of all the specimens tested in this study.

Table 5. HV30 hardness of the hardfaced layers and reference AISI D2 steel [kgf/mm^2].

	Specimen 1	Specimen 2	Specimen 3	Specimen 4	D2
Average values	663 ± 12	707 ± 8	712 ± 13	730 ± 6	640 ± 12

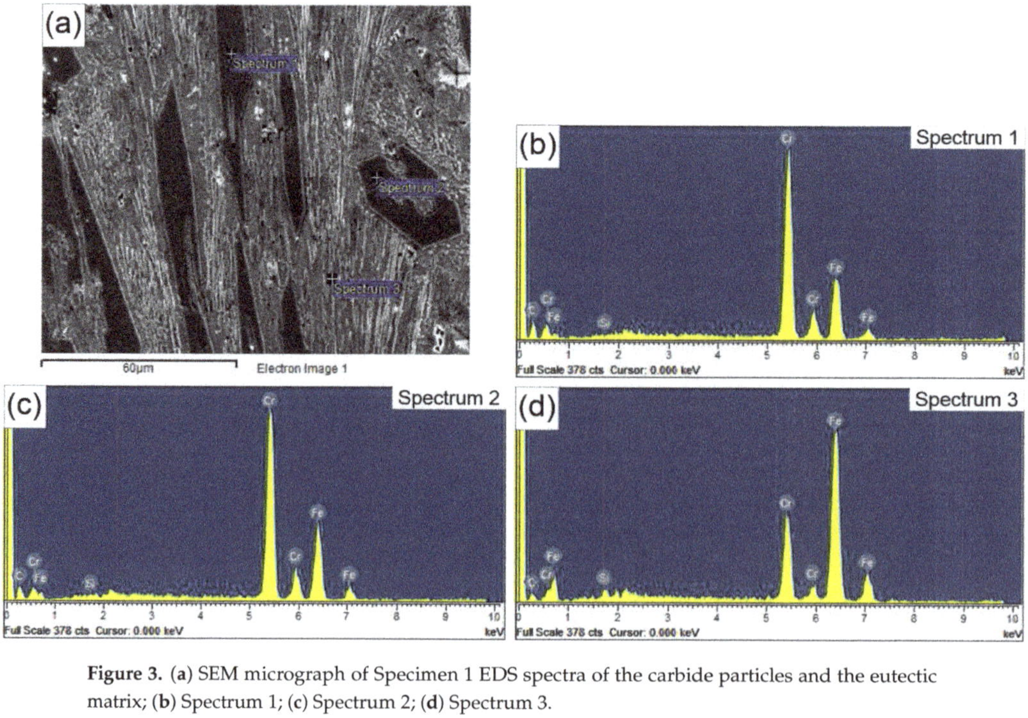

Figure 3. (**a**) SEM micrograph of Specimen 1 EDS spectra of the carbide particles and the eutectic matrix; (**b**) Spectrum 1; (**c**) Spectrum 2; (**d**) Spectrum 3.

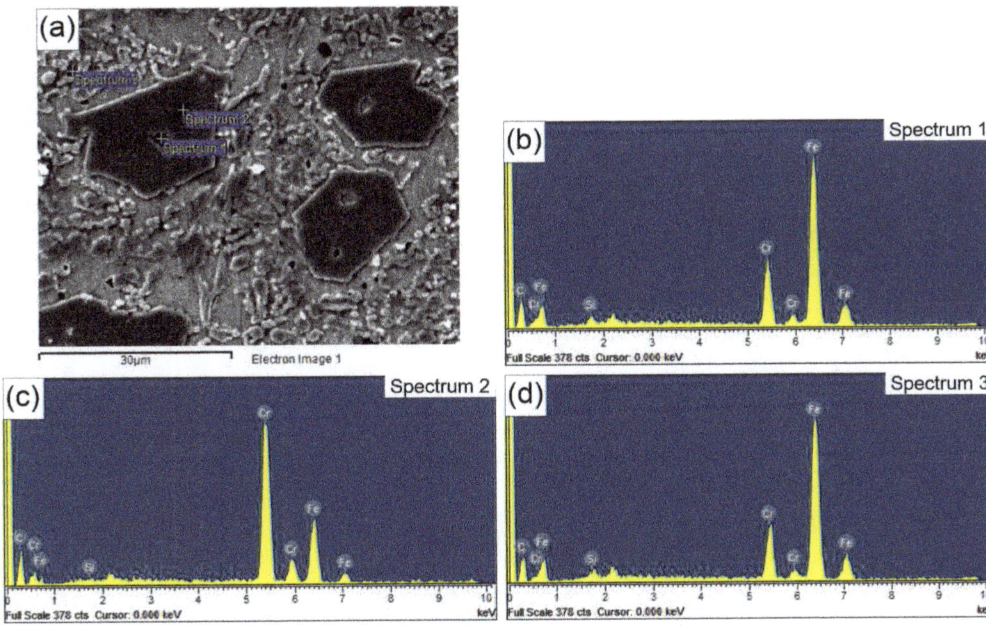

Figure 4. (**a**) SEM micrograph of Specimen 4 EDS spectra of the carbide particles and the eutectic matrix; (**b**) Spectrum 1; (**c**) Spectrum 2; (**d**) Spectrum 3.

Microhardness values for the carbides and eutectic matrix are given in Table 6. Although the same consumable was used, there are notable differences in microhardness, with the highest values obtained in the carbide and eutectic matrix regions of Specimen 4. In the case of M_7C_3 carbides, the hardness can be positively correlated with the carbide size. Larger carbides tend to exhibit slightly lower hardness compared to smaller ones, due to the possible presence of flaws, a less homogeneous structure, and a lower density of defects. However, in this study, the opposite result was observed, which can be explained by the large hollow central structure of the carbide filled with material, which is compositionally and morphologically similar to the eutectic matrix. This structure reduces the carbide's integrity, making it less resistant to deformation. This is supported by Figure 5, which depicts a hollow hexagonal M_7C_3 carbide that has cracked in several places. Additionally, a matrix-like structure can be observed in the center of the carbide, which is confirmed by the EDS results presented in Figure 3.

In Specimen D2, significantly lower microhardness values for both the carbides and matrix were obtained. This is consistent with the microstructure of the matrix, which consists of tempered martensite. Furthermore, the carbides are M_3C, which are not as hard as the M_7C_3 carbides found in the hardfaced layers. Moreover, the microhardness values of M_7C_3 primary carbides are approximately twice as high as those of the eutectic matrix. The carbide microhardness values for all specimens, presented in Table 6, are consistent with the results reported by Wang et al. [30].

The results of abrasive wear testing, specifically the weight loss rate, are shown in Figure 6. The data suggest that the highest abrasive wear loss rate was observed in Specimen 4. Its weight loss rate is significantly higher than that of Specimen 1, while the weight loss rates of Specimens 2 and 3 are similar. In contrast, the reference Specimen D2 exhibits a significantly lower abrasive wear loss rate compared to any of the hardfaced specimens (Specimens 1 to 4), especially Specimen 4. When compared to Specimen 4, reference Specimen D2 shows about half the weight loss rate. However, it is also evident that the standard deviation in Specimen D2 is the lowest of all the tested specimens, likely due to its relatively homogeneous microstructure, which consists mainly of tempered martensite with a smaller amount of M_3C carbides. Overall, the higher abrasive wear resistance of the hardfaced specimens can primarily be attributed to the increased amount of M_7C_3 carbides, compared to the M_3C carbides in the reference Specimen D2.

Table 6. HV0.01 microhardness of phases within the hardfaced layers [kgf/mm^2].

	Specimen 1	Specimen 2	Specimen 3	Specimen 4	D2
Carbides, average values	1601.9 ± 54.2	1683.9 ± 79.6	1735.2 ± 128.6	1837.4 ± 49.4	1052.4 ± 64.3
Matrix, average values	851.7 ± 61.8	910.8 ± 80.9	952.1 ± 121.0	997.4 ± 52.6	670.6 ± 50.6

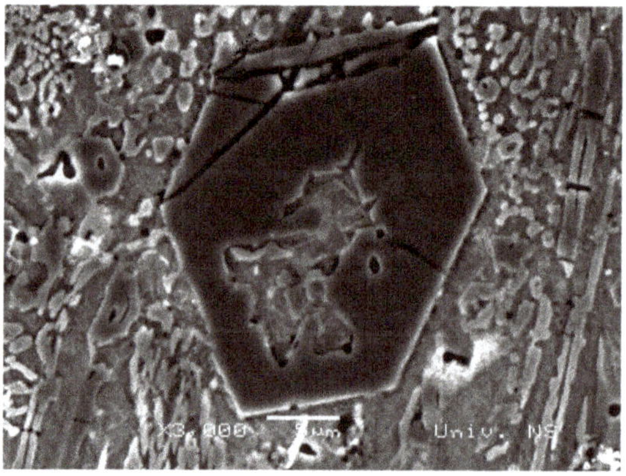

Figure 5. SEM micrograph of cracked M_7C_3 carbide in Specimen 1.

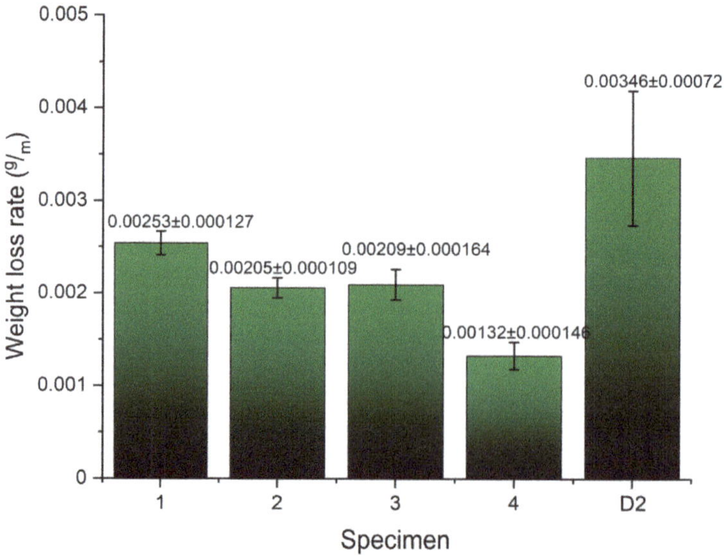

Figure 6. Weight loss rate of tested specimens.

The wear results are further supported by the surface roughness parameter measurements, shown in Figure 7. Based on these results, it can be deduced that the lowest roughness parameters were obtained in Specimen 4, which also exhibited the highest abrasive wear resistance, along with the highest hardness and microhardness. In contrast, Specimen D2 displayed the opposite behavior, exhibiting the highest roughness and the lowest wear resistance. A similar trend of increased roughness compared to Specimen 4 was observed in Specimen 1, which has a similar microstructure but slightly lower hardness and microhardness. The increased surface roughness can be attributed to the cutting action of the hard SiC particles, which abrade the specimen's surface consisting of carbides and a matrix of lower hardness.

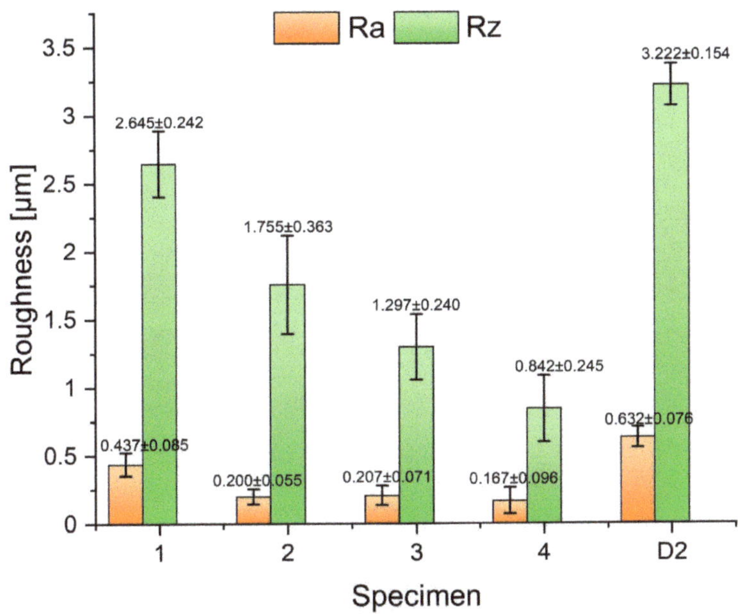

Figure 7. Roughness parameters Ra and Rz for wear tested specimens.

SEM analysis of the worn surfaces is shown in Figure 8a–e. It can be seen that, in Specimens 1 and 2 (Figure 8a,b), ploughing occurs along with mild delamination and the revealing of some pores, ranging from 1 to 3 μm in size, which influence the wear resistance of these hardfaced deposits. Ploughing is a typical mechanism of abrasion. In Specimen 1, a group of craters, likely pores, can also be observed. Porosity may result from a higher cooling rate of these specimens, where the pores form due to the incomplete escape of gases from the melt. In Specimens 3 and 4, ploughing with some delamination occurs, with polygonal carbides visible on the surface. These carbides appear larger in Specimen 4 compared to Specimen 3, which is consistent with the microstructural features of the deposited materials shown in Figure 2c,d. Additionally, in Specimens 3 and 4, small dark dots are present, possibly indicative of minor pores. In all Specimens 1 to 4, areas with flattened asperities are observed. Specimen D2 exhibits more uniform and seemingly more pronounced ploughing and delamination, which is similar to the results presented in [10]. Intensive ploughing is likely the result of intensive abrasion, in accordance with the lower hardness of the material and weight loss rate results compared to Specimens 1 to 4.

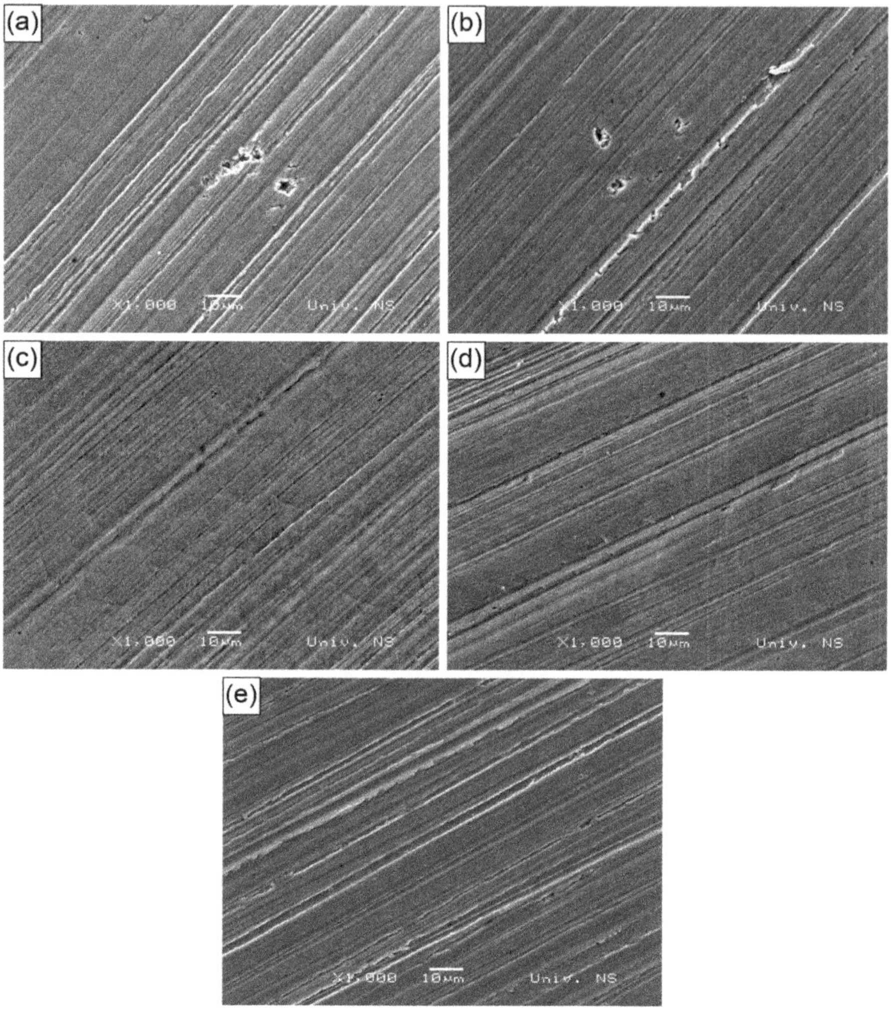

Figure 8. Worn surfaces: (**a**) Specimen 1; (**b**) Specimen 2; (**c**) Specimen 3; (**d**) Specimen 4; (**e**) Specimen D2.

4. Conclusions

The results obtained within the framework of this study, considering its limitations, may lead to the following conclusions:

- All hardfaced specimens exhibited higher abrasive wear resistance compared to the reference specimen made of D2 heat-treated, cold-work tool steel, due to a higher number of carbides and the overall bulk hardness of the material. An additional effect comes from the higher microhardness of both M_7C_3 carbides and the matrix in the hardfaced layers, compared to the microhardness of M_3C carbides and the martensitic matrix in D2 steel.
- The specimen hardfaced with higher current and lower speed exhibited the highest abrasive wear resistance against relatively hard SiC particles, as well as the lowest roughness parameters.

- High wear resistance is due to the relatively high microhardness of carbides, as well as a larger carbide size compared to other hardfaced specimens obtained with different parameters, i.e., a higher cooling rate.
- Slower cooling influenced the growth of larger M_7C_3 carbides, with smaller central hollow sections, which exhibit a higher microhardness compared to the specimen hardfaced with lower current and higher speed (i.e., with a lower cooling time).
- M_7C_3 carbides obtained with higher speed and lower current are relatively brittle and prone to cracking just after metallographic preparation, which negatively affects the wear resistance of the parent material. This section is not mandatory but can be added to the manuscript if the discussion is unusually long or complex.
- Weight loss rate results are supported by worn surface analysis. Besides ploughing, the hardfaced specimens exhibit minor delamination and pore formation as a result of incomplete gas escape from the melt. The control specimen was worn uniformly, with intensive ploughing and delamination.

In general, for abrasive wear, hardfacing proved to be more resistant than D2 steel, while in hardfacing, a slower cooling rate—achieved with higher current and slower hardfacing speed—can be beneficial.

Author Contributions: Conceptualization and methodology, S.B. and O.E.C.; validation, S.S. and M.R.; formal analysis, M.P., P.J. and D.L.Z.; investigation, D.L.Z. and M.R.; writing—original draft preparation, S.B.; writing—review and editing, D.L.Z. and S.S.; visualization, M.P. and P.J. All authors have read and agreed to the published version of the manuscript.

Funding: This research was financially supported by the Ministry of Science, Technological Development and Innovation of the Republic of Serbia under Contract numbers 451-03-65/2024-03/200156) through the project "Scientific and Artistic Research Work of Researchers in Teaching and Associate Positions at the Faculty of Technical Sciences, University of Novi Sad" (No. 01-3394/1), 451-03-65/2024-03/200108 (Faculty of Mechanical and Civil Engineering in Kraljevo, University of Kragujevac), 451-03-66/2024-03/200213 (Innovation Centre of the Faculty of Mechanical Engineering, Belgrade, University of Belgrade), 451-03-66/2024-03/200017 through the realization of research theme 1702414 ("Vinča" Institute of Nuclear Sciences—National Institute of the Republic of Serbia, University of Belgrade).

Institutional Review Board Statement: Not applicable.

Informed Consent Statement: Not applicable.

Data Availability Statement: The original contributions presented in this study are included in the article. Further inquiries can be directed to the corresponding authors.

Acknowledgments: The authors gratefully acknowledge research support by the project entitled "Materials Science, Joining and Allied Technologies in Production Engineering" from the Department of Production Engineering, Faculty of Technical Sciences Novi Sad, Serbia. Authors owe a special gratitude to Djordje Gacesa from company Tehnoalat, Novi Sad, Serbia for providing the roughness measurement equipment.

Conflicts of Interest: The authors declare no conflicts of interest. The funders had no role in the design of the study; in the collection, analyses, or interpretation of data; in the writing of the manuscript; or in the decision to publish the results.

References

1. Tandon, D.; Li, H.; Pan, Z.; Yu, D.; Pang, W. A Review on Hardfacing, Process Variables, Challenges, and Future Works. *Metals* **2023**, *13*, 1512. [CrossRef]
2. Singh, R.; Pandey, S.K. Experimental Investigation of Effect of Fe-Cr-C based Hardfacing on Wear Properties of Mild Steel. *Int. J. Eng. Res. Technol. (IJERT)* **2016**, *5*, 713–716. [CrossRef]

3. Gerard, B. *Fundamentals of Hardfacing by Arc Welding;* Welding Alloys: Florence, KY, USA, 2016; pp. 1–47. Available online: https://www.welding-alloys.com/wp-content/uploads/2021/12/Fundamentals_of_hardfacing_A5.pdf (accessed on 9 November 2024).
4. Kumar, S.; Mondal, D.P.; Jha, A.K.; Khaira, H.K. Improvement in high stress abrasive wear property of steel by hardfacing. *J. Mater. Eng. Perform.* **1999**, *8*, 711–715. [CrossRef]
5. Chakraborty, G.; Kumar, N.; Das, C.R.; Albert, S.K.; Bhaduri, A.K.; Dash, S.; Tyagi, A.K. Study on microstructure and wear properties of different nickel base hardfacing alloys deposited on austenitic stainless steel. *Surf. Coat. Technol.* **2014**, *244*, 180–188. [CrossRef]
6. Chang, C.-M.; Chen, Y.-C.; Wu, W. Microstructural and abrasive characteristics of high carbon Fe–Cr–C hardfacing alloy. *Tribol. Int.* **2010**, *43*, 929–934. [CrossRef]
7. Chatterjee, S.; Pal, T.K. Weld procedural effect on performance of iron basedhardfacing deposits on cast iron substrate. *J. Mater. Process. Technol.* **2006**, *173*, 61–69. [CrossRef]
8. Balos, S.; Janjatovic, P.; Dramicanin, M.; Labus Zlatanovic, D.; Pilic, B.; Hanus, P.; Jaworska, L. Microstructure, Microhardness, and Wear Properties of Cobalt Alloy Electrodes Coated with TiO_2 Nanoparticles. *Metals* **2019**, *9*, 1186. [CrossRef]
9. Garbade, R.R.; Dhokey, N.B. Overview on Hardfacing Processes, Materials and Applications. *IOP Conf. Ser. Mater. Sci. Eng.* **2021**, *1017*, 012033. [CrossRef]
10. Kaptanoglu, M.; Eroglu, M. Investigation of Hardfacing Deposits Obtained by Using Submerged Arc Welding Fluxes Containing High-Carbon Ferrochromium and Ferroboron. *Arch. Metall. Mater.* **2023**, *68*, 895–905. [CrossRef]
11. Correa, E.O.; Alcântara, N.G.; Valeriano, L.C.; Barbedo, N.D.; Chaves, R.R. The effect of microstructure on abrasive wear of a Fe–Cr–C–Nb hardfacing alloy deposited by the open arc welding process. *Surf. Coat. Technol.* **2015**, *276*, 479–484. [CrossRef]
12. Morsy, M.; El-Kashif, E. The effect of microstructure on high-stress abrasion resistance of Fe-Cr-C hardfacing deposits. *Zavar. I Zavarene Konstr.* **2014**, *3*, 119–129. [CrossRef]
13. Rajkumar, V.; Arjunan, T.V.; Rajesh Kannan, A. Metallurgical and mechanical investigations of Inconel 625 overlay welds produced by GMAW-hardfacing process on AISI 347 pipes. *Mater. Res. Express* **2019**, *6*, 076534. [CrossRef]
14. Abed, H.; Malek Ghaini, F.; Shahverdi, H.R. Characterization of Fe49Cr18Mo7B16C4Nb6 high-entropy hardfacing layers produced by gas tungsten arc welding (GTAW) process. *Surf. Coat. Technol.* **2018**, *352*, 360–369. [CrossRef]
15. Ahn, D.-G. Hardfacing technologies for improvement of wear characteristics of hot working tools: A Review. *Int. J. Precis. Eng. Manuf.* **2013**, *14*, 1271–1283. [CrossRef]
16. Liyanage, T.; Fisher, G.; Gerlich, A.P. Microstructures and abrasive wear performance of PTAW deposited Ni–WC overlays using different Ni-alloy chemistries. *Wear* **2012**, *274–275*, 345–354. [CrossRef]
17. Hou, Q.Y. Influence of molybdenum on the microstructure and properties of a FeCrBSi alloy coating deposited by plasma transferred arc hardfacing. *Surf. Coat. Technol.* **2013**, *225*, 11–20. [CrossRef]
18. Yao, J.; Ding, Y.; Liu, R.; Zhang, Q.; Wang, L. Wear and corrosion performance of laser-clad low-carbon high-molybdenum Stellite alloys. *Opt. Laser Technol.* **2018**, *107*, 32–45. [CrossRef]
19. Balakrishnan, M.; Balasubramanian, V.; Madhusudhan Reddy, G. Effect of Hardfacing Processes on Ballistic Performance of Q&T Steel Joints. *Mater. Perform. Charact.* **2014**, *3*, 265–284. [CrossRef]
20. Babu, S.; Balasubramanian, V.; Madhusudhan Reddy, G.; Balasubramanian, T.S. Improving the ballistic immunity of armour steel weldments by plasma transferred arc (PTA) hardfacing. *Mater. Des.* **2010**, *31*, 2664–2669. [CrossRef]
21. Venkatesh, B.; Sriker, K.; Prabhakar, V.S.V. Wear Characteristics of Hardfacing Alloys: State-of-the-art. *Procedia Mater. Sci.* **2015**, *10*, 527–532. [CrossRef]
22. Digambar, B.; Choudhary, D. A review paper on hardfacing processes, materials, objectives and applications. *Int. J. Sci. Res.* **2014**, *3*, 2400–2402.
23. Grilli, M.L.; Valerini, D.; Slobozeanu, A.E.; Postolnyi, B.O.; Balos, S.; Rizzo, A.; Piticescu, R.R. Critical Raw Materials Saving by Protective Coatings under Extreme Conditions: A Review of Last Trends in Alloys and Coatings for Aerospace Engine Applications. *Materials* **2021**, *14*, 1656. [CrossRef] [PubMed]
24. *Regulation (EU) 2024/1252 of the European Parliament and of the Council of 11 April 2024 Establishing a Framework for Ensuring a Secure and Sustainable Supply of Critical Raw Materials and Amending Regulations (EU) No 168/2013, (EU) 2018/858, (EU) 2018/1724 and (EU) 2019/1020;* European Commission: Brussels, Belgium, 2023.
25. Stubbington, P. *Hardfacing Alloy Design Consumable Review;* Lincoln Electric: Padstow, NSW, Australia, 2021.
26. Berns, H.; Fischer, A. Microstructure of Fe-Cr-C Hardfacing Alloys with Additions of Nb, Ti and B. *Metallography* **1987**, *20*, 401–429. [CrossRef]
27. Hobart Filler Metals Application Department. *Hardfacing Technical Welding Guide;* Hobart Brothers, LLC.: Troy, OH, USA, 2020.

28. Ito, S.; Kawai, H.; Oishi, Y.; Nogami, H. Structure and Solidification Process Observation of Fe-C-Cr Alloy Using Confocal Laser Scanning Microscope. *MATEC Web Conf.* **2021**, *333*, 11001. [CrossRef]
29. Buytoz, S. Microstructural properties of M7C3 eutectic carbides in a Fe–Cr–C alloy. *Mater. Lett.* **2006**, *60*, 605–608. [CrossRef]
30. Wang, J.; Xing, X.; Zhou, Y.; Liu, S.; Qi, X.; Yang, Q. Formation mechanism of ultrafine M_7C_3 carbide in a hypereutectic Fe-25Cr-4C-0.5Ti-0.5Nb-0.2N-2LaAlO_3 hardfacing alloy layer. *J. Mater. Res. Technol.* **2020**, *9*, 7711–7720. [CrossRef]

materials

Article

Numerical Fatigue Analysis of Dissimilar Lap Joints Fabricated by Dimple Spot Welding for Automotive Application

Paolo Livieri * and Michele Bortolan

Department of Engineering, University of Ferrara, via Saragat 1, 44121 Ferrara, Italy; michele.bortolan@unife.it
* Correspondence: paolo.livieri@unife.it

Abstract: This paper presents a numerical analysis of dimple spot welding (DSW) as an innovative joining technique for dissimilar materials, namely steel and aluminium alloys. Employing a finite element (FE) model, the study simulates the fatigue performance of DSW joints, considering crucial factors such as contact friction and cyclic loading conditions. While various numerical models are proposed, the simulation incorporating friction and fatigue loading appears to offer the highest accuracy. The research highlights that the fatigue behaviour of DSW joints can be effectively investigated through the non-local theory of the implicit gradient approach by utilising the fatigue curve of arc-welded structures composed of steel or aluminium alloys. Specifically, simulations incorporating friction and fatigue loading demonstrate that the steel spot weld does not represent the weakest point within the joints.

Keywords: fatigue; welded joints; implicit gradient; dimple spot welding

1. Introduction

The automotive industry is increasingly focused on developing lightweight vehicle structures to enhance fuel efficiency and reduce emissions. This shift has necessitated the exploration of alternative joining technologies, particularly for lightweight metals like aluminium alloys, which are gaining prominence in modern vehicle designs. Traditional methods, such as spot welding [1–3], are used for joining steel body panels, but their application to aluminium has proven challenging. Aluminium's high thermal conductivity, low melting range, and tendency to form an oxide layer complicate the spot welding process, often resulting in suboptimal joint integrity and reduced lifespan of welding electrodes [4]. To address these challenges, several alternative joining methods have been considered, including adhesive bonding [5], riveting [6], and clinching [7–10]. Among these, self-piercing riveting (SPR) has emerged as a promising technology for joining aluminium body panels. SPR operates as a cold forming process, allowing a semi-tubular rivet to be pressed into two sheets of material without the need for pre-drilled holes or precise alignment, thus simplifying assembly and enhancing the potential for joining dissimilar materials [11].

Recent studies have assessed the mechanical performance of SPR joints, revealing a complex interplay between static and fatigue strengths. For instance, while spot-welded joints often demonstrate higher static strength, SPR joints exhibit superior fatigue resistance. Research indicates that at 10^6 cycles, the fatigue strength of SPR joints can be twice that of their spot-welded counterparts [12,13]. This distinction makes SPR joints particularly suitable for applications subjected to cyclical loading.

The effectiveness of an SPR joint hinges on the quality of the mechanical interlock formed during the riveting process. Different factors influence this interlock, including die geometry, sheet thickness, rivet properties, and surface coatings [14]. Studies have shown that modifying die dimensions can significantly affect rivet flaring and joint strength. Similarly, the length and diameter of the rivet play crucial roles; longer and thicker rivets tend to absorb more energy, enhancing joint robustness [15].

Another critical aspect influencing joint strength is specimen configuration. Research demonstrates that joint strength varies significantly based on whether the configuration is optimised for shear or peel tests, with shear configurations generally exhibiting higher strength [16]. The design of the joint itself is also pivotal; evidence suggests that double-riveted joints do not inherently surpass the performance of single-riveted joints when normalised for stress, emphasising the importance of optimal rivet placement and spacing [17].

Despite the advantages of SPR, the diverse combinations of material grades and thicknesses present challenges in determining optimal rivet and die configurations. The need for tailored processes for numerous combinations increases complexity and cost in production [18].

The mechanical behaviour of SPR joints is further complicated by the loading conditions they face in service. Studies have highlighted how rivet parameters, sheet thickness, and edge distance significantly affect both static and fatigue performance. For instance, increasing edge distance can reduce joint distortion and enhance strength [19]. In summary, while the self-piercing riveting process offers a viable alternative to traditional joining methods for aluminium and dissimilar materials, ongoing research is essential needed to fully understand and optimise its mechanical performance. The exploration of SPR's capabilities and limitations is crucial as the automotive industry continues to embrace innovative materials and production techniques, paving the way for more sustainable and effective vehicle manufacturing solutions.

In this context, the innovative dimple spot welding (DSW) method developed by Sakaguchi et al. [20] represents an interesting advancement. DSW leverages existing resistance spot welding infrastructure to join aluminium alloys and steel sheets efficiently. This method involves creating a dimple in the steel and drilling a corresponding hole in the aluminium, allowing for a secure mechanical interlock when welded. Preliminary tests indicate that DSW joints exhibit static and fatigue strengths comparable to those achieved by SPR, thus offering a promising solution for integrating lightweight materials into automotive structures without necessitating costly equipment modifications.

Under this scenario, the aim of this paper is to numerically analyse dimple spot welding by means of the implicit gradient approach [21–23]. The analyses are performed by considering the characteristic fatigue strength of welded steel and aluminium alloy. The fatigue behaviour of dimple spot welds is analysed by investigating various models that incorporate the contact interaction between the two plates.

2. Theoretical Background

This paper explores a sophisticated approach to addressing notch effects in fatigue analysis by defining an effective non-local stress that captures the stress field surrounding critical points in a material. The foundation of this approach lies in the non-local theory, which suggests that the stress at a point cannot be understood solely by local values but should also incorporate the influence of surrounding points.

To formalise this concept, the effective stress $\sigma_{eff}(P)$ at any point P within a volume V is defined as a weighted average of equivalent stress σ_{eq}, represented mathematically as:

$$\sigma_{eff}(P) = \frac{\int\limits_V \Psi(P,Q)\sigma_{eq}(Q)dV}{\int\limits_V \Psi(P,Q)dV} \tag{1}$$

where Q is a variable point in the volume V, and Ψ is an isotropic weight function that diminishes with increasing distance $|P - Q|$. This function accounts for the stress gradient in the local field, enhancing the predictive capabilities of the model. The reference volume $\int\limits_V \Psi(P,Q)dV$ normalises the weighted average, ensuring that the effective stress is dimensionally consistent.

According to Peerlings at al. [24], this integral definition of effective stress σ_{eff}, can be approximated by transforming Equation (1) into an inhomogeneous Helmholtz equation:

$$\sigma_{eff} - c^2\nabla^2\sigma_{eff} = \sigma_{eq}\ in\ V \tag{2}$$

where ∇^2 denotes the Laplace operator, and c represents a characteristic length related to the material properties. This transformation is significant, as it allows the effective stress to be solved more easily than the original integral formulation given by Equation (1), particularly in three-dimensional domains where the critical point cannot be predetermined.

To apply this framework in practical scenarios, the authors consider Neumann-type boundary conditions, which facilitate the numerical solution of the Helmholtz equation. This strategy is further supported by the use of commercial finite element software (COMSOL Multiphysics^{TM}), where second-order tetrahedral elements are employed to mesh the three-dimensional models.

Moreover, the paper cites early contributions to the non-local theory by Kroner, Eringen, and Edelen [25], which laid the groundwork for understanding stress as a function of averaged strain within a representative volume of material. This theoretical background is crucial for modelling materials that exhibit strain-softening behaviour, as proposed by Pijaudier-Cabot and Bazant [26,27].

The paper emphasises that under fatigue loading, stress variations are critical for defining equivalent stress. In the case with proportional and uniaxial loading scenarios, the maximum principal stress variation $\Delta\sigma_1$ can be effectively used to characterise the equivalent stress range ($\Delta\sigma_{eq} = \Delta\sigma_1$), leading to equations that streamline the analysis:

$$\Delta\sigma_{eff} - c^2\nabla^2\ \Delta\sigma_{eff} = \Delta\sigma_{eq} \tag{3}$$

In particular, for welded joints, the characteristic length c is specified as 0.2 mm, whereas this becomes 0.15 for welded joints made of aluminium alloy.

Furthermore, the value of c can be calculated according to Equation (4) (see ref. [28]):

$$c = za_0 \tag{4}$$

where, for fatigue loading, the El-Haddad [29] critical length a_0 takes the form:

$$a_0 = \frac{1}{\pi}\left(\frac{\Delta K_{th}}{\Delta\sigma_0}\right)^2 \tag{5}$$

where $\Delta\sigma_0$ is the fatigue limit of the base material and ΔK_{th} is the threshold range of the Stress Intensity Factor. The parameter z is a constant that depends on the equivalent stress σ_{eq}. When the equivalent stress σ_{eq} is assumed to be equal to the maximum principal stress, z becomes 0.545.

Equations (4) and (5) allow us to assess the value of c when the relevant material properties of the parent material and the cracked specimen are known.

In Figure 1, typical spot weld joints subjected to tensile loadings are reported. Due to symmetry, only half the model is reported in the figure. The mesh is only fine near the weld where the stress becomes singular due to the sharp V-notch. The deformed shape in Figure 1b indicates that, at the two ends of the specimens, the rotation is neglected. The maximum principal stress tends to infinity at the notch tip whereas the effective stress σ_{eff} obtained by solving Equation (2) at the weld is not singular and assumes a finite value [21,22,24,30]. Hence, the value of $\Delta\sigma_{eff}$, related to the range of nominal stress $\Delta\sigma_{nom}$, can be used to evaluate the fatigue life of components. The location at which the maximum value of $\Delta\sigma_{eff}$ occurs will be the most likely site for fatigue nucleation.

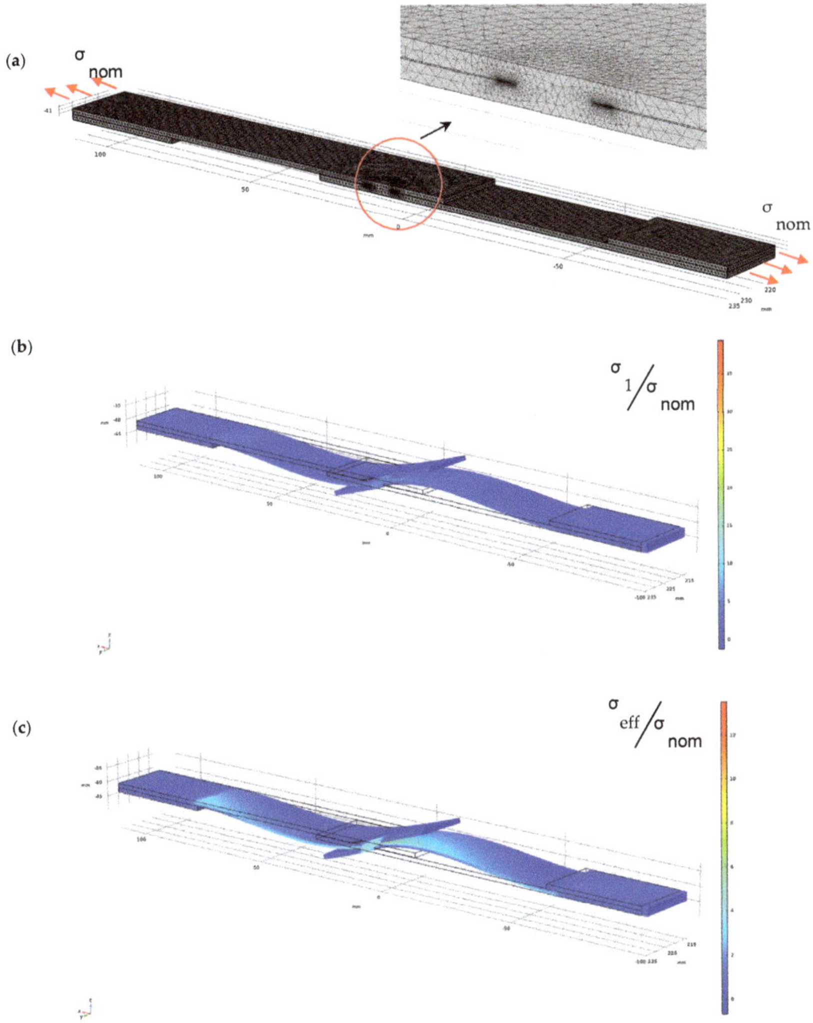

Figure 1. Effective stress σ_{eff} for spot weld joint evaluated by solving Equation (2), assuming that σ_{eq} coincides with the maximum principal stress σ_1. The plate is subjected to a nominal tensile stress σ_{nom}; (**a**) mesh; (**b**) maximum principal stress σ_1; and (**c**) effective stress σ_{eff}.

Figure 2 shows the fatigue curve for welded joints obtained from about 600 experimental data taken from literature, mainly in relation to the fatigue strength of cruciform and T joints subjected to tensile or bending loading. The fatigue strength, in terms of the range of effective stress $\Delta\sigma_{eff}$, depends only on the material: steels or aluminium alloy. The two scatter bands are separated and present different slopes: 3.0 for the weld made of steel and 3.7 for the weld made of aluminium alloy. For a complex welded joint, by solving Equation (4), the maximum effective stress range $\Delta\sigma_{eff}$ is the main physical parameter that, as introduced in Figure 2, gives us the fatigue life of components independently of the shape of the weld, and the size and type of loading (bending or tensile stress).

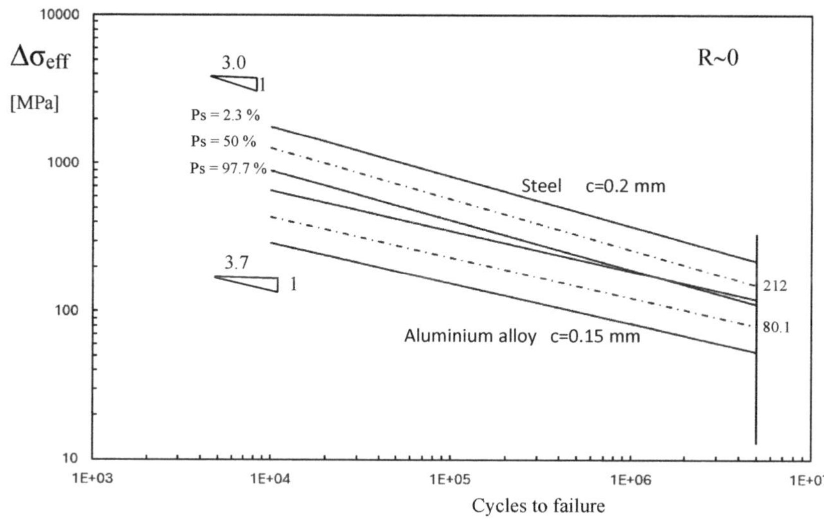

Figure 2. Scatter band of steel welded joints [21] in terms of maximum effective stress range (scatter bands related to mean values plus/minus 2 standard deviations; *Ps:* probability of survival).

3. Dimple Spot Welding

This section summarises the interesting experimental work on a novel type of connection between dissimilar materials, such as steel and aluminium alloys, proposed by Sakaguchi et al. [20].

3.1. Geometry

Dimple spot welding, as proposed by Sakaguchi et al. [20], is a process that uses a spot weld from two overlapping metal pieces and locks the displacement of an interposed sheet. The process begins by forming a dimple in a steel sheet and drilling a hole in an aluminium alloy plate. A backing steel plate is then prepared. The aluminium plate is positioned between the steel sheet and the backing plate, with the dimple from the steel sheet aligning through the hole in the aluminium. Resistance spot welding is used to join the steel sheet and backing plate, securely clamping the aluminium plate in place. A key advantage of this dimple spot welding method is that it utilises existing resistance spot welding equipment, allowing for seamless integration into current production lines without requiring modifications.

Figure 3 reports the geometry of the dimple spot welding analysed in [20]. The thickness of the two joined plates is 1.4 mm and 2.0 mm for steel and for aluminium alloy, respectively. A tab is present to align the load when remote tensile loading is applied. The overlapping area is a square of 40 × 40 mm, and the spot weld has a diameter of 5.5 mm.

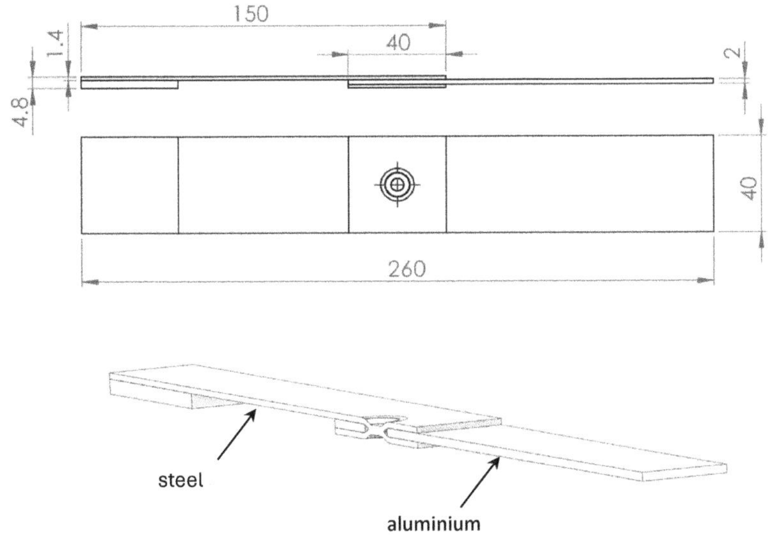

Figure 3. Specimen used in the experimental analysis [20].

3.2. Materials

Table 1 shows the material used in the four series of specimens for the plate made of steel and aluminium alloy. Table 2 presents the mechanical properties of the materials.

Table 1. Steel sheets and aluminium alloys considered in the dimple spot welding [20].

Specimen	Aluminium Alloy	Steel Sheet	Friction Coefficient
DSW-1	A6N01	SPC590	0.56
DSW-2	A7003	SPC590	0.45
DSW-3	A6N01	SPC980	0.42
DSW-4	A7003	SPC980	0.38

Table 2. Ultimate tensile strength, yield strength, and elongation of material used in dimple spot welding [20].

Material	Ultimate Tensile Strength (MPa)	Yield Strength (MPa)	Elongation (%)
A6N01	257	225	9.4
A7003	395	349	10.8
SPC590	590	444	24.8
SPC980	980	813	13.9

In order to calculate the material parameter c for the steel weld, a common approach is to use the value of 0.2 mm for the arc welding process, independently of the material properties [30,31]. Unfortunately, for the aluminium plate, the fatigue limit and the fatigue life curve of the base material as well as the threshold range of the stress intensity factor are not reported in reference [20]. The exact values can vary depending on factors such as the material heat treatment, environmental conditions, and the specific testing method. If we estimate these two important strength parameters from the tensile strength reported in Table 2 or from the average value reported in the literature, the size of c estimated by means of Equations (4) and (5) is around some tenths of a millimetre. However, due to the inherent uncertainty in this calculation, the typical value for welded joints made of aluminium alloy will be utilised. This is a strong exemplification of the problem. However, in this case we know the fatigue curve of the material and the characteristic length c that can address some important general information about fatigue performance of dimple spot welding.

3.3. Fatigue Strength

Fatigue tests were conducted at room temperature at constant amplitude and the nominal load ratio R was set equal to 0.1 [20]. The fatigue life comparison between dimple spot welding and self-piercing riveting joints shows that the DSW specimens, made from the same materials (SPC980 steel and A7003 aluminium alloy), exhibit fatigue lives approximately 10 times longer than self-piercing riveting specimens. DSW joints demonstrated superior fatigue strength, with all failures occurring in the aluminium plate. Sakaguchi et al. [20] underline that the fatigue life of DSW specimens depends on both the aluminium alloy strength and the steel sheet strength, with higher-strength materials increasing fatigue life. Figure 4 reports the comparison, in terms of fatigue strength, between the nominal stress applied to DSW specimens and that of SPR joints, which will be used in the next section for the numerical simulation. After failure, the analysis of the aluminium plate of DSW specimens shows a white discoloured area due to the repeated contact and friction observed at the top side while the crack nucleates in the opposite side of the hole where black debris was observed, as indicated in Figure 5. Sakaguchi et al. [20] justified this behaviour by observing that, in their FE analysis, the higher tangential stress area corresponded well to the actual site of crack initiation.

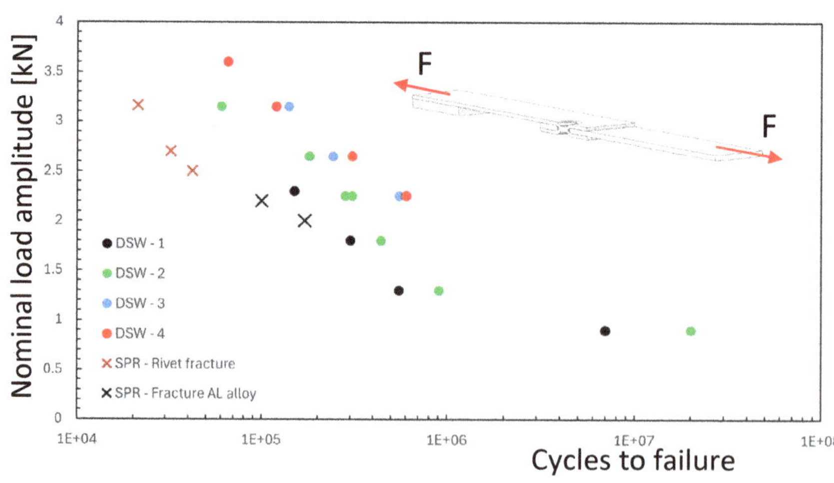

Figure 4. Comparison between the fatigue strength of dimple spot welding (DSW) joints and the self-piercing riveting (SPR) joints under tensile loading. Nominal load ratio R = 0.1 (F: nominal load; experimental results from reference [20]).

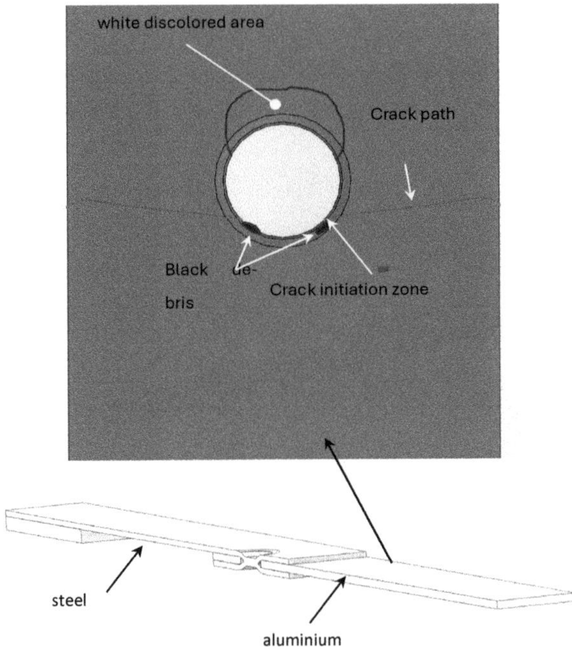

Figure 5. Typical fatigue failure at the aluminium plate.

4. Numerical Analysis of Dimple Spot Welding

4.1. FE Model

In order to simulate, with a tree-dimensional approach, the fatigue behaviour of dimple spot welding joints, a detailed FE analysis will be performed by means of two pieces of software: Ansys Workbench and Comsol Multiphysics. This is necessary because the complicated task of simulating the contact of the sheets, as reported in Figure 3, with or without the friction coefficients, will be undertaken using the Ansys code. Then, the FE results, such as the maximum principal stress, will be exported and used by Comsol Multiphysics as input data to solve the Helmholtz Equations (2) or (3).

The three-dimensional model was built by means of 3D Cad and reproduces the details of the geometry around the spot weld, as appears in reference [20]. Figure 6 shows the detail of the FE mesh used for the DSW joints. Due to the symmetry, only half the joint was modelled. At both ends, rotations were prevented by setting as zero the displacement perpendicular to the longitudinal plane (u_y in Figure 6a) and a force was applied at one end while restraining longitudinal displacements at the other (u_x in Figure 6a). Figure 6b shows the details of the mesh used in the FE analysis. Tetragonal elements were used with quadratic element order in the Ansys software, while in Comsol second-order element with Lagrange shape function was adopted. The size of the smallest element was in the order of 10^{-1} mm at the weld for the steel plate and all around the hole of the aluminium plate. A convergence analysis was performed to obtain a stable value of effective stress in the contact zone. Despite an oscillation value of the maximum principal stress in the zone where the plates are in contact (around the white discoloured area in Figure 5), the value of effective stress also resulted stable in that area. This is an advantage of the numerical implicit technique adopted in this study. The use of more accurate meshes would result in excessively long computation times. On the other hand, due to the uncertainty associated with the characteristic length *c*, the accuracy in the fatigue life prediction can only be accurate in the zone around the spot weld (see nest sections). As a general rule, convergence in finite element analysis is achieved when the minimum size of the elements in the region of highest stress is in the same order of magnitude as the characteristic length *c* [30].

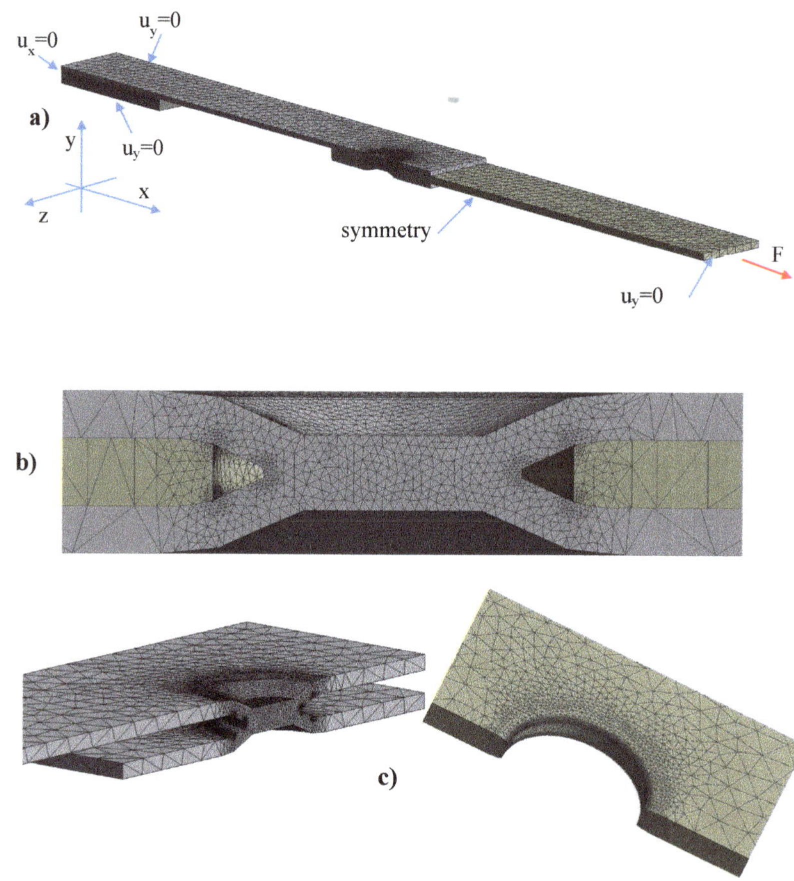

Figure 6. FE model and boundary conditions for the dimple spot welding; (**a**) Boundary conditions used in the model (u: displacement); (**b**) Mesh detail around the spot weld; (**c**) Mesh for spot weld and aluminium plate around the hole.

The residual stress due to the weld process was not considered in this study when no stress release was performed [31].

The material was considered as linear elastic. The Poisson's ratios of the steel sheets and aluminium alloys were set to 0.30 and 0.34, respectively. The Young modulus of 200 GPa was considered for the steel plate and was reduced to 70 GPa for the aluminium plate.

4.2. Results with Frictionless Simulation

The first analysis was performed without taking into account the friction coefficient between the two plates. The load was transmitted by the wedge effect between the steel and the aluminium plate. Figure 7 reports the values of the range of effective stress related to a load range of 4.5 kN applied at the remote section. This load, applied during the dimple spot welding, involves an average fatigue life of around $3 \cdot \times 10^5$ cycles, as appears in Figure 4 for DSW2. The results reported in Figure 7, as well as other similar results reported in the figure below, are focalised around the spot weld because it is the place of highest stress level for both the aluminium and the steel plate. For the steel plate, the maximum effective stress range is at the weld, whereas for the aluminium plate, the maximum is localised at the central part of the hole. Furthermore, the FE simulation results indicate that the contact zone between the two plates is localised around the white discoloured area observed in Figure 5.

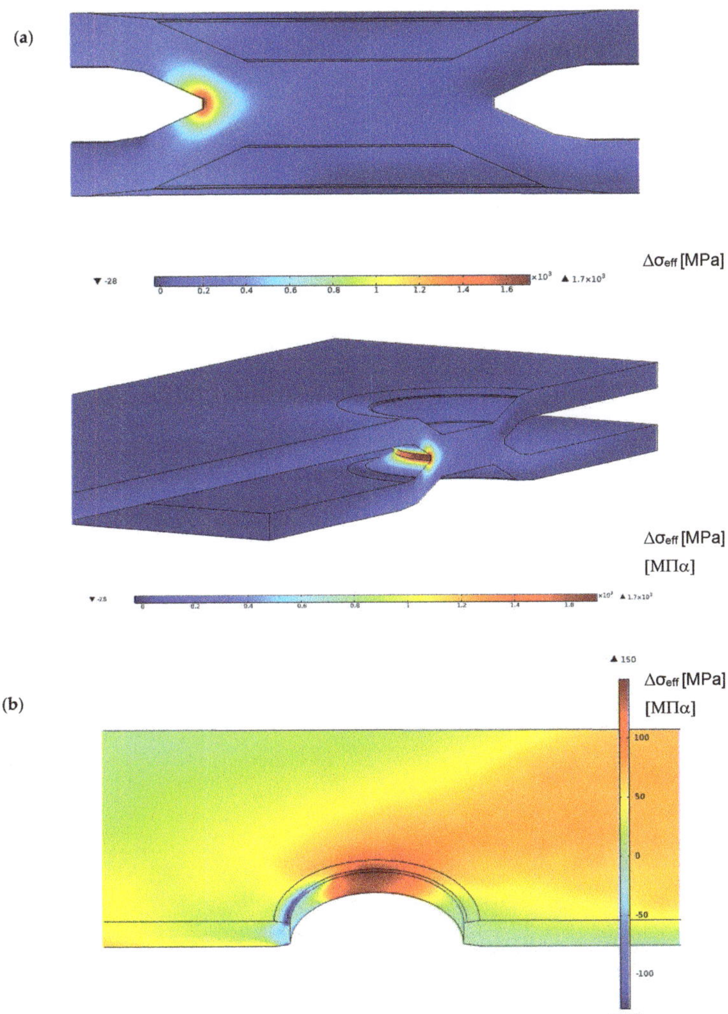

Figure 7. Range of effective stress in the dimple spot welding around the spot weld and the toe for a remote applied force range of 4.5 kN for the frictionless model (**a**) steel plate; (**b**) aluminium plate.

By considering series DSW2 as a reference, Figure 8 reports the experimental points in terms of effective stress range over the fatigue diagram of Figure 2. The experimental points of the aluminium plate fall into the scatter band of the aluminium alloy weld, but the steel spot weld shows a very high stress level. If this were true, the failure should occur in the steel plate and a fatigue crack should be visible at the weld, but this is in contrast with the experiments.

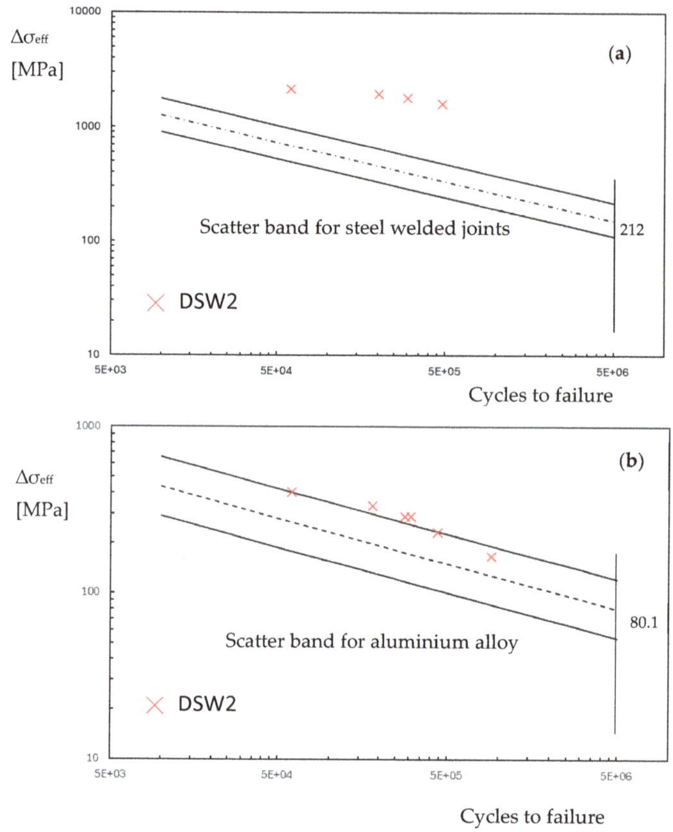

Figure 8. Fatigue life in terms of effective stress range in the dimple spot welding for the FE frictionless model (**a**) steel plate at the weld; (**b**) aluminium plate at the hole.

For the aluminium plate, the maximum effective stress, as shown in Figure 7, does not exactly agree with the points where the fatigue crack nucleates. Furthermore, the value of the characteristic length *c* and the Woehler curve of the aluminium plate in Table 2 are not known, and were approximated with those of the welded aluminium alloy. Nevertheless, in the spirit of a simplified analysis, the experimental points are located on the upper part of the welded aluminium alloy scatter band.

4.3. Results with Friction Simulation

In this section, a friction coefficient between the two plates of Figure 3 was introduced. Sakaguchi at al. [20] discussed the rule of friction coefficients ranging between 0.38 and 0.56, but in our simulations a reference value of 0.45 was considered independently of the couple of plates used (see Table 1 for the friction coefficients of each series). This constitutes a simplification of the contact problem, as a more accurate model would account for the influence of relative speed, surface topography, and contact pressures [32–34]. Figure 9 reports the effective stress related to a load range of 4.5 kN applied at the remote section, as shown in Figure 7. The trend of effective stress is similar to the case of the frictionless model but with lower values in the steel plate and higher values in the aluminium plate.

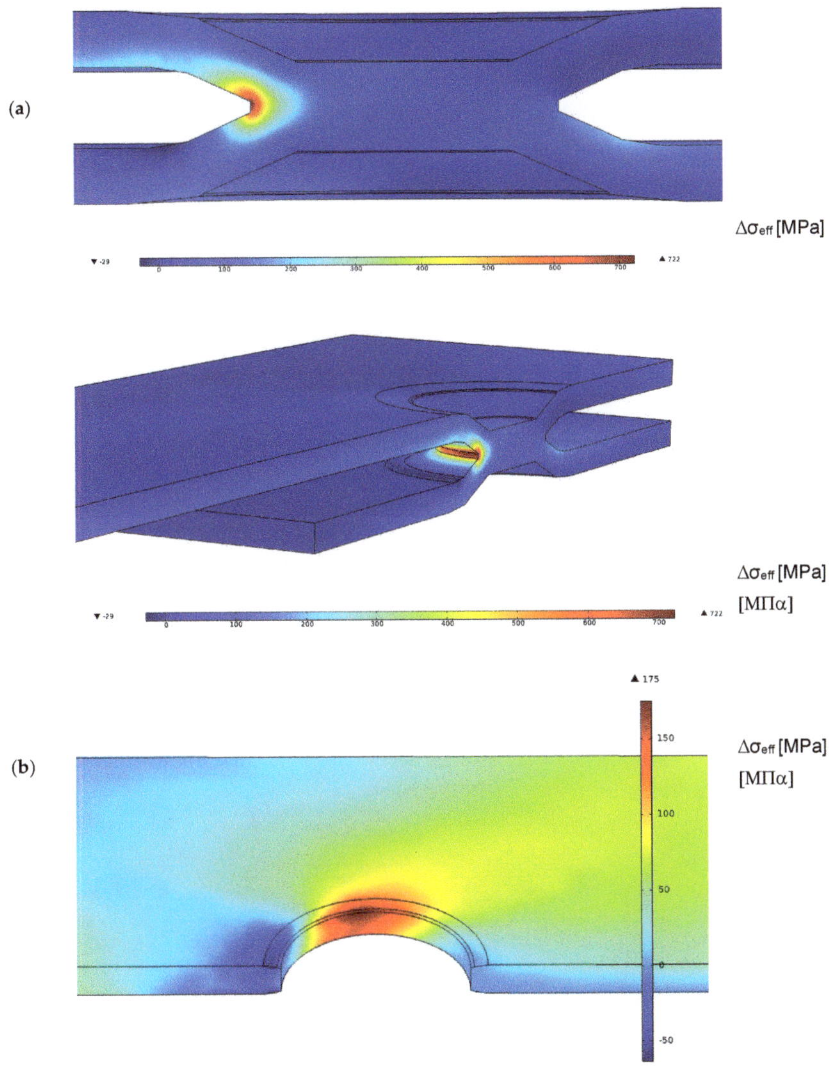

Figure 9. Range of effective stress in the dimple spot welding around the spot weld and the weld toe for a remote applied force range of 4.5 kN for the model with a friction coefficient of 0.45 (**a**) steel plate; (**b**) aluminium plate.

Similar to the frictionless simulation, assuming that the effective stress range corresponds to that of monotonic loading, it is possible to plot the experimental points of series DSW2 on the fatigue diagram in Figure 2 by considering the friction coefficient. The introduction of the friction coefficient does not substantially modify the fatigue behaviour of the steel plate, as reported in Figure 10.

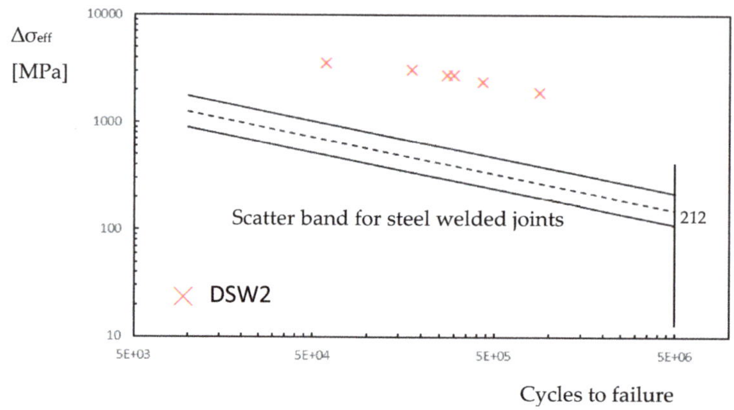

Figure 10. Fatigue life in terms of effective stress range in the dimple spot welding for the FE friction model (steel plate at the weld).

4.4. Results with Friction Simulation and Fatigue Loading Simulation

In this section, a friction coefficient of 0.45 was incorporated into the model, as in the previous section. However, the initial load cycles were simulated following the approach outlined in a previous study that numerically investigated the fatigue behaviour of double-riveted lap joints [35]. In reference [32], it was verified that the effect of the friction force tends to reduce the range of stresses at the net section during fatigue loadings. Therefore, in this study, the nominal load was considered as variable, as reported in Figure 11. Two complete cycles were simulated. The results at times 1 and 3 are quite similar, as are those at times 2 and 4, indicating that simulating additional cycles is unlikely to produce significant changes.

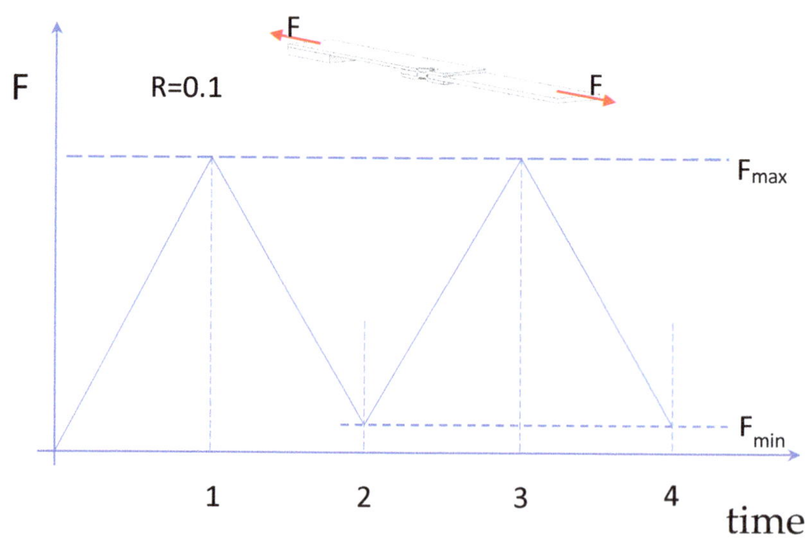

Figure 11. Load history at the nominal section for a nominal load ratio *R* of 0.1.

Figure 12 reveals that, despite a nominal load ratio R of 0.1, the effective stress range at point A is significantly lower than expected. In a frictionless scenario, the effective stress range is 90% of the maximum stress level. This discrepancy is less pronounced at point B in the aluminium plate.

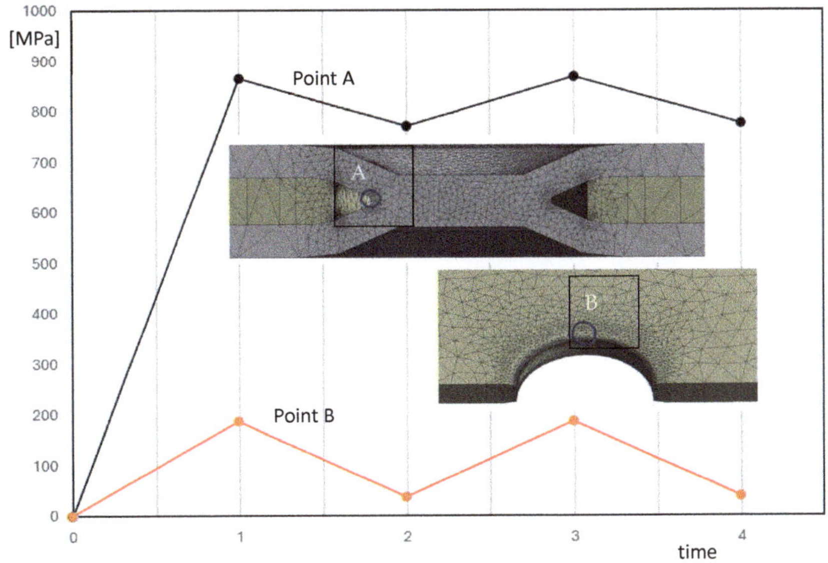

Figure 12. Effective stress range versus time at the weld (point A, steel plate) and at the border of the aluminium plate (point B) for the model with a friction coefficient of 0.45 by simulating the fatigue loading (ΔF = 2250 N, F_{max} = 2500 N, F_{min} = 250 N, nominal load ratio R = 0.1).

In order to obtain the fatigue damage of the weld, without taking into account a multiaxial fatigue criterion [36], at any points of the dimple spot welding, the difference between the maximum principal stress at time 3 and the maximum principal stress at time 4 is considered. Now, by integrating Equation (3), we obtain the range of effective stress to introduce in the fatigue curve of Figure 2. Figure 13 shows the range of effective stress due to the range of principal stress. The position of maximum effective stress moves from the weld to point C. In this region, the characteristic length may differ from that of the welded zone. However, due to the lack of precise knowledge regarding the correct value of *c* for the steel plate, and for the sake of simplicity, the fatigue resistance of welded steel structures is assumed to be equal to that of the weld throughout the plate. Figure 14 reports the fatigue points in terms of the effective stress range at the weld (point A) or at the plate (point C). In Figure 14, the points relative to C fall inside the scatter band but are not at all significative because of the uncertainty in the fatigue properties. On the contrary, the experimental points related to the fatigue strength of the weld show a low value of effective stress range, and this respects the condition that the fatigue failure does not occur in the steel plate. Finally, Figure 14 also includes the experimental data points for the other series, considering the effective stress at the weld toe.

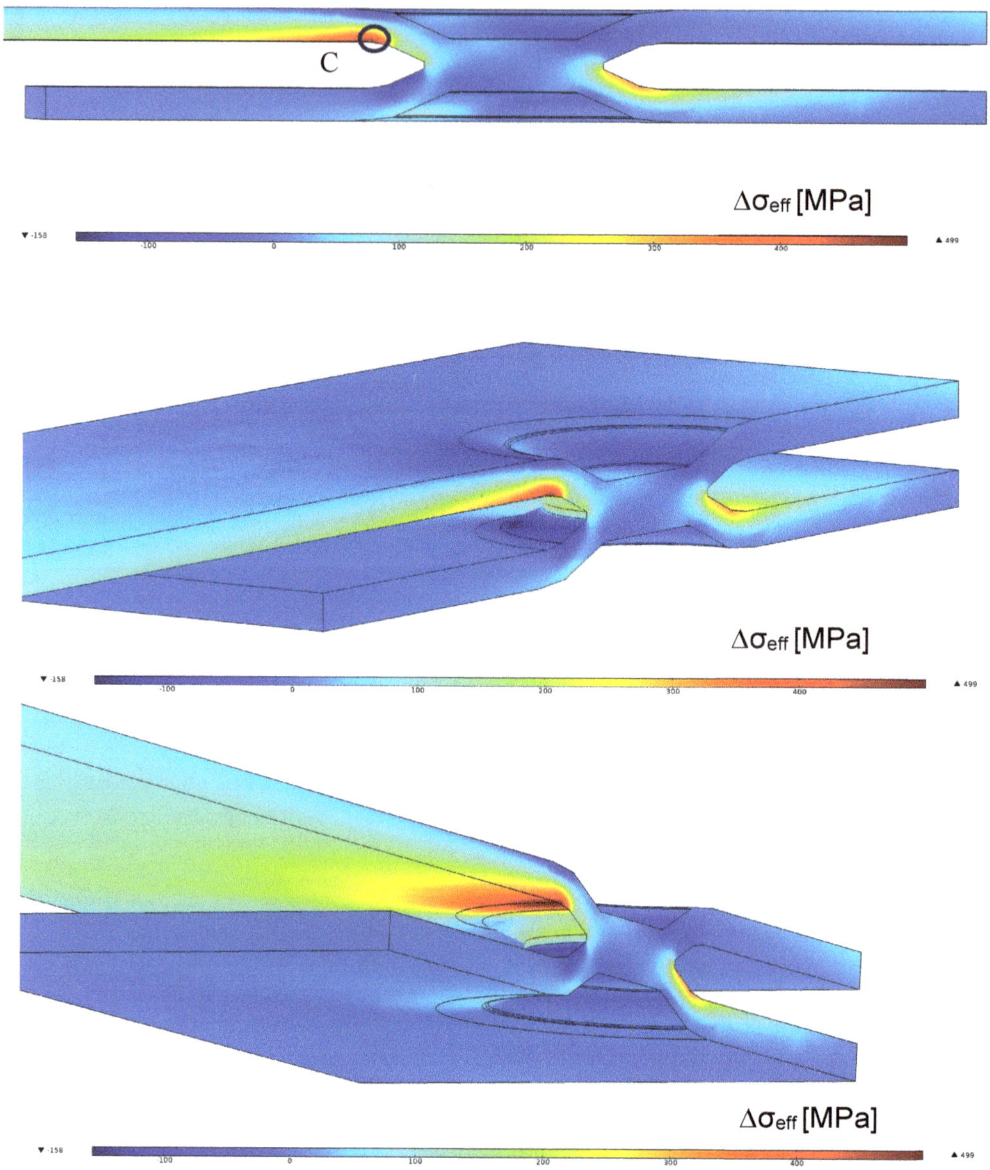

Figure 13. Range of effective stress in the dimple spot welding around the spot weld and the weld toe for a remote applied force range of $\Delta F = 4.5$ kN ($F_{max} = 5$ kN, $F_{min} = 500$ N, nominal load ratio R = 0.1) for the model with a friction coefficient of 0.45 by simulating the fatigue loading (steel plate).

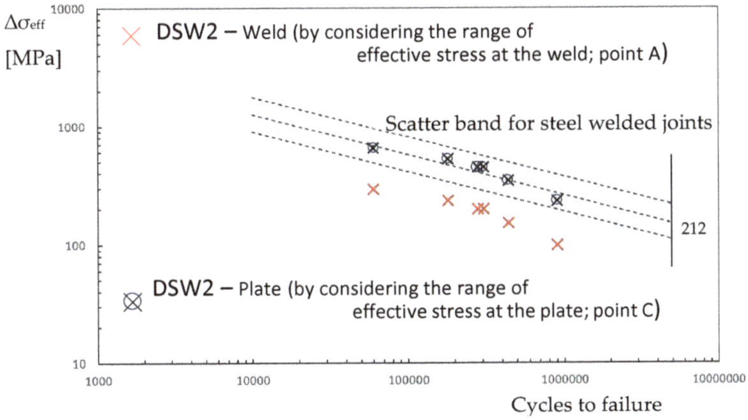

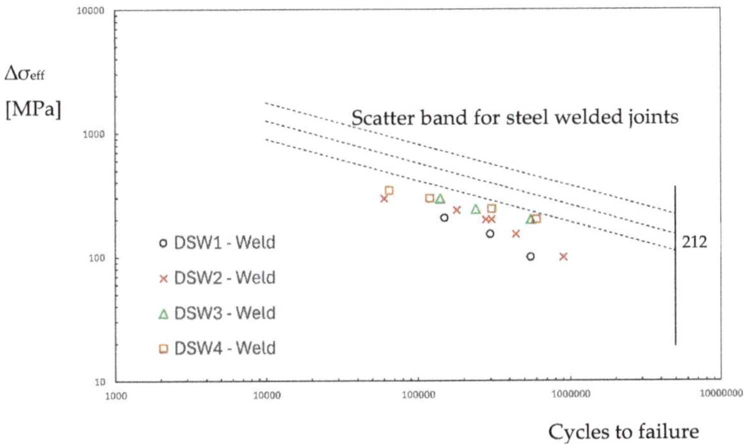

Figure 14. Fatigue life in terms of effective stress range in the dimple spot welding for the FE with friction simulation and fatigue lading simulation (steel plate).

5. Conclusions

This paper analysed the stress distribution and fatigue life under cyclic loading of dimple spot weld (DSW) used to connect two different plates made of steel and aluminium alloy, respectively.

The frictionless model results in unusually high stress on the steel plate, which deviates from experimental observations showing that fatigue cracks initiate in the aluminium plate. This discrepancy implies that, in practical DSW applications, additional factors such as friction likely play a role in altering stress distribution and crack locations.

By adding a friction coefficient in the FE model with a monotonic loading, the analysis reveals a decrease in stress concentration at the weld, but around the aluminium plate hole the magnitude of the effective peak stress increases.

Through cyclic loading simulations, the fatigue analysis underscores the critical importance of both friction and load history in accurately predicting fatigue life. Friction significantly reduces the range of effective stress at the weld.

To enhance the accuracy of numerical predictions in future work, incorporating a non-linear material model, such as elasto-plastic, and a more detailed friction model, should be considered.

Author Contributions: Conceptualisation, P.L.; Formal analysis, M.B.; Investigation, P.L. and M.B.; Writing—original draft, P.L.; Writing—review and editing, M.B. All authors have read and agreed to the published version of the manuscript.

Funding: EU funding—NextGenEU-M4C2, Inv.1.1—CUP: F53D23001710006—project title: Hybrid joints between dissimilar metal materials for lightweight structures of environmentally-friendly vehicles: experimental characterisation, theoretical modelling and CAE-based simulation chain from joint manufacturing to fatigue design by local approaches.

Institutional Review Board Statement: Not applicable.

Informed Consent Statement: Not applicable.

Data Availability Statement: The original contributions presented in this study are included in the article. Further inquiries can be directed to the corresponding author.

Conflicts of Interest: The authors declare no conflict of interest.

References

1. Karabulut, B.; Rossi, B. On the applicability of the hot spot stress method to high strength duplex and carbon steel welded details. *Eng. Fail. Anal.* **2021**, *128*, 105629. [CrossRef]
2. Zhou, F.; Chen, Y.; Wang, W.; Xie, W.; Ying, T.; Yue, G. Investigation of hot spot stress for welded tubular K-joints with stiffeners. *J. Constr. Steel Res.* **2022**, *197*, 107452. [CrossRef]
3. Sonsino, C.; Baumgartner, J.; Breitenberger, M. Equivalent stress concepts for transforming of variable amplitude into constant amplitude loading and consequences for design and durability approval. *Int. J. Fatigue* **2022**, *162*, 106949. [CrossRef]
4. Mallick, P. *Materials, Design and Manufacturing for Lightweight Vehicles*; Elsevier: Amsterdam, The Netherlands, 2020. [CrossRef]
5. Gerstmann, T.; Awiszus, B. Hybrid joining: Numerical process development of flat-clinch-bonding. *J. Mater. Process. Technol.* **2021**, *277*, 116421. [CrossRef]
6. Galinska, A.; Galinski, C. Mechanical joining of fibre reinforced polymer composites to metals—A review. Part II: Riveting, clinching, non-adhesive form-locked joints, pin and loop joining. *Polymers* **2020**, *12*, 1681. [CrossRef]
7. Eshtayeh, M.; Hrairi, M.; Mohiuddin, A. Clinching process for joining dissimilar materials: State of the art. *Int. J. Adv. Manuf. Technol.* **2016**, *82*, 179–195. [CrossRef]
8. Schramm, B.; Martin, S.; Steinfelder, C.; Bielak, C.R.; Brosius, A.; Meschut, G.; Tröster, T.; Wallmersperger, T.; Mergheim, J. A Review on the Modeling of the Clinching Process Chain—Part I: Design Phase. *J. Adv. Join. Process.* **2022**, *6*, 100133. [CrossRef]
9. Schramm, B.; Friedlein, J.; Gröger, B.; Bielak, C.; Bobbert, M.; Gude, M.; Meschut, G.; Wallmersperger, T.; Mergheim, J. A Review on the Modeling of the Clinching Process Chain—Part II: Joining Process. *J. Adv. Join. Process.* **2022**, *6*, 100134. [CrossRef]
10. Schramm, B.; Harzheim, S.; Weiß, D.; Joy, T.D.; Hofmann, M.; Mergheim, J.; Wallmersperger, T. A Review on the Modeling of the Clinching Process Chain—Part III: Operational Phase. *J. Adv. Join. Process.* **2022**, *6*, 100135. [CrossRef]
11. Ma, Y.; Lou, M.; Li, Y.; Lin, Z. Effect of rivet and die on self-piercing rivetability of AA6061-T6 and mild steel CR4 of different gauges. *J. Mech. Work. Technol.* **2018**, *251*, 282–294. [CrossRef]
12. Hahn, O.; Meschut, G.; Peetz, A. Mechanical properties of punchriveted and adhesive-bonded aluminum sheets. *Weld. Cut.* **1999**, *51*, E130–E134.
13. Fu, M.; Mallick, P. Fatigue of self-piercing riveted joints in aluminum alloy 6111. *Int. J. Fatigue* **2003**, *25*, 183–189. [CrossRef]
14. Fu, M.; Mallick, P. *Effect of Process Variables on the Static and Fatigue Properties of Self-Piercing Riveted Joints in Aluminum Alloy 5754*; Society of Automotive Engineers: Warrendale, PA, USA, 2001.
15. Pickin, C.; Young, K.; Tuersley, I. Joining of lightweight sandwich sheets to aluminium using self-pierce riveting. *Mater. Des.* **2006**, *28*, 2361–2365. [CrossRef]
16. Haque, R.; Beynon, J.H.; Durandet, Y.; Kirstein, O.; Blacket, S. Feasibility of measuring residual stress profile in different self-pierce riveted joints. *Sci. Technol. Weld. Join.* **2012**, *17*, 60–68. [CrossRef]
17. Dong, H.; Chen, S.; Song, Y.; Guo, X.; Zhang, X.; Sun, Z. Refilled friction stir spot welding of aluminum alloy to galvanized steel sheets. *Mater. Des.* **2016**, *94*, 457–466. [CrossRef]

18. Meschut, G.; Merklein, M.; Brosius, A.; Drummer, D.; Fratini, L.; Füssel, U.; Gude, M.; Homberg, W.; Martins, P.; Bobbert, M.; et al. Review on mechanical joining by plastic deformation. *J. Adv. Join. Process.* **2022**, *5*, 100113. [CrossRef]
19. Li, D.; Han, L.; Thornton, M.; Shergold, M. Influence of rivet to sheet edge distance on fatigue strength of self-piercing riveted aluminium joints. *Mater. Sci. Eng. A* **2012**, *558*, 242–252. [CrossRef]
20. Sakaguchi, M.; Kurokawa, Y.; Nakamura, F.; Hashimura, T. Fatigue strength of steel–aluminum alloy dissimilar lap joints fabricated by dimple spot welding for automotive application. *Fatigue Fract. Eng. Mater. Struct.* **2024**, *47*, 939–951. [CrossRef]
21. Livieri, P.; Tovo, R. Fatigue strength of aluminium welded joints by a non-local approach. *Int. J. Fatigue* **2020**, *143*, 106000. [CrossRef]
22. Livieri, P.; Tovo, R. Overview of the geometrical influence on the fatigue strength of steel butt welds by a nonlocal approach. *Fatigue Fract. Eng. Mater. Struct.* **2019**, *43*, 502–514. [CrossRef]
23. Livieri, P.; Tovo, R. Optimization of Welded Joints under Fatigue Loadings. *Metals* **2024**, *14*, 613. [CrossRef]
24. Peerlings, R.H.J.; de Borst, R.; Brekelmans, W.A.M.; de Vree, J.H.P. Gradient enhanced damage for quasi-brittle material. *Int. J. Numer. Methods Eng.* **1996**, *39*, 3391–3403. [CrossRef]
25. Eringen, C.A.; Edelen, D.G.B. On nonlocal elasticity. *Int. J. Eng. Sci.* **1972**, *10*, 233–248. [CrossRef]
26. Pijaudier-Cabot, G.; Bažant, Z.P. Nonlocal Damage Theory. *J. Eng. Mech.* **1987**, *10*, 1512–1533. [CrossRef]
27. Bažant, Z.P. Imbrigate continuum and its variational derivation, Journal of the Engi-neering Mechanics Division. *ASCE* **1984**, *110*, 1693–1712.
28. Tovo, R.; Livieri, P. An implicit gradient application to fatigue of complex structures. *Eng. Fract. Mech.* **2007**, *75*, 1804–1814. [CrossRef]
29. El Haddad, M.H.; Topper, T.H.; Smith, K.N. Fatigue crack propagation of short cracks. *ASME J. Eng. Mater. Technol.* **1979**, *101*, 42–45. [CrossRef]
30. Tovo, R.; Livieri, P. A numerical approach to fatigue assessment of spot weld joints. *Fatigue Fract. Eng. Mater. Struct.* **2010**, *34*, 32–45. [CrossRef]
31. Radaj, D.; Sonsino, C.M.; Fricke, W. *Fatigue Assessment of Welded Joints by Local, Approaches*, 2nd ed.; Woodhead Publishing: Boca Raton, FL, USA; CRC Press: Cambridge, UK, 2006.
32. Saelzer, J.; Berger, S.; Iovkov, I.; Zabel, A.; Biermann, D. Modelling of the friction in the chip formation zone depending on the rake face topography. *Wear* **2021**, *477*, 203802. [CrossRef]
33. Karupannasamy, D.K.; de Rooij, M.B.; Schipper, D.J. Contact pressure-dependent friction characterization by using a single sheet metal compression test. *Wear* **2013**, *308*, 222–231. [CrossRef]
34. Karupannasamy, D.; de Rooij, M.; Schipper, D. Multi-scale friction modelling for rough contacts under sliding conditions. *Wear* **2013**, *308*, 222–231. [CrossRef]
35. Livieri, P. Numerical Analysis of Double Riveted Lap Joints. *Lubricants* **2023**, *11*, 396. [CrossRef]
36. Cristofori, A.; Livieri, P.; Tovo, R. An application of the implicit gradient method to welded structures under multiaxial fatigue loadings. *Int. J. Fatigue* **2009**, *31*, 12–19. [CrossRef]

materials

MDPI

Article

Punch Edge Topological Design for Reduction of Work Hardening Damage in Shearing of Non-Oriented Electrical Steel Sheets

Ryoma Okada [1], Kentaro Ito [1], Tatsuya Funazuka [2], Tatsuhiko Aizawa [3] and Tomomi Shiratori [2,*]

[1] Graduate School of Science and Engineering, University of Toyama, Toyama 930-8555, Japan; m23c1422@ems.u-toyama.ac.jp (R.O.); m23c1411@ems.u-toyama.ac.jp (K.I.)
[2] Academic Assembly Faculty of Engineering, University of Toyama, Toyama 930-8555, Japan; funazuka@eng.u-toyama.ac.jp
[3] Surface Engineering Design Laboratory, Shibaura Institute of Technology, Tokyo 144-0045, Japan; taizawa@sic.shibaura-it.ac.jp
* Correspondence: shira@eng.u-toyama.ac.jp

Abstract: A new shearing tool is necessary to reduce the iron loss of motor cores by minimizing the work hardening damage on the sheared non-oriented electrical steel sheets. The punch edge topology and the clearance between the punch and the die were controlled to investigate their influence on the sheared surface condition and the work hardening damage of steel sheets. A non-oriented electrical steel sheet with the thickness of 500 μm was used and sheared at the speed of 5 mm/s. After that, the sheared surface was investigated. In particular, hardness mapping was utilized to quantitatively analyze the work-hardened area of the sheared steel sheets and the dissipation of the plastic work. Among the four punch edge topological configurations explored, the nano-grooved punch employed straight along the shearing direction reduced the damage dealt to the sheared steel sheets and the plastic dissipation work to one-third compared to conventional punches.

Keywords: shearing; punching; non-oriented electrical steel

1. Introduction

In the pursuit of SDGs with carbon neutrality, electric vehicles with highly efficient motors are in high demand to reduce energy consumption [1]. The motor cores are made from non-oriented electrical steels with magnetic isotropy. The energy dissipation by the loss of iron induced in them has become one of the most essential issues in related engineering processes. This loss is mainly attributed to the shearing-induced work hardening damage deal to the electrical steel sheets [2–4]. The plastic strains and residual stresses caused by the work hardening process pose a significant risk of worsening the magnetic domain structure. This deterioration induces hysteresis loss and eddy current loss [5]. In particular, the hysteresis loss caused by plastic straining results in magnetic anisotropy. Magnetization processes, characteristic of the damaged zone width, are influenced by the value of the applied magnetic field strength. A link can be established between the damaged zone's magnetic and mechanical properties because an increased hardness value and bad magnetic properties usually characterize the areas most affected by cutting [6]. Hence, the shearing process must be improved to minimize the affected zone on the non-oriented electrical steel sheets in the shearing process and to reduce the iron loss.

In the shearing tool design, the clearance between the punch and the die plays a role in the shearing performance [7,8]. A larger clearance induces a bending moment applied to the work material, enhances work hardening, and deteriorates its magnetic properties. On the other hand, a smaller clearance is preferable in reducing work hardening, causing little degradation of magnetic properties [9,10]. However, shearing in low clearance often induces a secondary fracture surface and shortens the tool life. The punch edge configuration influences the shearing process. As stated in [11,12], a mechanically ground edge yields a burnished surface in shearing; the sheared surface quality deteriorates with its abrasive wearing. The work hardening area is exaggerated to be three times larger than the work sheet thickness when using a mechanically ground punch edge [2]. When using a carbide punch with an edge sharpened via argon-ion beam polishing, the work hardening area is much reduced in the shearing of AISI304 steel sheets [13,14]. As reported recently in [15–18], carbide and nitrided punches with periodically nano-grooved edges via short-pulse laser texturing improved the sheared surface quality by reducing the abrasive and adhesive wear of the punch. However, there has been no quantitative evaluation of the optimization of the clearance and punch edge conditions in the shearing of a non-oriented electrical steel sheet.

The present paper aims to quantitatively analyze the effect of the clearance and tool edge conditions on the minimization of work hardening in the shearing of non-oriented electrical steel sheets to identify a shearing condition that will significantly reduce iron loss. Four types of shearing punches are employed to analyze, experimentally, the effect of the clearance and punch-edge topological configurations on the work hardening damage incurred in the shearing of the non-oriented steel sheets. Using the specified experimental procedure, a die set with three different clearances is utilized. Four punches are prepared such that each of the experiments has different edge configurations, namely, a mechanically ground edge, an ion-milled edge, and a periodically nano-grooved edge with two groove orientations (20° and 90°).

From the experiments, it was found that the nano-grooved punches minimize the work hardening damage to work sheets in the shearing process. In addition, the nano-grooved edge, either straight along the shearing direction or at an angle of 90° against its edge, minimized the plastic dissipation work through the localization of the stress concentration and through control of local shearing.

2. Methods and Materials

2.1. Shearing Process

The screw servo stamping system (Precise Stamping Laboratory, DT-J311; Tokyo, Japan) was utilized to punch out the work materials, as shown in Figure 1a. The stroke was measured using the laser displacement meter (KEYENCE, SI-F01; Tokyo, Japan) and data acquisition system (KEYENCE, NR-600; Tokyo, Japan). Non-oriented Fe-3Si electrical steel sheets, with a thickness of 500 μm (NIPPON STEEL Corp., 50H310; Tokyo, Japan), were used as the work material. The WC (Co) die in Figure 1b and the work piece were inserted into a die holder. An SKD11 punch was inserted into the center hole of the punch guide and fastened to the die holder. The punch diameter was 5 mm. Three types of clearances were prepared to investigate the effect of clearance (δ) on work hardening; δ = 25 μm, i.e., 5% of work piece thickness (h), as a normal condition; δ = 5 μm, i.e., 1% of h; and δ = 2.5 μm.

The loading speed was constant at 5 mm/s. Two shearing experiments were conducted: short-shot at a 30% reduction in thickness and fully punching out.

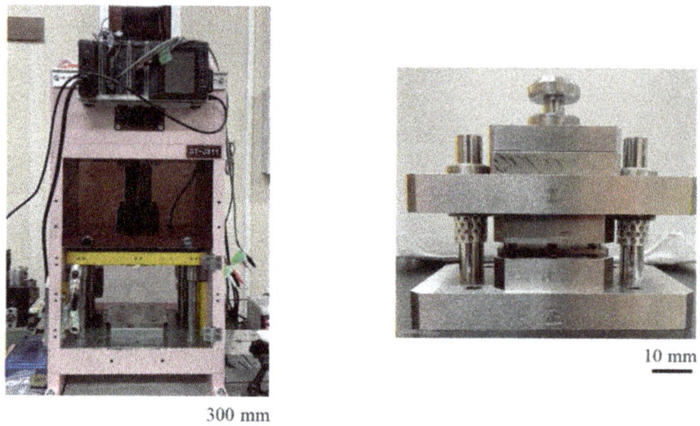

(**a**) Servo-press machine (**b**) Die set

Figure 1. Experimental setup for the dry punching of non-oriented electrical steel sheets: (**a**) a CNC (computer numerical control) stamper; (**b**) a die set.

2.2. Punch Edge Configurations

Four types of punches were used in this experiment, as listed in Figure 2: a normal punch with a mechanically ground edge; an ion-beam-treated punch with an edge sharpened by argon-ion irradiation; Nano20° and Nano90° punches with nano-grooved edges, with each width being 300 nm on their side surface at orientations of 20° and 90°, respectively, against their edges; and a femtosecond laser with a wave length of 515 nm, a pulse width of 200 fs, and a repetition rate of 20 MHz were utilized for nano-grooving [17].

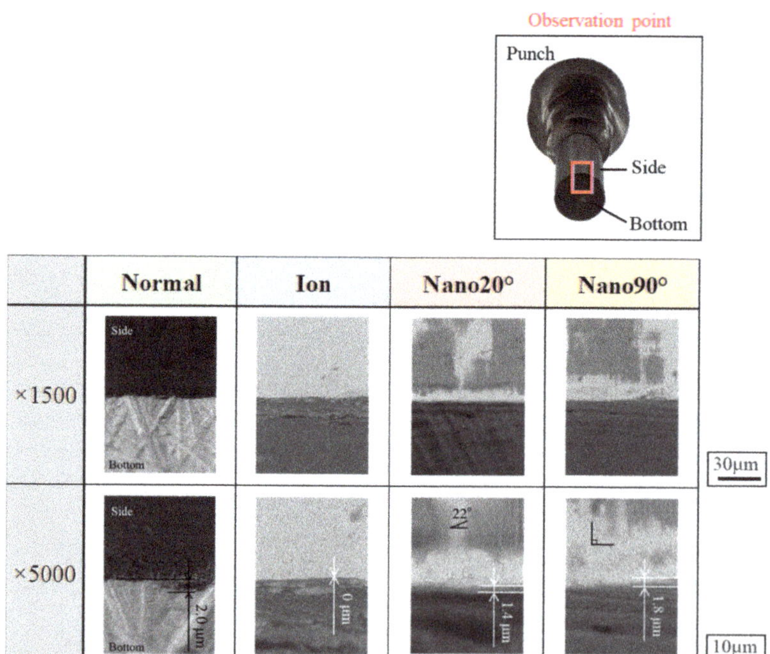

Figure 2. Four types of shearing punches with different edge configurations. Low magnification 1000x SEM images are listed in the upper row, and high magnification 5000x SEM images are listed in the lower row.

2.3. Evaluation on Quality of Sheared Electrical Steel Sheets

The area near the right side of sheared hole was observed by SEM (JCM-7000, JEOL; Tokyo, Japan). The fractured surface area ratio was calculated using image processing software (mediBang Paint Pro ver. 2023.06, Photoshop ver. 26.3). The boundary of the fractured area was estimated via image processing. One pixel of the image was approximately 1 μm; the digitization error was 0.2%.

2.4. Mechanical Characterization

The work hardening zone of sheared electrical steel sheets was determined from the Vickers hardness map after cutting and polishing near the right side of the punched hole. The Vickers hardness tester (HM-100, Mitutoyo; Kawasaki, Japan) was used for mapping. The load was applied by 0.1 N for 10 s. The mapping scheme is shown in Figure 3. The hardness map was measured at three points with an interval of 25 μm on the first zone, at three points with an interval of 50 μm on the second zone, at two points with an interval of 75 μm on the third zone, and at three points with an interval of 100 μm on the forth zone. This hardness map is depicted by ten color-coded levels in each 25 HV. The personal error in measurement was suppressed to be lower than 15 HV, or 6% of the reference hardness of 250 HV for the electrical steel sheet without work hardening damages.

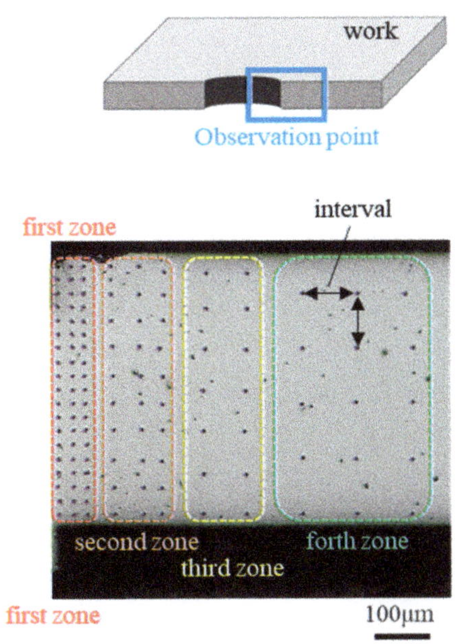

Figure 3. Hardness mapping to evaluate the affected zone of the sheared electrical steel sheets using four types of punches.

2.5. Estimate on the Plastic Work

To obtain material properties, a 1/2-scale JIS 13B tensile test specimen was used for tensile testing. The stress–strain curves were obtained at a strain rate of 0.1 mm/s.

The plastic work (W) induced in the sheared sheets is estimated using the measured hardness map. This W is defined by

$$W = \int_V w \times dv \quad (1)$$

where w is the work density, dv is the representative volume of each measuring zone in the hardness mapping, and V is the total volume of the plastically deforming work materials, excluding the elastically deforming volume. The stress–strain curve of the non-oriented electrical steel sheet is modeled by the rigid plastic model; e.g., $\sigma = \sigma_y + K \times \varepsilon$, where σ is the stress, K is the work hardening modulus, σ_y is the yield stress, and ε is the plastic strain. These K and σ_y values are given by the uniaxial tensile test. Then, the plastic work density is represented by the following.

$$w = \frac{1}{2}\left(\sigma_y + \sigma\right) \times \varepsilon = \frac{1}{2K} \times \left(\sigma + \sigma_y\right) \times \left(\sigma - \sigma_y\right) \tag{2}$$

In the case of the hardness mapping in Figure 4, V was discretized into M representative elements, including each measuring point of hardness. Then, the plastic work is calculated by

$$\mathrm{W} = \sum_{m=1}^{M} \left(\frac{1}{2K}\right) \times \left({\sigma_m}^2 - {\sigma_y}^2\right) \times 2\pi r_m \, \Delta \mathrm{r} \Delta \mathrm{z} \tag{3}$$

where $\mathrm{r_m}$ is a radius of m-th measuring point from the symmetric axis, and Δr and Δz are pitches in the radial and axial directions, respectively, in Figure 4.

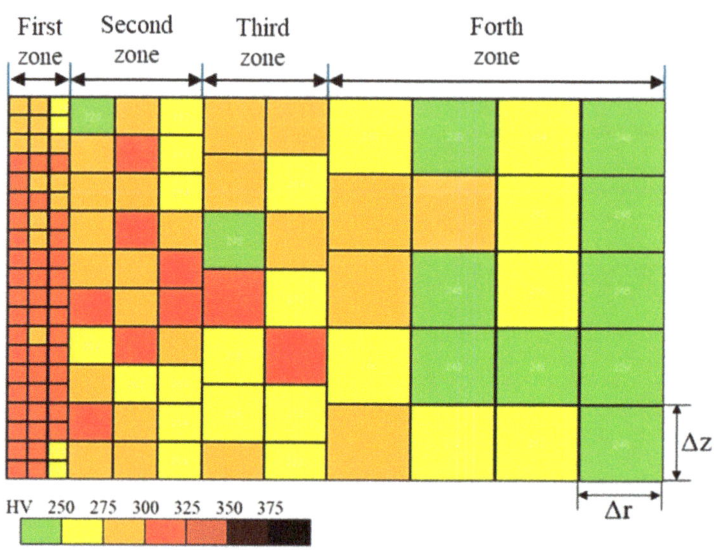

Figure 4. Discretization of the work hardening area induced in the right-hand side of the sheared steel sheet. The m-th square element has the m-th measuring point of hardness.

3. Experimental Results

3.1. Sheared Surface Observation

The sheared surface conditions attained using four punches were evaluated via SEM observation. Figure 5 compares four sheared surface conditions where δ = 25 µm or at CL25 µm. The fracture surface extends heterogeneously. In the case of CL5 µm in Figure 6, the fractured surface area ratio decreases and the burnished surface area ratio increases. Less-fractured surface areas are detected at CL2.5 µm in Figure 7.

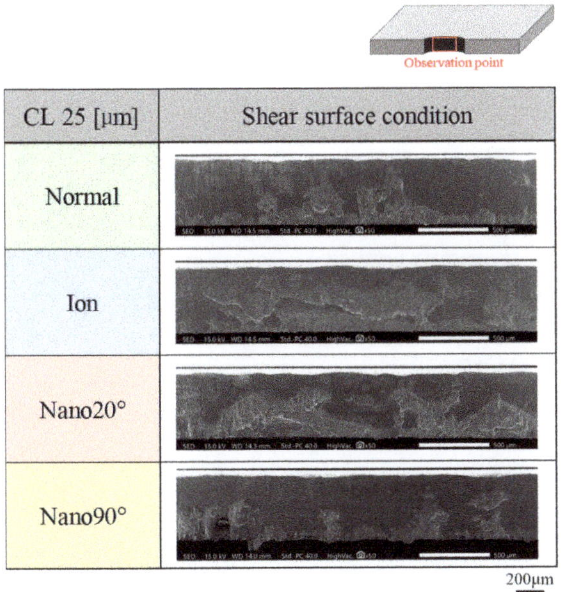

Figure 5. Comparison of sheared surface conditions among four punches at a clearance of CL25 μm.

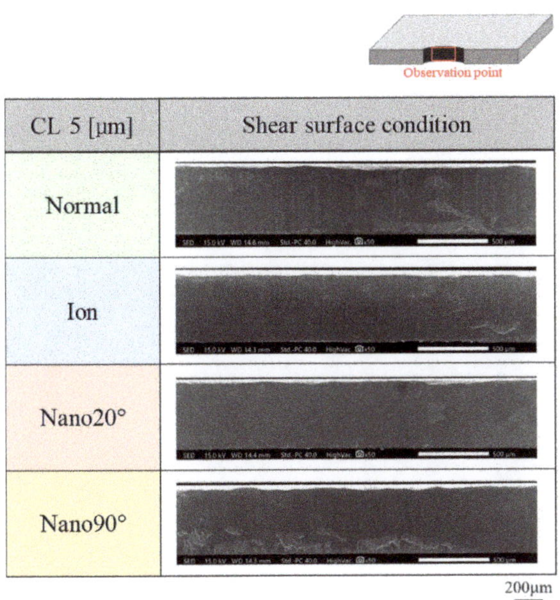

Figure 6. Comparison of sheared surface conditions among four punches at a clearance of CL5 μm.

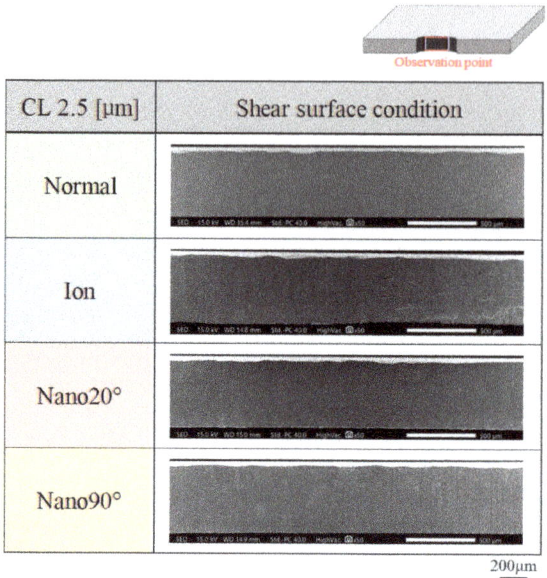

Figure 7. Comparison of sheared surface conditions among four punches at a clearance of CL2.5 μm.

The variation of these fractures and the burnished surface area ratios attained by changing the clearance are listed in Figure 8, and the four punches are compared together with the measured shear droop ratio. At CL25 μm, the fractured surface area ratio exceeds 27% for every condition. Reducing the clearance down to CL5 μm, the fractured surface area ratio significantly decreases to 3% in the case of Nano20°. At CL 2.5 μm, no fractured surface areas are detected when using the Nano20° and Nano90° punches.

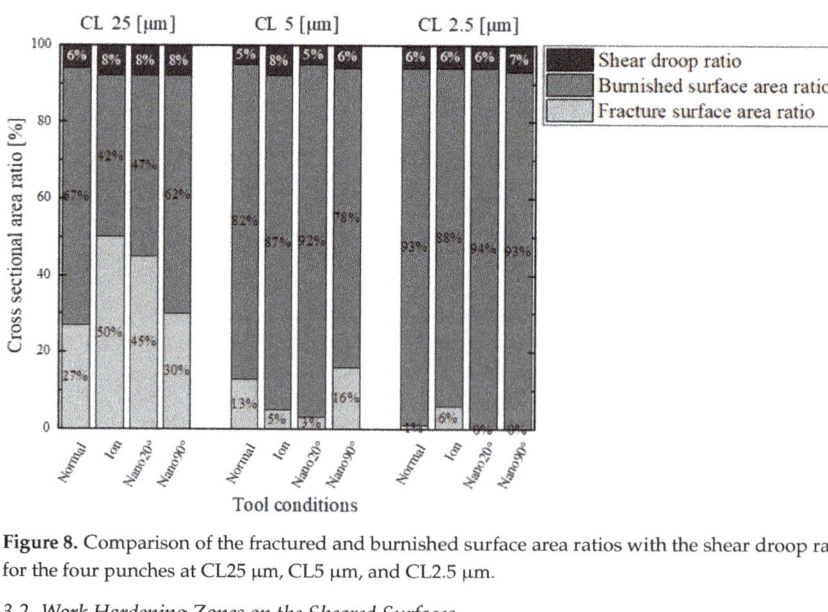

Figure 8. Comparison of the fractured and burnished surface area ratios with the shear droop ratio for the four punches at CL25 μm, CL5 μm, and CL2.5 μm.

3.2. Work Hardening Zones on the Sheared Surfaces

The work hardening zone is defined as the area with a hardness greater than 250 HV in the hardness map. Each zone is compared at each clearance for each punch condition in

Figure 9. The elastic zone is represented by the green color. Based on the hardness map in Figure 9, Figure 10 shows the affected area ratio of six work hardening zones and an elastic zone.

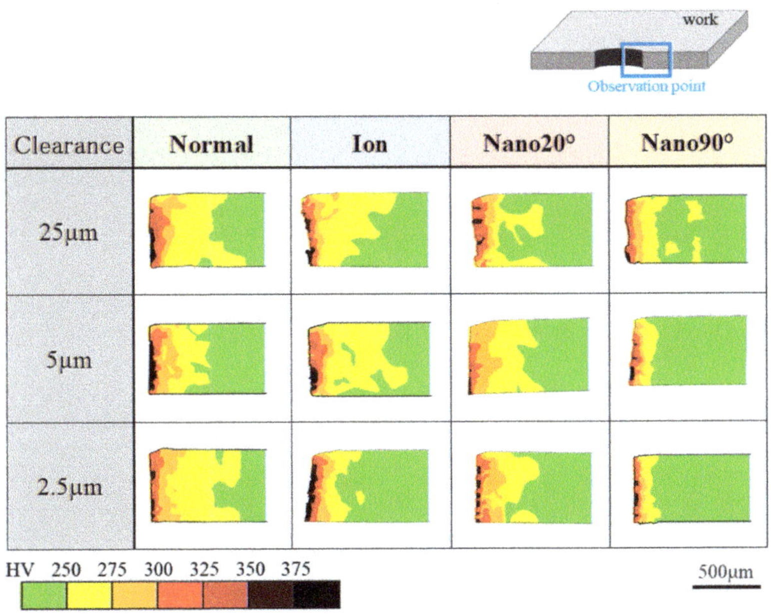

Figure 9. Hardness comparison of the work hardening zones for each clearance of CL25 μm, CL5 μm, and CL2.5 μm for the four punches.

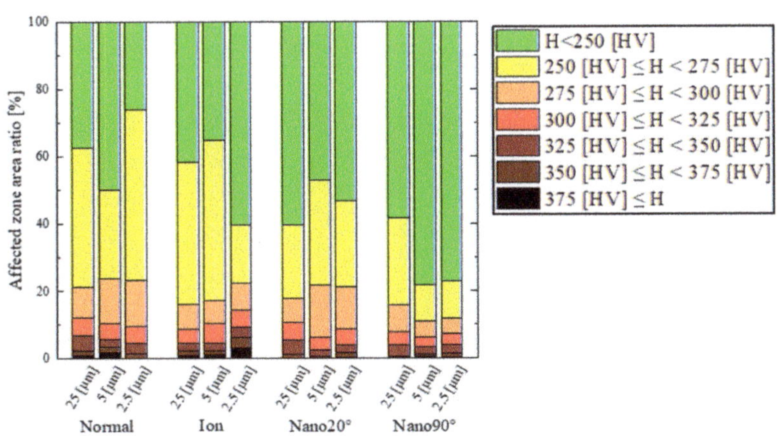

Figure 10. Hardness contribution of six work hardening and elastic zone ratios on the whole cross-sectional area at each clearance for the four punching conditions.

When using a normal punch, a larger work hardening zone is detected irrespective of the clearance. It should be noted that since the red-colored zone with hardness greater than 300 HV is detected at the nearest to edge, work hardening concentrates in the vicinity of the punch edge.

In the case of the ion punch, the work hardening zone decreases with the reduction in clearance. The work hardening zone with a hardness greater than 375 HV is detected in the vicinity of sheared surface. It is enhanced with the reduction in clearance. This is because of the severe plastic flow around the sharpened punch edge.

When using the Nano20° punch, a large work hardening zone fluctuates by itself on the cross-section, irrespective of clearance. It should be noted that the highest hardened zone, i.e., the zone with a hardness greater than 375 HV, is only detected pointwise in the vicinity of the sheared surface.

In the case of the Nano90° punch, the work hardening area greater than 250 HV, shown in yellow color, is concentrated in the vicinity of the hole and is minimized when the clearance is lower than 5 μm.

As summarized in Figure 10, the fraction of each work hardening zone in the whole cross-sectional area is compared at each clearance and with each punching condition. When using the normal punch, the yellowed colored zone fraction becomes the maximum (51%) at CL2.5 μm. In case of the Nano90° punch, the yellowed colored zone fraction becomes the minimum (11%) at CL5 μm and CL2.5 μm. This fraction is less than the minimum fraction using the ion beam (i.e., 17% at CL2.5 μm) by 35%, and it is half of the minimum fraction attained using Nano20° (i.e., 22% at CL25 μm).

Table 1 lists the contribution of the elastic zone and the medium and high work hardening zones to the total cross-sectional area. When using the Nano90° punch, the cross sectional area is determined to be elastic; the medium work hardening zone is limited by 15%; and the high work hardening zone is reduced to less than 5%. Most of the errors come from the measured area at the first zone in Figure 3, with the pitch of 25 μm. The maximum error area was estimated to be 4%.

Table 1. Comparison of the elastic zone (H < 250 HV; green-colored zone in Figure 9), the medium work hardening zone (250 HV < H < 300 HV; yellow- and orange-colored zones in Figure 9), and the high work hardening zone (300 HV < H < 375 HV and H < 375 HV; red, dark-red-, brown-, and black-colored zones in Figure 9) for each clearance and punch.

Tool Conditions	Clearance	Elastic Zone	Medium Work Hardening Zone	High Work Hardening Zone
Normal	25 μm	37.5 [%]	50.4 [%]	12.1 [%]
	5 μm	50.0 [%]	39.7 [%]	10.3 [%]
	2.5 μm	26.1 [%]	64.3 [%]	9.6 [%]
Ion	25 μm	41.8 [%]	49.7 [%]	8.5 [%]
	5 μm	35.2 [%]	54.5 [%]	10.3 [%]
	2.5 μm	60.5 [%]	25.3 [%]	14.2 [%]
Nano20°	25 μm	60.5 [%]	28.8 [%]	10.7 [%]
	5 μm	47.1 [%]	46.9 [%]	6.0 [%]
	2.5 μm	53.3 [%]	38.0 [%]	8.7 [%]
Nano90°	25 μm	58.5 [%]	33.5 [%]	8.0 [%]
	5 μm	78.3 [%]	15.5 [%]	6.2 [%]
	2.5 μm	77.2 [%]	15.6 [%]	7.2 [%]

3.3. Measurement of Work Hardening Zone for Short-Shot Specimens

A short-shot shearing experiment using 30% of the full stroke was performed to describe the transients of the work hardening process compared to fully punching-out. Three shots (N1, N2, and N3) were made with a constant clearance of CL2.5 μm in Figure 11.

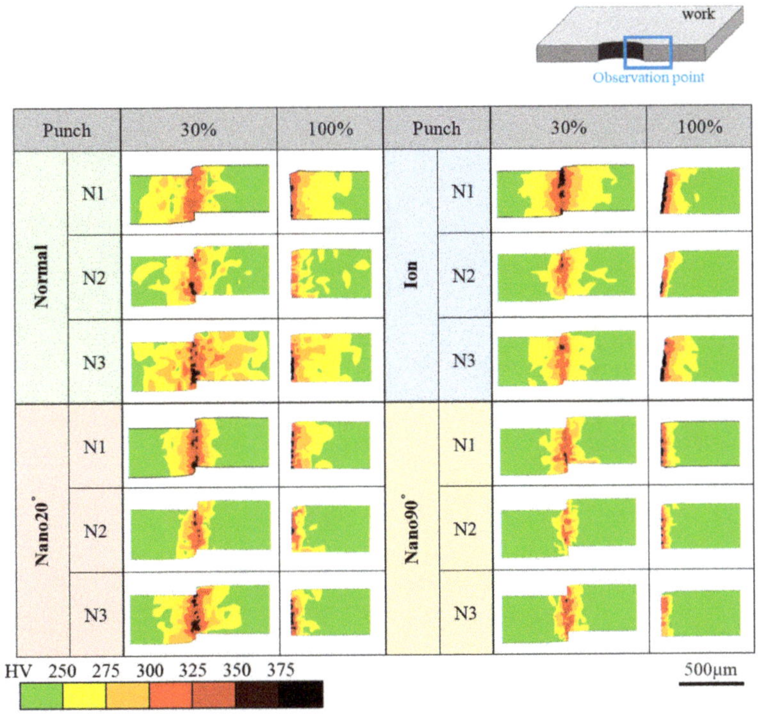

Figure 11. Hardness map of transients of the work hardening zone from a sheared state (30%) to a fully sheared state for the four punching conditions. Three shots—N1, N2, and N3—were performed in each shearing experiment.

When shearing with the normal punch, the medium work hardening zone fluctuates by itself among the three shots. This fluctuation is also seen on the cross-section after fully punching out. This implies that the metal flow of the work materials is significantly affected by the heterogeneous edge profile of the normal punch.

In the case of the ion punch, the medium work hardening zone is narrowed, with lower fluctuation among the three shots after short-shot when compared to normal punching. This zone of the short-shot specimen is reduced after fully punching out. This is because a part of the work hardening zone has been determined to be elastic in the unloading process after fully punching out. It should be noted that the highest work hardening zone increased after punching out. This implies that the shear strain localizes along the shearing line due to the sharpened edge of the ion punch.

In the case of the Nano20° punch, the medium work hardening zone becomes larger in the N3 test than in the N1 and N2 tests in the short-shot specimen. This zone is much reduced after punching out. This reveals that most of the medium work hardening zone is elastic in the unloading process after fully punching out. In fact, the highest work hardening zone is commonly seen in the vicinity of the sheared surface, even after punching out. This proves that the shearing process using the Nano20° punch affects strain concentration in a similar manner to shearing via the ion punch.

When using the Nano90° punch, the medium work hardening zone area is greatly narrowed; there are not the high work hardening zones even in the vicinity of the sheared surface. In addition, no significant fluctuation of the work hardening zones is detected after the short-shot and fully punching out experiments.

Figure 12 compares the work hardening zone area fractions after short-shot (30%) and fully punching out (100%) experiments for the four punches.

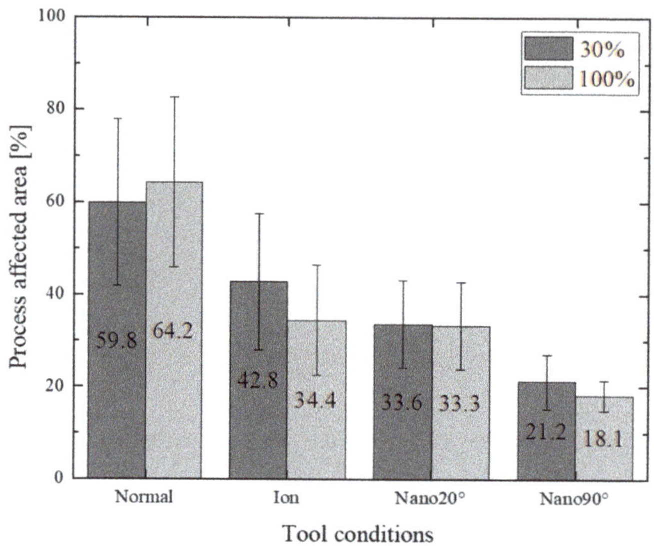

Figure 12. Effect of tool conditions on the work hardening damage fraction at the short-shot after fully punching out.

The normal punch induces the largest work hardening zones due to its heterogeneous edge profile. In the case of the ion punch, the total work hardening zone is reduced by 64.4% to 43.8% after fully punching out due to its homogeneously sharpened edge. When using the Nano90° punch, far fewer work hardening zones are induced to be 20% along the shearing line. This difference in the work hardening process among the four punches comes from the difference in the metal flow pattern during shearing. The plastic work dissipation might well be a measure by which to describe the punch edge profile's effect on the shearing process.

4. Discussion

The tool edge configuration and the clearance have influence on the work hardening damage inflicted on the sheared electrical steel sheets. When using the mechanically ground normal punch, the widest work hardening volume is depicted in Figures 9 and 11, irrespective of the clearance. This is because the work plastic flow around the heterogeneous punch edge enhances the work hardening. Figure 12 also shows the largest deviation in the work hardening zone area fractions for three shots. The heterogeneous punch edge also exaggerates the mechanical fluctuation in the work hardening behavior from experiment to experiment.

When using the ion-sharpened punch, the severe work hardening zone volume reaches its maximum for the four punches in the case of CL2.5 μm. The fracture surface area ratio also reaches the maximum value in the case of CL25 μm. This implies that the work plastic flow localizes at the sharpened punch edge, results in the most severe work hardening state, and induces the onset of fracture. This local work hardening behavior is also observed when using the Nano20° punch. However, the work hardening zone volume is much smaller in cases of greater clearance than 5 μm. The severe work hardening zone volume does not increase via a decrease in the clearance. The fracture surface area ratio is always smaller than that attained using the ion-punch. This reveals that the nano-grooves in the direction of 20° control the local work plastic flow around the sharped edge and reduce the work hardening damage.

When using the Nano90° punch, the local work hardening zone volume reduces significantly, even in case of a clearance under 5 μm. It becomes smaller than that by 69% when using the normal punch, and smaller than that by 51% when using the Nano20° punch. This drastic suppression of work hardening damages comes from the nano-groove structure with a groove orientation change from 20° to 90°.

When using the Nano20° punch, the surface of the workpiece interacts with the nano-periodic structure formed by the punch tip during shearing. When using the Nano90° punch, on the other hand, the workpiece plastically flows without interaction between its surface and the nano-periodic structure. Therefore, a state of local stress concentration continues until shearing is completed, and the work hardening zone area is considered to be narrow and concentrated in the vicinity of the shearing edge.

The plastic dissipation work (W) provides a measurement by which to analyze the tool edge topology effect on the work hardening damage. As listed in Table 2, W, calculated by Equation (3), reduces monotonously by changing the (normal punch) → (ion-sharpened punch) → (Nano20° punch) → (Nano90° punch). The deviation (ΔW) of W among three strokes in the shearing process represents the effect of homogeneity in the punch edge configuration to the local plastic flow. The monotonous decrease in ΔW, with (normal punch) → (Ion) → (Nano20°) → (Nano90°), proves that local plastic flow is sufficiently well controlled to suppress local work hardening using the sharpened and nano-grooved punch edge.

Table 2. Comparison of the plastic dissipation work (W).

	Normal	**Ion**	**Nano20°**	**Nano90°**
N1	0.040 [J]	0.063 [J]	0.031 [J]	0.027 [J]
N2	0.042 [J]	0.034 [J]	0.019 [J]	0.008 [J]
N3	0.104 [J]	0.034 [J]	0.046 [J]	0.018 [J]
Average	0.062 [J]	0.044 [J]	0.032 [J]	0.018 [J]

In comparison to the normal punch, the plastic work (W) is halved when using the Nano20° punch and reduced to one-third when using the Nano90° punch. Let us discuss the effect of the nano-grooved punch edge configurations on W through a comparison to the ion-milled punch. When using this punch, the shear strain exceeds a certain level and induces ductile separation at its sharp and homogeneous edge. The nano-grooves work to enhance the strain concentration at the nano-grooved edges and to onset the shear-slipping process of the work materials from the edges. The difference in the shearing and burnishing behavior between the ion-milled and nano-grooved punches comes from the stress concentration, which is determined by the indentation of the nano-grooves to induce the separation of the work materials. The plastic work is minimized by using these nano-grooved punches.

Regarding the shear processing of non-oriented electrical steel sheets using a nano-grooved punch, a case has been reported in the research of Aizawa et al., in which plasma nitriding was applied to the alloy tool steel SKD11 [17]. This study reports that a non-oriented electrical steel sheet with a thickness of 0.15 mm was sheared along the entire burnished surface. However, the mechanism by which shear stress concentrates on nano-grooved structures has not been considered. In the piercing of amorphous electrical steel sheet, five stacked sheet piercings have been attempted by imparting a nano-periodic structure to the sides of a diamond punch at an angle of 60° from the bottom of the punch [19]. As a result, it was reported that the nano-periodic structure has the effect of suppressing the attraction of the work material to the die side. It is thought that even with a brittle amorphous electrical steel sheet with a thin thickness of 0.025 mm and an elongation of only 0.2%, shear stress is concentrated in the nano-periodic structure at the punch tip, thereby reducing the radial effect of the hole. These reports also suggest that the addition of nano-periodic structures to the punch may be a factor in upgrading the sheared surface as it helps to concentrate the shear stress on the punch edge.

The difference in local shearing and burnishing behavior between Nano20° and Nano90° is considered next. In the case of Nano20°, the locally sheared work material starts to flow along the skewed orientation by 20° on the punch side surface and to separate from the nano-grooves. This work material separation from nano-grooves distributes the shear stresses and induces the broadening of the work hardening zones in the radial direction. On the other hand, no work separation occurs from the 90°-nano-grooved edge to preserve the shear stress concentration at the nano-grooved edges against the work materials.

The plastic dissipation work is minimized by this local control of the material flow around the nano-grooved edges. This indicates that the continuous contact between the nanometer-sized workpiece and the tool is a significant factor underpinning shear stress concentration.

5. Conclusions

Four types of shearing punches are employed to experimentally analyze the effect of clearance and punch-edge topological configurations on the work hardening damage by shearing the non-oriented steel sheets. This conclusion is summarized in the following:

(1) When using a nano-grooved punch, the shear strain concentrates at the edge to induce ductile separation at its sharp and homogeneous edge.
(2) The nano-grooves work to enhance the strain concentration at the nano-grooved edges and to onset the shear-slipping of the work materials from the edges.
(3) When using the Nano90° punch, the fractured surface areas, work hardening zone, and plastic work are minimized.
(4) The addition of nano-grooves to the punch provides an engineering means of upgrading the sheared surface by helping to concentrate shear stress on the punch edge.

Author Contributions: Conceptualization, T.A. and T.S.; methodology, T.F. and R.O.; investigation, R.O. and K.I.; resources, R.O., T.S. and T.F.; writing—original draft preparation, R.O. and K.I.; writing—review and editing, T.A., T.F. and T.S.; supervision, T.A. and T.S. All authors have read and agreed to the published version of the manuscript.

Funding: The present study was financially supported, in part, by the Charitable Trust Ame Hisaharu Toyama Prefecture University, etc., Research Grant Fund.

Institutional Review Board Statement: Not applicable.

Informed Consent Statement: Not applicable.

Data Availability Statement: The original contributions presented in the study are included in the article, further inquiries can be directed to the corresponding author.

Acknowledgments: The authors would like to thank for Ohmura S. for his help in the experiments.

Conflicts of Interest: The authors declare no conflicts of interest.

References

1. Enomoto, Y.; Soma, K. Development of the Amorphous Motor Balancing both Resource Saving and Energy Saving. *J. Soc. Mech. Eng.* **2014**, *117*, 753–756.
2. Zaizen, Y.; Omura, T.; Fukumura, M.; Senda, K.; Toda, H. Evaluation of stress distribution due to shearing in non-oriented electrical steel by using synchrotron radiation. *AIP Adv.* **2016**, *6*, 055926. [CrossRef]
3. Zaizen, Y.; Omura, T.; Senda, K.; Fukumura, M.; Oda, Y.; Toda, H. Influence of Grain Size of Non-oriented Electrical Steel Sheets on Iron Loss Deterioration by Shearing Process. *IEEJ Trans. FM.* **2018**, *138*, 576–581. [CrossRef]
4. Toda, H.; Zaizen, Y.; Namikawa, M.; Shiga, N.; Oda, Y.; Morimoto, S. Iron Loss Deterioration by Shearing Process in Non-Oriented Electrical Steel with Different Thickness and Its Influence on Estimation of Motor Iron Loss. *IEEJ J. Ind. Appl.* **2013**, *3*, 55–61.
5. Tani, Y.; Daikoku, A.; Nakano, M.; Arita, H.; Yamaguchi, S.; Toide, Y. Magnetic Power Loss Characteristics of Non-oriented Electrical Steel Sheets under Stress. *J. Magn. Soc. Jpn.* **2006**, *30*, 196–200. [CrossRef]
6. Gheorghe, P.; Veronica, M.; Aurora, A.; Iosif, V.N.; Horia, G. Mechanical and Magnetic Properties Variation in Non-Oriented Electrical Steels with Different Cutting Technology: A Review. *Materials* **2024**, *17*, 1345.
7. Hirota, K. Precision Blanking with Negative Clearance. *J. Jpn. Soc. Technol. Plast.* **2014**, *55*, 185–189.
8. Yoshida, Y. Deformation Behavior of Die and Tools in Shearing. *J. Jpn. Soc. Technol. Plast.* **2017**, *58*, 903–907.
9. Yiwei, Z.; Qiusheng, Y.; Jiabin, L.; Wei, X. Effects of shearing processing on the quality and magnetic properties of electric steel sheet. *Ain Shams Eng. J.* **2024**, *15*, 102395.
10. Ján, F.; Samuel, D.; Ivan, P.; Peter, K.; František, K.; Ján, S. Correlation between Cutting Clearance, Deformation Texture, and Magnetic Loss Prediction in Non-Oriented Electrical Steels. *Materials* **2021**, *14*, 6893.
11. Weiss, H.A.; Leuning, N.; Steentjes, S.; Hameyer, K.; Andorfer, T.; Jenner, S.; Volk, W. Influence of shear cutting parameters on the electromagnetic properties of non-oriented electrical steel sheets. *J. Magn. Magn. Mater* **2017**, *421*, 250–259. [CrossRef]
12. Ján, S.; L'uboš, K.; Stanislaw, K. FEM Modeling of Shear Cutting of Electrical Steel Sheets under Various Technological Conditions. *Acta Mech. Slovaca* **2018**, *22*, 24–30.
13. Fujimoto, K.; Yang, M.; Hotta, M.; Koyama, H.; Nakano, S.; Morikawa, K.; Cairney, J. Fabrication of dies in micro-scale for micro-sheet metal forming. *J. Mater. Process. Technol.* **2006**, *177*, 639–643. [CrossRef]
14. Xu, J.; Guo, B.; Shan, D.; Wang, C.; Wang, Z. Surface quality improvements of WC–Co micro-punch finished by ion beam irradiation for micro-punching process of metal foil. *Surf. Coat. Technol.* **2013**, *235*, 803–810. [CrossRef]
15. Aizawa, T.; Inohara, T. Pico- and Femtosecond Laser Micromachining for Surface Texturing. In *Micromachining*; Stanimirović, Z., Stanimirović, I., Eds.; IntechOpen: Belgrade, Serbia, 2019; Volume1.
16. Aizawa, T.; Inohara, T.; Wasa, K. Nano-Texturing onto Tool-Surface by the Femtosecond Laser Processing. In Proceedings of the 4th WCMNM, Mumbai, India, 21–23 September 2021.
17. Aizawa, T.; Yoshino, T.; Suzuki, Y.; Inohara, T. Micro-/Nano-Texturing onto Plasma-Nitrided Tool Surface by Laser Printing for CNC Imprinting and Piercing. *Micromachines* **2022**, *13*, 265. [CrossRef] [PubMed]
18. Aizawa, T.; Inohara, T.; Yoshino, T.; Shiratori, T.; Suzuki, Y. Laser Treatment CVD Diamond Coated Punch for Ultra-Fine Piercing of Metallic Sheets. In *Engineerng Applications of Diamond*; Mallik, A., Ed.; IntechOpen: Hasselt, The Netherlands, 2021; Volume 1, pp. 63–84.
19. Aizawa, T.; Shiratori, T.; Kira, Y.; Yoshino, T.; Suzuki, Y. Femtosecond Laser Trimming with Simultaneous Nanostructuring to Fine Piercing Punch to Electrical Amorphous Steel Sheets. *Micromachines* **2021**, *12*, 568. [CrossRef] [PubMed]

materials

MDPI

Article

Effects of Prior Heat Treatment and Induction Hardening on the Properties of JIS SUJ3 Bearing Steel

Shao-Quan Lu [1,2], Liu-Ho Chiu [1,2,*], Pei-Jung Chang [2,3] and Chung-Kwei Lin [1,2,4,*]

1 Department of Mechanical and Materials Engineering, Tatung University, Taipei 104327, Taiwan; d11101001@o365.ttu.edu.tw
2 Research Center of Digital Oral Science and Technology, College of Oral Medicine, Taipei Medical University, Taipei 110301, Taiwan; peronchang@tmu.edu.tw
3 Graduate Institute of Manufacturing Technology, National Taipei University of Technology, Taipei 106344, Taiwan
4 School of Dental Technology, College of Oral Medicine, Taipei Medical University, Taipei 110301, Taiwan
* Correspondence: lhchiu@gm.ttu.edu.tw (L.-H.C.); chungkwei@tmu.edu.tw (C.-K.L.)

Abstract: Bearing steels are frequently used in highly loaded components, such as roller bearings, due to their excellent hardenability and wear resistance. Microstructure, hardness, and residual stress distribution of the bearings significantly affect the wear resistance of the parts. In the present study, experiments investigated the effects of austenitizing temperature (850, 900, and 950 °C), with or without cryogenic treatment, and induction hardening treatment (9 and 12 kW) on the microstructure, microhardness, the amount of retained austenite, surface residual stress, and wear behavior of JIS SUJ3 steel. The experimental results revealed that the austenitized specimens' microstructure consisted of martensite, retained austenite, and dispersed granular alloy carbide exhibiting high hardness. After cryogenic or induction hardening treatment, the surface residual stress of austenitized specimens exhibited compressive stress rather than its original tensile stress state. The induction hardening treatment can significantly increase the microhardness of austenitized specimens, followed by quenching. Furthermore, the induction-hardened surface possessed less retained austenite. For practical industrial applications, a prior austenitizing heat treatment at 950 °C followed by hardening with an induction power of 12 kW was the optimal parameter for JIS SUJ3 bearing steel. The maximum microhardness and surface residual stress were 920 $HV_{0.3}$ and −1083 MPa, respectively, while the lowest weight loss was 0.5 mg after the 10,000-revolution wear test.

Keywords: bearing steel; prior heat treatment; induction hardening; microhardness; retained austenite; residual stress; wear

1. Introduction

With the rapid development of industrial machinery, the requirements for wear resistance and fatigue resistance of rolling parts such as bearing rings, ball screws, and other mechanical components have increased to ensure their proper function, and the selection of materials is a crucial issue. The most commonly used material in manufacturing precision and motion components is high chromium bearing steel [1]. The damages inflicted on the bearing steel are mainly due to wear and fatigue caused by the conditions of their usage. Thus, to ensure superior comprehensive mechanical properties and high dimensional stability, the bearing steel should exhibit high hardness and good wear resistance, which directly impact its lifespan. JIS SUJ3 steel is a high-quality alloy steel that has excellent wear resistance after hardening treatment, with a hardness of over 700 HV and a tensile strength of 1200 MPa.

Precision and motion parts made of bearing steel are usually required with good wear resistance to ensure their proper function. A major factor that affects wear behaviors is the hardness of the materials used to manufacture these mechanical parts. Other factors that affect resistance to mechanical damage include heat treatment procedures. The heat treatment of bearing steel is often conducted by quenching and tempering. The bearing steel, however, often forms retained austenite after quenching. This high content of retained austenite is due to the high carbon content in the bearing steel, as reported by Rivero et al. [2] and Roy et al. [3,4]. The transformation of the retained austenite into martensite can occur when external environmental factors such as temperature and load are introduced. This can negatively impact the stability of precision and motion parts in their usage [5–7].

High-temperature tempering, cryogenic treatment, and shot peening treatment are typical methods for alleviating the above-mentioned concerns about retained austenite. High-temperature tempering treatment causes a significant decrease in hardness, while shot peening treatment can only eliminate the retained austenite up to a limited depth in the subsurface region [8–11], and the resulting surface roughness does not meet the requirement of a bearing part. Unlike the other two methods, cryogenic treatment further induces phase transformation of the retained austenite through heat exchange generated at low temperatures, as reported by Bensely et al. [12], Senthilkumar et al. [13] and Mahdi et al. [14]. Its advantages not only include reducing the amount of retained austenite but also significantly relaxing residual stress [15,16]. Thus, cryogenic treatment is a preferable method. However, the cost and production time of cryogenic treatment are relatively high due to its technical threshold.

To reduce production time and costs to further reduce the consumption of energy, induction hardening is a good choice to consider. Shorter processing times make induction hardening treatment more affordable. Furthermore, induction hardening treatment also significantly improves surface properties, especially surface residual stress. Hu et al. [17], Gao et al. [18], and Hayama et al. [19] investigated the surface properties after induction quenching, and their results revealed that induction hardening treatment creates compressive residual stress and prolonged service life when compared to those mechanical parts with a tensile stress surface [20–22]. Moreover, the parts treated by induction hardening respond to less grain growth, possibly higher hardness, and a lower amount of retained austenite due to its rapid cooling rate.

Numerous studies explored the relationship between the heat treatment (quenched, tempered, and cryogenic treatment) and the resulting properties (hardness and the retained austenite content) and the wear behavior of bearing steel. The effects of prior treatment and induction hardening treatment on the dry wear behavior of bearing steel were not investigated in detail. The aim of the present study is to investigate the influence of different parameters of prior heat treatment (austenitization and cryogenic treatment) on the properties of induction-hardened JIS SUJ3 steel. In this study, JIS SUJ3 steel was selected for investigation. Firstly, the specimens were austenitized at various temperatures (850, 900, and 950 °C), quenched, processed with or without cryogenic treatment, and tempered to achieve a hardness of 62 ± 1 HRC to meet the requirements for industrial use in precision and motion parts. Subsequently, induction hardening with a power of 9 and 12 kW was applied to the prior heat-treated specimens. The microhardness, the amount of retained austenite, and the surface residual stress of the specimens were examined to evaluate the effects of different treatment parameters. In addition, block-on-roller type dry wear testing was used to compare the wear resistance. The properties and wear behavior of JIS SUJ3 steel specimens with different treatments were investigated for practical industrial applications.

2. Materials and Methods

2.1. Heat Treatment of Specimens

The specimens used in the experiment were made of JIS SUJ3 steel, a bearing steel manufactured by Gloria Material Technology Corporation (Tainan, Taiwan). The steel bar was spheroidized and machined to form cubic specimens with a dimension of 12.7 × 12.7 × 12.7 mm with chemical compositions listed in Table 1. The size of the specimens was designed to fix the specimen with a sample holder for wear testing according to the ASTM G77-17 specification [23].

Table 1. Chemical compositions (wt.%) of the JIS SUJ3 steel.

	C	Si	Mn	P	S	Cr	Ni	Mo	Cu	Fe.
JIS SUJ3 Spec.	0.95–1.10	0.40–0.70	0.90–1.15	≤0.025	≤0.025	0.90–1.20	≤0.25	≤0.08	≤0.30	Bal.
Specimens	1.05	0.326	1.02	0.0106	0.0051	1.14	0.0023	0.0014	0.0865	Bal.

The machined specimens were then heat-treated in a salt bath using three different austenitization temperatures. First, the specimens were preheated at a heating rate of 50 °C/min to 600 °C. Then, the specimens were austenitized at 850, 900, and 950 °C for 30 min and quenched to 100 °C with oil. After quenching, some specimens were immersed in liquid nitrogen at −196 °C for 2 h (cryogenic treatment). At the final stage, the quenched and the cryogenic-treated specimens were tempered at 200 °C for 1 h and cooled with water. Schematic illustrations of the temperature-time diagrams for QT and QCT treatments are shown in Figures S1a and S1b, respectively.

The effects of the austenitizing and cryogenic treatments were confirmed by examining surface hardness. The specimens, subjected to various conditions, exhibited a surface hardness of 62 ± 1 HRC, as determined using a Rockwell hardness tester (FUTURE-TECH FR-3e, Kawasaki, Japan) with an applied load of 150 kg.

2.2. Induction Hardening of Austenitized Specimens

The austenitized SUJ3 specimens immediately underwent induction hardening treatment. A digital controllable induction hardening system consists of an inductor coil, an induction power supply, and a circulating cooling water system. It is also equipped with a controllable mobile stage. The inductor was made of copper and applied with a frequency of 200 kHz. After austenitization, induction power levels of 9 and 12 kW were immediately applied, and specimens were quenched by moving them at a rate of 2.5 mm/s. After the induction hardening treatment, specimens were tempered again at 200 °C for 1 h. Table 2 summarizes all the parameters of austenitizing, cryogenic and induction hardening treatment, and the respective specimen codes. Three specimens for each sample code were prepared to ensure the reproducibility and stability of the experiments.

Table 2. Prior heat treatment and induction hardening process parameters and code.

Austenitizing Temperature	Cryogenic Treatment	Induction Power (kW)	Code
850 °C 900 °C 950 °C	No	No	850 QT 900 QT 950 QT
850 °C 900 °C 950 °C	Yes	No	850 QCT 900 QCT 950 QCT
850 °C 900 °C 950 °C	No	9	850 QT-9 900 QT-9 950 QT-9
850 °C 900 °C 950 °C	Yes	9	850 QCT-9 900 QCT-9 950 QCT-9
850 °C 900 °C 950 °C	No	12	850 QT-12 900 QT-12 950 QT-12
850 °C 900 °C 950 °C	Yes	12	850 QCT-12 900 QCT-12 950 QCT-12

2.3. *Characterizations of Austenitized and Induction-Hardened Specimens*

After the austenitizing, cryogenic, and induction hardening treatments, the specimens were investigated to reveal the microhardness, the percentage of retained austenite, and the surface residual stress.

2.3.1. Microhardness

The microhardness (HV) values as a function of depth (from surface to 6 mm) were determined by a Matsuzawa Vickers hardness tester (MXT50, Akita, Japan) to reveal the effects of various process parameters. Three specimens for each sample code ($n = 3$) were prepared and tested five times using a load of 300 g (2.94 N) and holding for 10 s. The microhardness (mean ± standard deviation) in the middle of the three tested specimens was used for comparison to reveal the effects of austenitizing, cryogenic, and induction hardening treatments with different parameters.

2.3.2. X-Ray Diffraction

A μ-X360s portable X-ray diffractometer (Pulstec Industrial Co., Ltd., Shizuoka, Japan) was used to determine the retained austenite content and surface residual stress using the two-dimensional cosα method [24–26]. A Cr target operating at 30 kV and 1.5 mA was used for X-ray diffraction with a diffraction angle set at 0° and 35° to determine the percentage of retained austenite and residual stress, respectively. The X-ray diffraction parameters are summarized in Table 3. The amount of retained austenite and residual stress was performed at three different locations for each treated specimen. Before the experiments, the X-ray diffractometer was aligned and validated using a stress-free iron powder sample for calibration. Equation (1) was used to calculate the retained austenite content (γ%) [27]. In the equation, I and R are the integrated intensity and theoretical relative intensity, respectively. The Greek letters γ and α indicate austenite and ferrite, respectively.

$$\gamma\% = (I\gamma/R\gamma)/[(I\gamma/R\gamma) + (I\alpha/R\alpha)] \times 100\% \quad (1)$$

Table 3. The parameters of the X-ray diffractometer.

Diffractometer Parameters	Specification/Values
Target	Cr
Diffraction plane (h,k,l)	α Fe (211)
Bragg angle for diffraction (2θ)	156.5°
X-ray tube current	1.5 mA
X-ray tube voltage	30 kV
Irradiation time	15 s
Collimator diameter	2 mm
Collimator distance	51 mm

2.4. Dry Wear Behavior of Austenitized and Induction-Hardened Specimens

To investigate the effect of different parameters on the wear behavior of the specimens, a multi-functional abrasion tester (PLINT TE53, Phoenix Tribology Ltd., Kingsclere, Berkshire, UK) was used to conduct the dry wear test. A schematic illustration of the experimental setup is shown in Figure S2a. The wear tester is a computer-controlled block-on-roller contact instrument in which a cubic block (specimen) is in contact with a roller, as shown in Figure S2b. In the case of the wear test, the specimen experiences one contact cycle per revolution at a constant contact normal load (92 N) with a roller rotation speed of 250 rpm. A SUJ2 steel roller with a diameter of 60 mm and a hardness of 62 ± 1 HRC was used for the wear test that was performed under dry friction conditions with a controlled temperature of 25 ± 1 °C and a relative humidity of 70 ± 5%. The test sample was ultrasonically cleaned with alcohol, and the weight loss was measured using a microbalance every 2000 revolutions. Each sample experienced a total revolution of 10,000 cycles for the dry wear test. All tests were repeated three times. After the wear test, the surface morphology of the worn specimen was examined by field emission scanning electron microscopy (Hitachi SU8000, Tokyo, Japan).

3. Results and Discussion

3.1. Heat Treatment of the JIS SUJ3 Specimen

3.1.1. The Effect of Austenitizing Temperatures on the JIS SUJ3 Specimen

Figure 1 shows the microstructure of the specimen after the austenitization treatment (at 850, 900, and 950 °C) and quenching to 100 °C. The microstructure of the as-received SUJ3 specimen was shown in Figure 1a, where the spheroidal Fe_3C cementite particles were uniformly distributed within the ferrite matrix. Figure 1b–d show the microstructure of the SUJ3 specimen after the austenitization treatment at 850, 900, and 950 °C, respectively. The ferrite matrix transformed into a martensite matrix can be observed after quenching. The proportion of the cementite decreased with increasing austenitizing temperature. However, the amount of the retained austenite increased with increasing austenitizing temperature. This is due to the higher austenitizing temperature increasing the dissolution ratio of the cementite. Moreover, high carbon content causes the martensitic transformation temperature (Ms) to decrease, thereby inducing an increase in the volume of the retained austenite. A higher magnification of these SEM images is available in Figure S3a–d for more detailed investigations.

After austenitizing and quenching treatments, a Vickers hardness tester was used to check the hardness of the SUJ3 specimen. The hardness of the specimen was 715 ± 5, 725 ± 8 and 728 ± 15 $HV_{0.3}$ after austenitization at 850, 900, and 950 °C, respectively. A slight increase in microhardness with increasing austenitizing temperature was observed but without significant difference. To further examine the condition of the amount of retained austenite, a portable X-ray diffractometer was used, and the measuring principle of the single incident angle method (cosα method) was applied [24–26]. Figure 2 shows the resulting X-ray diffraction patterns. The SUJ3 specimen, after undergoing austenitizing and quenching treatments, exhibited a distinct martensitic phase (PDF No. 44-1290) and a couple of very weak retained austenite diffraction peaks (PDF No. 65-4150). More detailed information concerning these two powder diffraction files (PDF No. 44-1290 and PDF No. 65-4150) is available in Tables S1 and S2. The peak intensities of austenite increased with increasing austenitizing temperature. The amount of retained austenite of the specimen austenitized at 850 °C was 3.4 ± 0.3%. By increasing the temperature to 900 °C, the amount of austenite retained increased to 10.4 ± 0.7% and further increased to 15.5 ± 0.6% at 950 °C. It should be pointed out that while carbon content can significantly affect the Ms temperature, the amount of retained austenite increased due to the cementite dissolved in the lattice (carbon concentration of martensite).

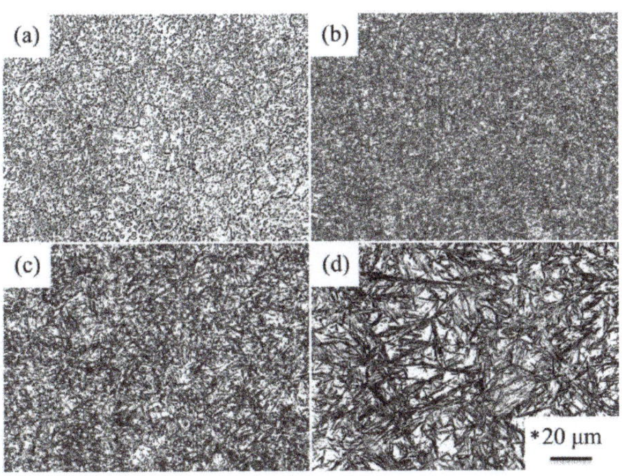

Figure 1. Microstructure of the (**a**) as-received SUJ3 specimen, (**b**) 850 QT, (**c**) 900 QT, and (**d**) 950 QT specimens. * The scale bars were the same for these OM images.

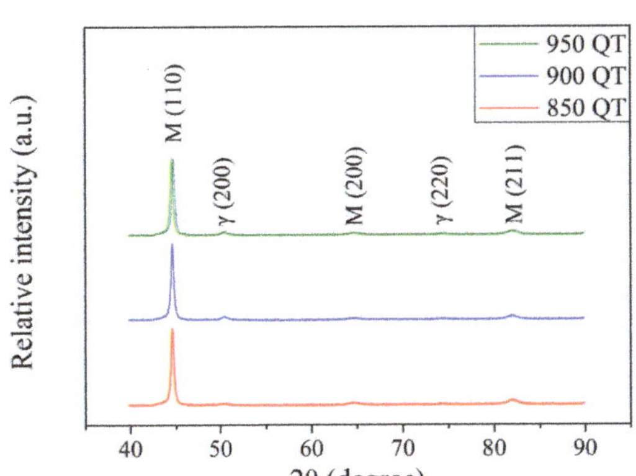

Figure 2. XRD patterns of SUJ3 specimens after austenitizing at various temperatures.

In addition to the retained austenite, the martensitic transformation was accompanied by volume expansion, and the boundary effect resulted in a tensile surface residual stress. The residual stress on the surface of the specimen austenitized at 850 °C was 53 ± 12 MPa. It increased gradually to 78 ± 17 and 115 ± 15 MPa at an austenitizing temperature of 900 and 950 °C, respectively. Table 4 summarizes the microhardness, the retained austenite content, and the surface residual stress of the SUJ3 specimen austenitized at various temperatures.

Table 4. The microhardness, the percentage of retained austenite and surface residual stress profile of SUJ3 specimen after austenitizing and quenching treatments.

Sample Code	Microhardness ($HV_{0.3}$)	Retained Austenite (%)	Surface Residual Stress (MPa)
850 QT	715 ± 5	3.4 ± 0.3	53 ± 12
900 QT	725 ± 8	10.4 ± 0.7	78 ± 17
950 QT	728 ± 15	15.5 ± 0.6	115 ± 15

3.1.2. The Effect of Cryogenic Treatment on the JIS SUJ3 Specimen

In industrial applications, the cryogenic treatment is commonly used to achieve dimensional stabilization after quenching. Figure 3 shows the microstructure of the austenitized specimens followed by cryogenic treatment. Figure 3a–c show the microstructure after the cryogenic treatment of the SUJ3 specimen with austenitizing temperatures of 850, 900, and 950 °C, respectively. A major difference observed was the proportion of the retained austenite, which was smaller than those in Figure 1b–d. It should be mentioned that martensitic transformation of the retained austenite occurred due to the heat exchange [8–11] and resulted in a decrease in the percentage of retained austenite after cryogenic treatment. However, the cryogenic treatment had no obvious effect on the microhardness of the SUJ3 specimen. After the cryogenic treatment, the microhardness of the specimen was 718 ± 3, 723 ± 5, and 730 ± 8 $HV_{0.3}$ with austenitizing temperatures of 850, 900, and 950 °C, respectively. These SEM images are magnified in Figure S4a–c for better observations.

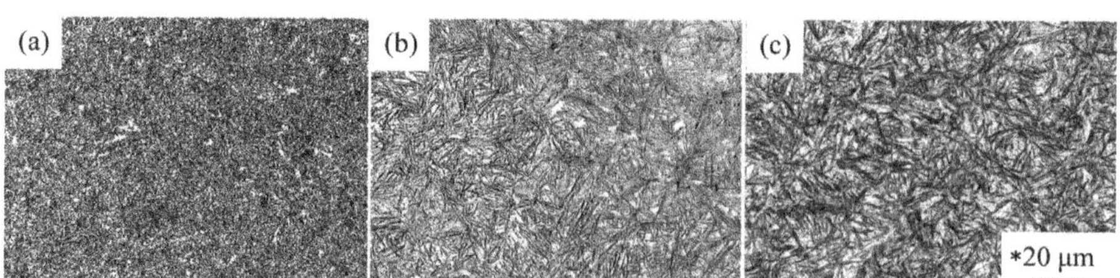

Figure 3. Microstructure of (**a**) 850 QCT, (**b**) 900 QCT, and (**c**) 950 QCT specimens. * The scale bars are the same for these OM images.

The XRD patterns of the SUJ3 specimen after cryogenic treatment at various austenitizing temperatures are shown in Figure 4. The peak intensities of austenite decreased after the cryogenic treatment. A similar observation was reported by Chen et al. [6] and Pöhl et al. [7]. The amount of retained austenite of the specimen decreases to 3.5 ± 0.2, 5.2 ± 0.3 and 7.4 ± 0.3% after cryogenic treatment for samples austenitized at 850, 900, and 950 °C, respectively. Moreover, the residual stress on the surface of the specimen changed from a tensile stress to a compressive stress of −61 ± 11, −92 ± 8 and −132 ± 14 MPa after cryogenic treatment for 850, 900, and 950 °C austenitized samples, respectively.

Bensely et al. [12] and Senthilkumar et al. [13] reported similar phenomena. The cryogenic treatment is accompanied by volume contraction. Therefore, the surface residual stress appears as compressive residual stress. Table 5 summarizes the important results of the SUJ3 specimens austenitized at 850, 900, and 950 °C and subsequently underwent a cryogenic treatment.

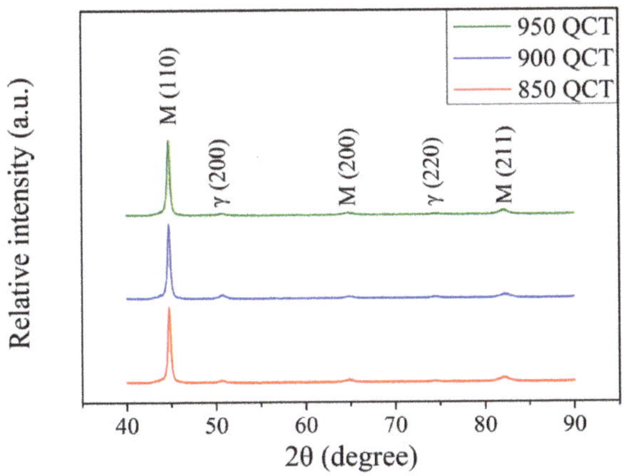

Figure 4. XRD patterns of the SUJ3 specimen after austenitizing at different temperatures and cryogenic treatment.

Table 5. The microhardness, the percentage of retained austenite, and the surface residual stress profile of the SUJ3 specimen at given conditions.

Sample Code	Microhardness ($HV_{0.3}$)	Retained Austenite (%)	Surface Residual Stress (MPa)
850 QCT	718 ± 3	3.5 ± 0.2	−61 ± 11
900 QCT	723 ± 5	5.2 ± 0.3	−92 ± 8
950 QCT	730 ± 8	7.4 ± 0.3	−132 ± 14

3.2. Induction Hardening of the JIS SUJ3 Specimen

3.2.1. The Effect of 9 kW Induction Hardening Treatment on the JIS SUJ3 Specimen

In Section 3.1, the effects of austenitizing temperature and cryogenic treatment were investigated. The higher the austenitizing temperature, the higher the amount of retained austenite and surface residual tensile stress. Cryogenic treatment reduces the amount of retained austenite and converts the residual stress into a compressive state. Different austenitizing temperatures and cryogenic treatment were examined, and the effects of induction hardening treatment were investigated further. The mechanical properties of the induction-hardened specimen as a function of depth were analyzed using the Vickers microhardness test with an applied load of 300 g. Figure 5 shows the microhardness in the subsurface region of the induction-hardened specimen with different prior austenitizing temperatures. After the induction hardening treatment, the microhardness increased significantly at the subsurface region up to a relatively deep depth of 0.8 mm. The microhardness exhibited a maximum of 845 ± 11, 865 ± 8, and 890 ± 12 $HV_{0.3}$ with a prior austenitizing temperature of 850, 900, and 950 °C, respectively. It decreased continuously to a minimum of 400 ± 9, 410 ± 6, and 420 ± 6 $HV_{0.3}$ at a depth of 2.0 mm. Then, it increased gradually to 725 ± 20 $HV_{0.3}$ at depths of 4.0 mm and deeper. A similar microhardness profile was also observed in the literature [17–19]. The hardened layer (up to 0.8 mm) was attributed to the induction-hardened treatment that included the heat transfer, temperature gradient, and proportion of cementite dissolved in the lattice (carbon concentration of martensite) due to the rapid heating rate and cooling rate.

Figure 6 shows the microhardness in the subsurface region of the induction-hardened specimen with different prior austenitizing temperatures and cryogenic treatment. Compared to those without cryogenic treatment (Figure 5, results of QT-9 samples), specimens with cryogenic treatment exhibited a similar microhardness of 840–890 $HV_{0.3}$ up to a depth of 0.8 mm. The microhardness decreased continuously until reaching a minimum at a depth of 2.0 mm when treated with a prior cryogenic treatment. Thereafter, it increased gradually to 725 ± 20 $HV_{0.3}$ at depths of 4.0 mm and deeper. Thus, the cryogenic treatment does not significantly increase or decrease the microhardness in the subsurface region. However, raising the prior austenitizing temperature (up to 950 °C) increased the maximum microhardness and affected the depth of hardening in the induction-hardened specimen.

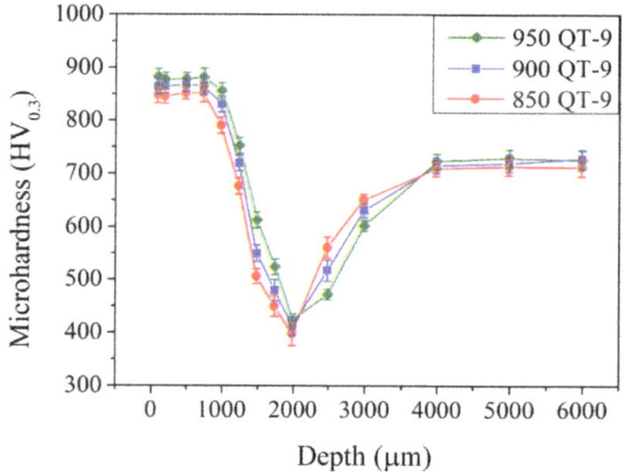

Figure 5. Microhardness profiles of specimens after various austenitizations followed by 9 kW induction hardening.

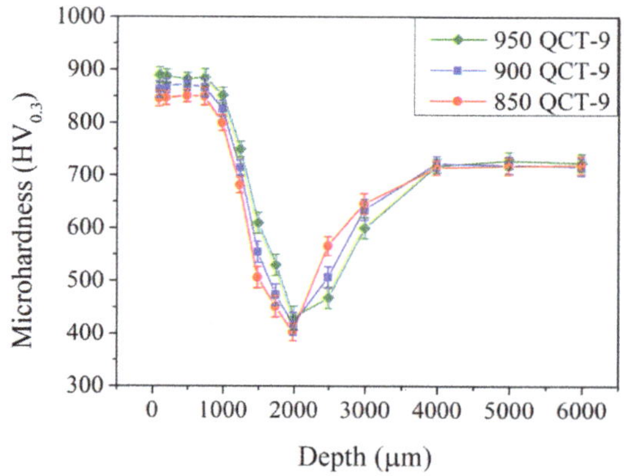

Figure 6. Microhardness profiles of specimens after various austenitizations followed by cryogenic treatment that were induction-hardened by 9 kW.

The XRD patterns of induction-hardened specimen with an induction power of 9 kW and the different prior austenitizing temperatures are shown in Figure 7. Compared to specimens without an induction hardening treatment (Figure 2), samples that underwent induction hardening treatment exhibited lower the peak intensities of austenite. It can be noted that the rapid cooling rate of induction hardening treatment is due to water sprayed during the process. The faster the quenching treatment cooling rate, the lower the amount of retained austenite. This shows a similar trend as reported by Babasafari et al. [28] and Cao et al. [29]. Figure 8 shows the XRD patterns of the induction-hardened specimen with different prior austenitizing temperatures and cryogenic treatment. Similar to the results of Figure 7 (QT-9 samples), the peak intensities of austenite remained at a relatively low level. The amount of retained austenite of induction-hardened specimens without prior cryogenic treatment was 5.3 ± 0.7, 7.3 ± 0.5 and 9.4 ± 0.9% at austenitizing temperatures of 850, 900 and 950 °C, respectively. The corresponding residual stress on the surface of the specimen was −951 ± 13, −977 ± 17 and −974 ± 9 MPa. The amount of retained austenite in the induction-hardened specimens with prior cryogenic treatment was 4.9 ± 0.4, 6.5 ± 0.6 and 8.7 ± 0.7% at austenitizing temperatures of 850, 900 and 950 °C, respectively. The corresponding residual stress on the surface of the specimen was −900 ± 20, −922 ± 15 and −915 ± 19 MPa. These results indicate that prior cryogenic treatment does not significantly affect the amount of retained austenite and residual stress after induction hardening treatment. Table 6 summarizes the amount of retained austenite and surface residual stress of the SUJ3 specimens austenitized at 850, 900, and 950 °C and followed by a 9 kW induction hardening treatment.

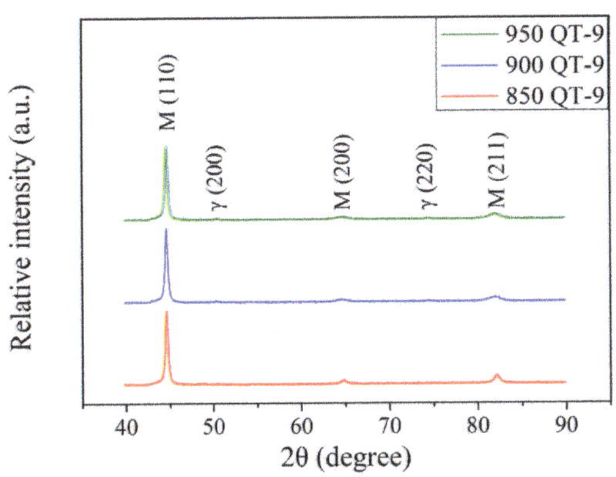

Figure 7. XRD patterns of specimens after various austenitizations followed by 9 kW induction hardening.

Table 6. The percentage of retained austenite and surface residual stress of 9 kW induction-hardened SUJ3 specimens.

Sample Code	Retained Austenite (%)	Surface Residual Stress (MPa)
850 QT-9	5.3 ± 0.7	−951 ± 13
850 QCT-9	4.9 ± 0.4	−900 ± 20
900 QT-9	7.3 ± 0.5	−977 ± 17
900 QCT-9	6.5 ± 0.6	−922 ± 15
950 QT-9	9.4 ± 0.9	−974 ± 9
950 QCT-9	8.7 ± 0.7	−915 ± 19

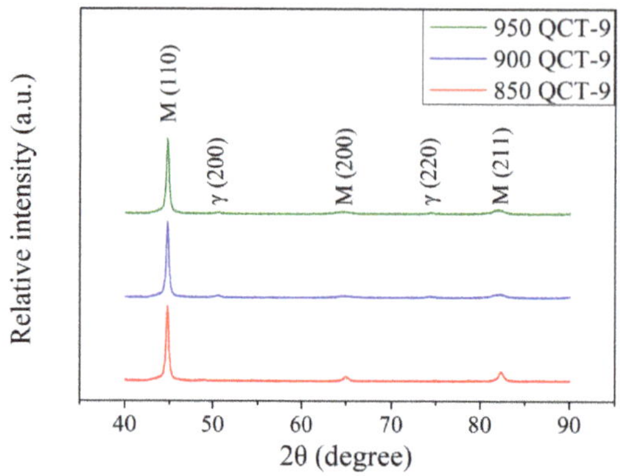

Figure 8. XRD patterns of specimens after various austenitizations followed by cryogenic treatment that were induction-hardened by 9 kW.

3.2.2. The Effect of 12 kW Induction Hardening Treatment on the JIS SUJ3 Specimen

Figure 9 shows the microhardness in the subsurface region of the induction-hardened specimen with different prior austenitizing temperatures. After induction hardening, the microhardness increased significantly at the subsurface region up to a relatively deep depth of 1.0 mm. The microhardness exhibited a maximum at a depth of 0.1–1.0 mm and was 885 ± 6, 900 ± 5, and 920 ± 10 $HV_{0.3}$ with prior austenitizing temperatures of 850, 900, and 950 °C, respectively. It decreased continuously to a minimum of 380 ± 12, 400 ± 9, and 420 ± 8 $HV_{0.3}$ at a depth of 2.5 mm. Then, it increased gradually to 725 ± 20 $HV_{0.3}$ at depths of 5.0 mm and deeper. As reported by Hu et al. [17] and Gao et al. [18], increasing the power of induction-hardening treatment increases microhardness in the induction-hardened region. In the present study, increasing the power of induction hardening treatment (up to 12 kW) increased the microhardness and the affected depth of the specimen.

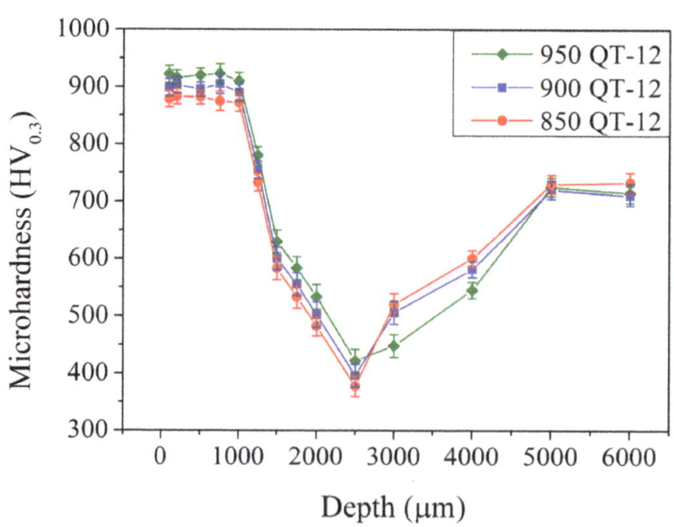

Figure 9. Microhardness profiles of specimens after various austenitizations followed by 12 kW induction hardening.

Figure 10 shows the microhardness in the subsurface region of the induction-hardened specimens with various prior austenitization and cryogenic treatments. Compared to those without a prior cryogenic treatment (Figure 9, QT-12 samples), the microhardness remained 880–930 $HV_{0.3}$ up to a depth of 1.0 mm and decreased continuously to a minimum microhardness at a depth of 2.5 mm when treated with a prior cryogenic treatment. Thereafter, it increased gradually to 725 ± 20 $HV_{0.3}$ at depths of 5.0 mm and further. The prior cryogenic treatment does not significantly increase or decrease the microhardness in the subsurface region.

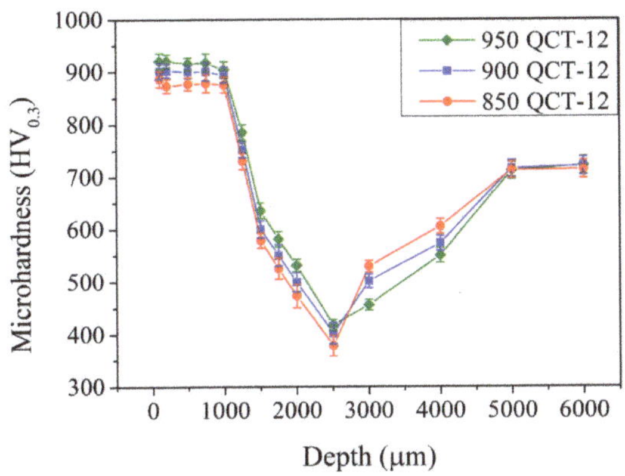

Figure 10. Microhardness profiles of specimens after various austenitizations followed by cryogenic treatment that were induction-hardened by 12 kW.

The XRD patterns of the 12 kW induction-hardened specimen with different prior austenitizing temperatures are shown in Figure 11. After 12 kW induction hardening, there was an increase in the peak intensities of austenite when compared to specimens treated at a power of 9 kW (Figure 7). This phenomenon is similar to those reported by Kaiser et al. [30] and Fisk et al. [31]. It should be pointed out that the increase in the power resulted in the specimen reaching a higher austenitizing temperature during the induction hardening treatment. Figure 12 shows the XRD patterns of 12 kW induction-hardened specimens with different prior austenitizing temperatures and cryogenic treatment. The peak intensities of austenite remained at a relatively low level. Compared to those without the prior cryogenic treatment (Figure 11, QT-12 samples), the effect of cryogenic treatment on induction-hardened specimens is not significant. The amount of retained austenite of the induction-hardened specimen without prior cryogenic treatment was 6.2 ± 0.3, 8.9 ± 0.4, and 11.8 ± 0.3% at austenitizing temperatures of 850, 900, and 950 °C, respectively. The corresponding surface residual stress of the specimen was −1068 ± 3, −1073 ± 11, and −1083 ± 5 MPa. The amount of retained austenite of the induction-hardened specimen with prior cryogenic treatment was 5.8 ± 0.2, 8.5 ± 0.4, and 11.3 ± 0.4% at austenitizing temperatures of 850, 900, and 950 °C, respectively. The corresponding residual stress on the surface of the specimen was −1031 ± 10, −1027 ± 6, and −1055 ± 9 MPa. These results indicate that prior cryogenic treatment does not significantly affect the amount of retained austenite and residual stress after the induction hardening treatment. The induction-hardened specimens were considered to be in a re-hardened state. Therefore, the benefits of cryogenic treatment are not obvious. The percentage of retained austenite and surface residual stress of the 12 kW induction-hardened SUJ3 specimens with various prior heat treatments are summarized in Table 7.

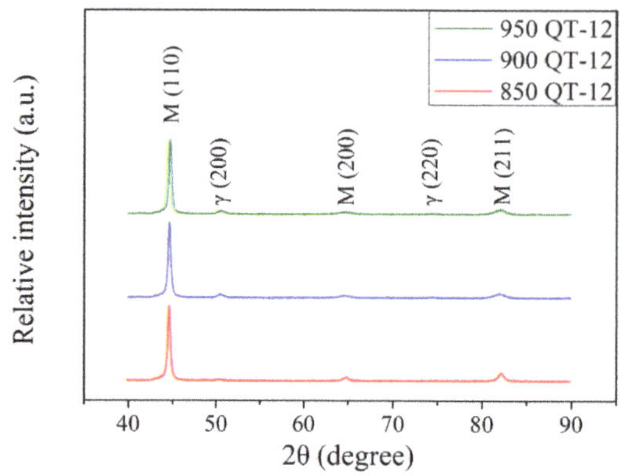

Figure 11. XRD patterns of specimens after various austenitizations followed by 12 kW induction hardening.

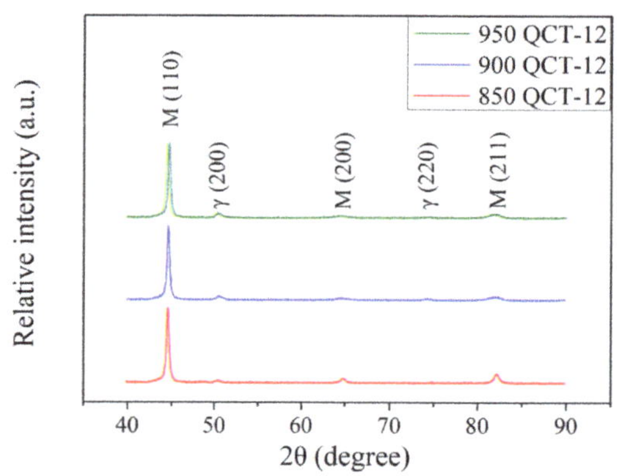

Figure 12. XRD patterns of specimens after various austenitizations followed by cryogenic treatment that were induction-hardened by 12 kW.

Table 7. The percentage of retained austenite and surface residual stress of 12 kW induction-hardened SUJ3 specimens.

Sample Code	Retained Austenite (%)	Surface Residual Stress (MPa)
850 QT-12	6.2 ± 0.3	−1068 ± 3
850 QCT-12	5.8 ± 0.2	−1031 ± 10
900 QT-12	8.9 ± 0.4	−1073 ± 11
900 QCT-12	8.5 ± 0.4	−1027 ± 6
950 QT-12	11.8 ± 0.3	−1083 ± 5
950 QCT-12	11.3 ± 0.4	−1055 ± 9

3.3. The Effect of Austenitizing Temperature, Cryogenic Treatment, and Induction Hardening Treatment on Dry Wear Behavior of the JIS SUJ3 Specimen

As demonstrated in the above sections, the effects of different austenitizing temperatures, cryogenic treatment, and induction hardening treatment were investigated. The microhardness, retained austenite content, and surface residual stress were measured and are summarized in Figure S5. Within the limitations of the present work, data on the depth of residual compressive stress and the changes in the specimen's diameters after different heat treatments are not available and will be addressed elsewhere.

To investigate the dry wear behavior of specimens, a block-on-roller wear test with an applied contact load of 92 N and a roller rotation rate of 250 rpm was performed at room temperature. The cumulative weight losses of SUJ3 specimens with different prior austenitizing temperatures and induction hardening treatments with and without a prior cryogenic treatment are shown in Figures 13 and 14, respectively. Cumulative weight loss increased with a prolonged revolution cycle. With induction hardening, the weight loss of the specimen significantly decreased. By increasing the power applied to the induction hardening treatment, weight loss can be further decreased. Similar behavior has been observed by Hu et al. [17] and Kaiser et al. [30]. It should be pointed out that the induction hardening treatment can induce significant compressive residual stress due to the rapid cooling rate. As a result, the wear resistance was improved.

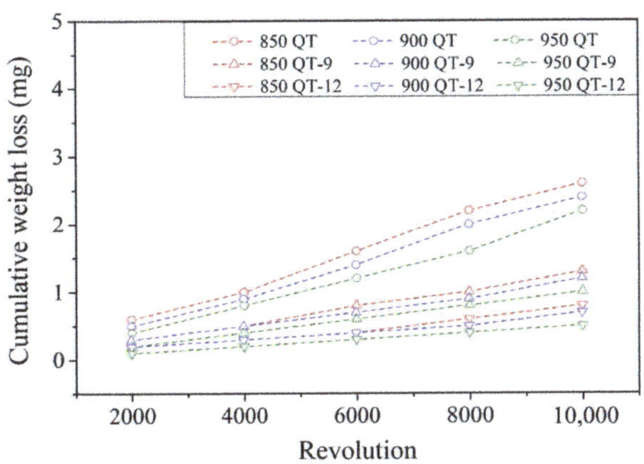

Figure 13. The cumulative weight loss profile of specimens without a prior cryogenic treatment.

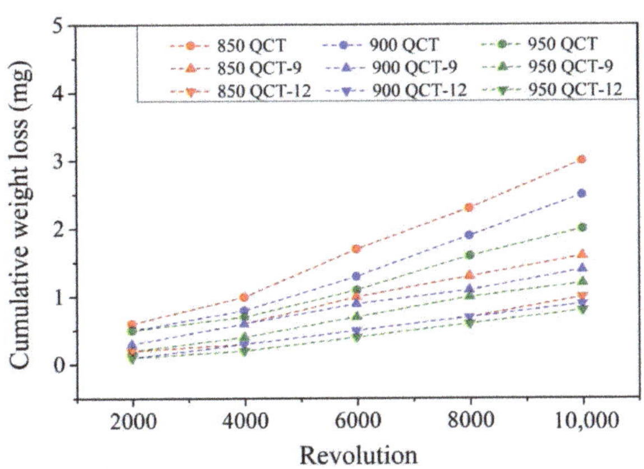

Figure 14. The cumulative weight loss profile of specimens with prior cryogenic treatment.

Figure 15 shows the weight loss of the JIS SUJ3 specimen under different conditions after the 10,000-cycle dry wear test. As shown by the orange bars in Figure 15, the weight loss of heat-treated specimens without prior cryogenic treatment was 2.60 ± 0.11, 2.40 ± 0.10, and 2.20 ± 0.10 mg with austenitizing temperatures of 850, 900, and 950 °C, respectively. It can be decreased further with induction hardening. The weight loss of the 9 kW induction-hardened specimen without prior cryogenic treatment was 1.30 ± 0.09, 1.20 ± 0.05, and

1.00 ± 0.07 mg with an austenitizing temperature of 850, 900, and 950 °C, respectively. Then, it reduced to 0.80 ± 0.05, 0.70 ± 0.07, and 0.50 ± 0.03 mg at an induction power of 12 kW. It can be noted that the increase in austenitizing temperature and power applied for induction hardening treatment resulted in lower weight loss, which decreased after the dry wear test due to the increase in microhardness. However, specimens with prior cryogenic treatment do not exhibit a significant decrease in weight loss. It should be pointed out that the prior cryogenic treatment has lower compressive residual stress compared to those without a prior cryogenic treatment. Therefore, the wear resistance was not improved.

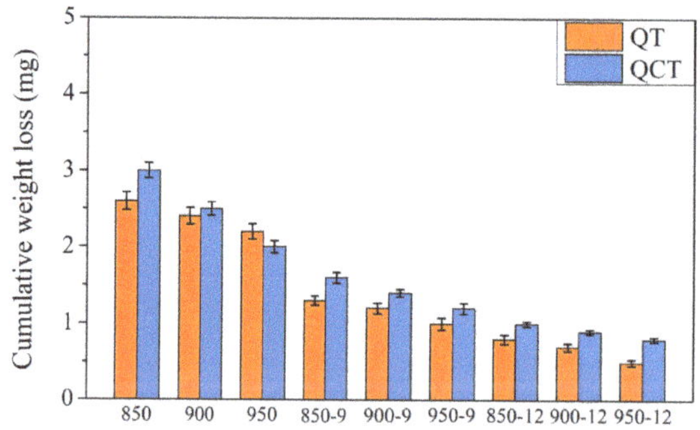

Figure 15. Comparison of weight loss among the specimens after the 10,000-cycle dry wear test.

The surface morphologies of the SUJ3 specimens with or without prior cryogenic treatment with different austenitizing temperatures are shown in Figure 16. The worn surface of the specimens was distributed by groove, scraped abrasive particles, and partial peeling. This includes abrasive wear and adhesive wear. Figure 17 shows the surface morphology of the induction-hardened SUJ3 specimens with and without prior cryogenic treatment with an austenitizing temperature of 950 °C. Compared to those without the induction hardening treatment (Figure 16c,f), it shows a smoother worn surface due to the lower weight loss.

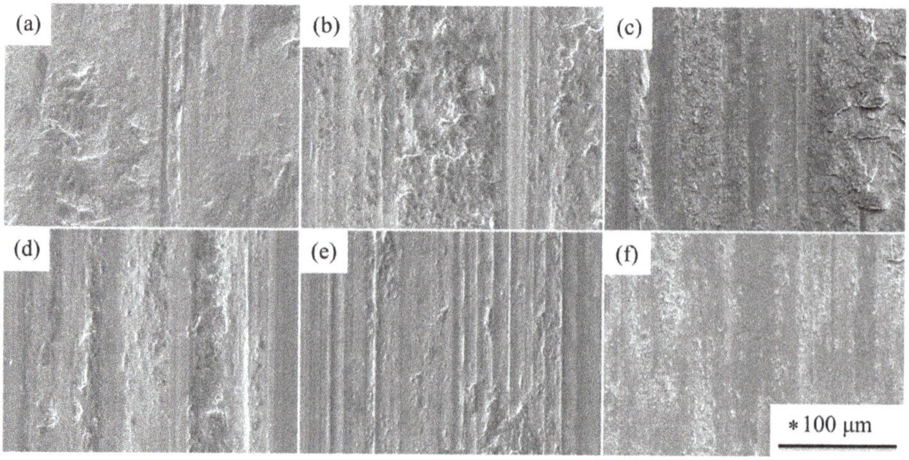

Figure 16. Surface morphology of worn specimens treated with the given conditions: (**a**) 850 QT, (**b**) 900 QT, (**c**) 950 QT, (**d**) 850 QCT, (**e**) 900 QCT, and (**f**) 950 QCT. * The scale bars were the same for these SEM images.

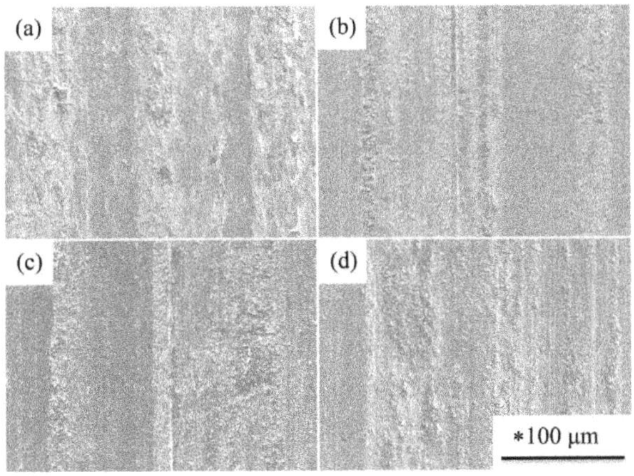

Figure 17. Surface morphology of worn specimens treated with the given conditions: (**a**) 950 QT-9, (**b**) 950 QCT-9, (**c**) 950 QT-12 and (**d**) 950 QCT-12. * The scale bars were the same for these SEM images.

4. Conclusions

In the present study, the JIS SUJ3 specimens underwent austenitizing treatment with different temperatures. Alternative cryogenic and induction hardening treatments were also investigated. The microhardness, surface residual stress, amount of retained austenite, and wear behaviors were investigated, and the following conclusions were drawn:

1. The residual tensile stress on the surface of the austenitized specimen was converted into compressive stress accompanied by cryogenic treatment, and a decrease in the amount of retained austenite can be observed.
2. By induction hardening, the microhardness and the residual compressive stress of austenitized and quenched specimens were significantly increased.
3. The prior austenitizing temperature and induction power can affect the induction hardening results. The induction hardening effect increased with increasing the prior austenitizing temperature and induction power. A prior austenitizing temperature of 950 °C and induction power of 12 kW produced optimal results. The maximum compressive residual stress reached −1083 MPa (on the surface), and the microhardness (in the induction-hardened subsurface region) was also the highest (920 ± 10 $HV_{0.3}$).
4. The cryogenic treatment can decrease the amount of retained austenite. However, the induction-hardened specimens with a prior cryogenic treatment did not show a lower or even higher weight loss after the dry wear test. Thus, for bearing parts used in dry wear conditions, induction hardening the pre-austenitized specimen without prior cryogenic treatment was the optimal choice.

Supplementary Materials: The following supporting information can be downloaded at https://www.mdpi.com/article/10.3390/ma18081797/s1. Figure S1: Temperature time diagrams for (a) QT and (b) QCT treatments. Figure S2: (a) Schematic diagram of abrasion tester (block-on-roller type); (b) experimental setup of the block (specimen) and roller on the abrasion tester. Figure S3: Microstructure images with a higher magnification of the (a) as-received SUJ3 specimen, (b) 850 QT, (c) 900 QT, and (d) 950 QT specimens. Figure S4: Microstructure images with a higher magnification of (a) 850 QCT (b) 900 QCT, and (c) 950 QCT specimens. Figure S5: (a) The microhardness (the original HV before heat treatment was shown as the red line in the figure), (b) the amount of retained austenite, and (c) the surface residual stress profile of samples after different heat treatment methods. Table S1.

Powder diffraction file of martensitic phase (PDF No. 44-1290). Table S2. Powder diffraction file of austenite (PDF No. 65-4150).

Author Contributions: Conceptualization, S.-Q.L. and L.-H.C.; methodology, S.-Q.L. and L.-H.C.; software, P.-J.C. and C.-K.L.; investigation, S.-Q.L. and L.-H.C.; validation, P.-J.C. and C.-K.L.; data curation, S.-Q.L. and P.-J.C.; funding acquisition, L.-H.C.; writing—original draft, S.-Q.L. and P.-J.C.; writing—review and editing, C.-K.L. All authors have read and agreed to the published version of the manuscript.

Funding: This study was supported by the National Science and Technology Council, Taiwan, NSTC-113-2622-E-036-001.

Institutional Review Board Statement: Not applicable.

Informed Consent Statement: Not applicable.

Data Availability Statement: The original contributions presented in this study are included in the article. Further inquiries can be directed to the corresponding author.

Conflicts of Interest: The authors declare no conflicts of interest.

References

1. Fernandes, A.P.; Gallego, J.; Picon, C.A.; Filho, G.T.; Casteletti, L.C. Wear and corrosion of niobium carbide coated AISI 52100 bearing steel. *Surf. Coat. Technol.* **2015**, *279*, 112–117. [CrossRef]
2. Rivero, I.V.; Ruud, C.O. Deviation of residual stress patterns in 52100 bearing steel due to inherent microstructural transformations after rolling contact. *Mater. Charact.* **2004**, *53*, 381–393. [CrossRef]
3. Roy, S.; Sundararajan, S. Effect of retained austenite on spalling behavior of carburized AISI 8620 steel under boundary lubrication. *J. Mater. Res. Technol.* **2019**, *119*, 238–246. [CrossRef]
4. Roy, S.G.; Ooi, T.C.; Sundararajan, S. Effect of retained austenite on micropitting behavior of carburized AISI 8620 steel under boundary lubrication. *Materialia* **2018**, *3*, 192–201. [CrossRef]
5. Su, Y.Y.; Chiu, L.H.; Chuang, T.L.; Huang, C.L.; Wu, C.Y. The Study of Evaluating the Amount of Retained Austenite in Steels by Micrographic and X-ray Diffraction Analysis. *Met. Heat Treat.* **2011**, *111*, 34–44.
6. Chen, W.; He, X.F.; Yu, W.C.; Shi, J.; Wang, M.Q.; Yao, K.F. Characterization of the microstructure and hardness of case-carburized gear steel. *Micron* **2021**, *144*, 103028. [CrossRef]
7. Pöhl, F. Local deformation and transformation behavior of retained austenite in 18CrNiMo7-6 after high-carbon carburizing treatment. *Mater. Charact.* **2020**, *167*, 110446. [CrossRef]
8. Lu, S.Q.; Chiu, L.H.; Chang, P.J.; Lin, C.K. Effects of Shot Peening Pressure, Time, and Material on the Properties of Carburized Steel Shafts. *Materials* **2024**, *17*, 4124. [CrossRef]
9. Wu, J.Z.; Liu, H.J.; Wei, P.T.; Zhu, C.C.; Lin, Q.J. Effect of shot peening coverage on hardness, residual stress and surface morphology of carburized rollers. *Surf. Coat. Technol.* **2020**, *384*, 125273. [CrossRef]
10. Qu, S.G.; Duan, C.F.; Hu, X.F.; Jia, S.Y.; Li, X.Q. Effect of shot peening on microstructure and contact fatigue crack growth mechanism of shaft steel. *Mater. Chem. Phys.* **2021**, *274*, 125116. [CrossRef]
11. Lin, Q.J.; Liu, H.J.; Zhu, C.C.; Chen, D.F.; Zhou, S.S. Effects of different shot peening parameters on residual stress, surface roughness and cell size. *Surf. Coat. Technol.* **2020**, *398*, 126054. [CrossRef]
12. Bensely, A.; Venkatesh, S.; Mohan Lal, D.; Nagarajan, G.; Rajadurai, A.; Junik, K. Effect of cryogenic treatment on distribution of residual stress in case carburized En 353 steel. *Mater. Sci. Eng. A* **2008**, *479*, 229–235. [CrossRef]
13. Senthilkumar, D.; Rajendran, L.; Pellizzari, M.; Siiriainen, J. Influence of shallow and deep cryogenic treatment on the residual state of stress of 4140 steel. *J. Mater. Process. Technol.* **2011**, *211*, 396–401. [CrossRef]
14. Mahdi, K.; Meshinchi Asl, K.; Khomamizadeh, F. Effect of cryogenic treatment on microstructure, mechanical and wear behaviors of AISI H13 hot work tool steel. *Cryogenics* **2011**, *51*, 55–61.
15. Kara, F.; Küçük, Y.; Özbek, O.; Özbek, N.A.; Gök, M.S.; Altaş, E.; Uygur, İ. Effect of cryogenic treatment on wear behavior of Sleipner cold work tool steel. *Tribol. Int.* **2023**, *180*, 108301. [CrossRef]
16. Ramesh, S.; Bhuvaneshwari, B.; Palani, G.S.; Lal, D.M.; Mondal, K.; Gupta, R.K. Enhancing the corrosion resistance performance of structural steel via a novel deep cryogenic treatment process. *Vacuum* **2019**, *159*, 468–475. [CrossRef]
17. Hu, Y.; Qin, Q.; Wu, S.; Zhao, X.; Wang, W. Fatigue resistance and remaining life assessment of induction-hardened S38C steel railway axles. *Int. J. Fatigue* **2021**, *144*, 106068. [CrossRef]

18. Gao, J.W.; Pan, X.N.; Han, J.; Zhu, S.P.; Liao, D.; Li, Y.B.; Dai, G.Z. Influence of artificial defects on fatigue strength of induction hardened S38C axles. *Int. J. Fatigue* **2020**, *139*, 105746. [CrossRef]
19. Hayama, M.; Kikuchi, S.; Tsukahara, M.; Misaka, Y.; Komotori, J. Estimation of residual stress relaxation in low alloy steel with different hardness during fatigue by in situ X-ray measurement. *Int. J. Fatigue* **2024**, *178*, 107989. [CrossRef]
20. Li, B.Z.; Jiang, X.H.; Yang, J.G.; Liang, S.Y. Effects of depth of cut on the redistribution of residual stress and distortion during the milling of thin-walled part. *J. Mater. Process. Technol.* **2015**, *216*, 223–233. [CrossRef]
21. Rego, R.; Löpenhaus, C.; Gomes, J.; Klock, F. Residual stress interaction on gear manufacturing. *J. Mater. Process. Technol.* **2018**, *252*, 249–258. [CrossRef]
22. Kikuchi, S.; Minamizawa, K.; Arakawa, J.; Akebono, H.; Takesue, S.; Hayakawa, M. Combined effect of surface morphology and residual stress induced by fine particle and shot peening on the fatigue limit for carburized steels. *Int. J. Fatigue* **2023**, *168*, 107441. [CrossRef]
23. *ASTM G77-17*; Standard Test Method for Ranking Resistance of Materials to Sliding Wear Using Block-on-Ring Wear Test. ASTM International: West Conshohocken, PA, USA, 2017.
24. Ao, S.S.; Li, C.J.; Huang, Y.F.; Luo, Z. Determination of residual stress in resistance spot-welded joint by a novel X-ray diffraction. *Measurement* **2020**, *161*, 107892. [CrossRef]
25. Tanaka, K. The cosα method for X-ray residual stress measurement using two-dimensional detector. *Mech. Eng. Rev.* **2019**, *6*, 378. [CrossRef]
26. Ramírez-Rico, J.; Lee, S.; Ling, J.; Noyan, I.C. Stress measurement using area detectors: A theoretical and experimental comparison of different methods in ferritic steel using a portable X-ray apparatus. *J. Mater. Sci.* **2016**, *51*, 5343–5355. [CrossRef]
27. *ASTM E975-13*; Standard Practice for X-Ray Determination of Retained Austenite in Steel with Near Random Crystallographic Orientation. ASTM International: West Conshohocken, PA, USA, 2013.
28. Babasafari, Z.; Pan, A.V.; Pahlevani, F.; Hossain, R.; Sahajwalla, V.; Du Toit, M.; Dippenaar, R. Effects of austenizing temperature, cooling rate and isothermal temperature on overall phase transformation characteristics in high carbon steel. *J. Mater. Res. Technol.* **2020**, *9*, 15286–15297. [CrossRef]
29. Cao, K.; Wang, Z.G.; Liu, P.F.; Zhao, A.M. Effect of austenitizing temperature and cooling rate on Ms temperature of Fe–Ni–Cr cast iron. *J. Mater. Res. Technol.* **2023**, *26*, 796–806. [CrossRef]
30. Kaiser, D.; Damon, J.; Mühl, F.; de Graaff, B.; Kiefer, D.; Dietrich, S.; Schulze, V. Experimental investigation and finite-element modeling of the short-time induction quench-and-temper process of AISI 4140. *J. Mater Process. Technol.* **2020**, *279*, 116485. [CrossRef]
31. Fisk, M.; Lindgren, L.E.; Datchary, W.; Deshmukh, V. Modelling of induction hardening in low alloy steels. *Finite Elem. Anal. Des.* **2018**, *144*, 61–75. [CrossRef]

materials

MDPI

Article

Influence of Nanocomposite PVD Coating on Cutting Tool Wear During Milling of 316L Stainless Steel Under Air Cooling Conditions

Jarosław Tymczyszyn [1,*], Artur Szajna [1] and Grażyna Mrówka-Nowotnik [2]

[1] Department of Manufacturing Techniques and Automation, Faculty of Mechanical Engineering and Aeronautics, Rzeszow University of Technology, 12 Al. Powstancow Warszawy Street, 35-959 Rzeszow, Poland; a.szajna@prz.edu.pl

[2] Department of Material Science, Faculty of Mechanical Engineering and Aeronautics, Rzeszow University of Technology, 12 Al. Powstancow Warszawy Street, 35-959 Rzeszow, Poland; mrowka@prz.edu.pl

* Correspondence: j.tymczyszyn@prz.edu.pl

Abstract: This study examines the impact of PVD coatings on cutting tool wear during the milling of 316L stainless steel under air cooling conditions. In the experiment, a carbide milling cutter coated with a nanocomposite nACo3 (AlTiSiN) coating was used. The coating was deposited using a next-generation device, the PLATIT π411PLUS, which features one central and three lateral rotating cathodes. The nanocomposite nACo3 coating obtained with this method exhibits exceptionally high structural density and excellent mechanical properties. The new generation of the nACo3 coating demonstrates improved surface properties and a lower friction coefficient compared to previous generations. The findings indicate that PVD nACo3 coatings significantly enhance wear resistance, extending tool life while maintaining acceptable surface quality. The optimal cutting time was determined to be approximately 90 min, after which a sharp increase in surface roughness and tool wear was observed. After 120 min of machining, substantial deterioration of surface quality parameters was recorded, suggesting increasing cutting forces and cutting edge degradation. SEM and EDS analyses revealed the presence of adhered material on the tool and sulfide inclusions in the microstructure of 316L stainless steel, which influenced the machining process. The nACo3 coating demonstrated high thermal and wear resistance, making it an effective solution for machining difficult-to-cut materials. This study suggests that selecting appropriate cutting parameters, tool geometry, protective coatings, and cooling strategies can significantly affect tool longevity and machining quality. The novelty of this research lies in the application of innovative nanocomposite PVD coatings during the milling of 316L stainless steel under air cooling conditions. These studies indicate potential future research directions, such as the use of minimum quantity lubrication (MQL) or cryogenic cooling as methods to reduce tool wear and improve post-machining surface quality.

Keywords: 316L stainless steel; wear; cutting forces; surface roughness; nACo3

1. Introduction

Due to the continuous improvement of semifinished product manufacturing technologies, the development of cutting tool materials, and advancements in modern manufacturing machinery, dry machining is becoming an increasingly desirable production method from both ecological and economic perspectives. The elimination of cutting fluids reduces production costs and has a significant impact on environmental protection [1]. Dry machining offers numerous advantages but also presents certain challenges. The primary benefits include improved cleanliness of machined surfaces, reduced costs associated with chip recycling (due to the absence of oil contamination), and lower expenses related to cutting fluids [2]. To fully exploit the potential of dry machining, extensive research and development efforts are required to minimize excessive tool wear [3–8].

As early as 2001, within the framework of the National Nanotechnology Initiative in the United States, efforts were initiated to develop coatings and nanostructured materials specifically designed for cutting tools used in dry machining [9].

Tungsten carbide, widely utilized in the production of cutting tools, is the most commonly employed material for machining operations. During machining, cutting tools are exposed to various wear mechanisms. For instance, in dry machining of commonly used 1.0503 steel, the cutting zone temperature can reach between 600 and 950 °C, depending on the cutting speed, while even higher temperatures are observed when machining stainless steels. Under these conditions, cutting tools are particularly susceptible to wear caused by adhesion and diffusion processes, especially when temperatures approach 900 °C (for tungsten carbides) or 1000 °C (for tungsten carbides with titanium) [10].

The application of PVD coatings significantly enhances the performance of cutting tools, particularly in the machining of stainless steel. Carbide tools coated with AlCrN exhibit nearly twice the tool life and improved surface finish compared to uncoated tools [11].

TiAlN coatings, especially when applied in multiple layers, demonstrate the lowest wear rates [12].

The integration of coating design and substrate material is crucial for optimizing tool performance. Both CVD (TiCN + Al_2O_3) and PVD (AlTiN) coatings on cemented carbide substrates are effective in machining austenitic stainless steel, with coating thickness and substrate material playing a significant role in tool longevity and wear resistance [13].

Post-processing of PVD-coated surfaces, such as drag finishing and abrasive jet machining, can improve coating surface quality by reducing droplets formed during the PVD deposition process (which involves evaporating solid material in a vacuum and depositing it onto the component surface) and enhancing wear resistance [14].

Overall, PVD coatings substantially improve cutting tool durability and the machining quality of 316L stainless steel under various material removal conditions.

The exceptional properties of 316L stainless steel, including its biocompatibility and distinctive physical, mechanical, and biological characteristics, have led to its increased application across various industries, particularly in the biomedical sector over the past 50 years. 316L stainless steel is widely used in welding, where it provides optimal corrosion resistance. However, machining stainless steel using conventional methods presents certain challenges. The machining of 316L stainless steel has been investigated from various perspectives to enhance efficiency and minimize tool wear. Wire electrical discharge machining (WEDM) has proven effective for complex geometries [15]. The application of minimum quantity lubrication (MQL) has resulted in reduced tool wear compared to dry machining [16]. The study presented in [17] examined the influence of tungsten carbide (WC) on wear resistance, cutting temperature, and surface quality during the milling of 316L stainless steel.

The literature contains numerous studies on cutting forces and temperatures, primarily in the context of turning [18,19]. Umbrello et al. [20] present a study on the influence of different sets of material constants in the Johnson–Cook (J–C) constitutive equation on the finite element modeling of orthogonal turning of 316L stainless steel, examining their impact on experimental and predicted cutting forces, chip morphology, temperature distributions, and residual stresses. The authors observed that all considered process outputs, particularly residual stresses, are highly sensitive to the J–C material constants. Arrazola et al. [21] proposed a novel approach to friction identification in machining through finite element modeling. This approach involves applying a variable friction coefficient at the tool–chip interface, which enhances the agreement between numerical results and feed force measurements.

Uysal et al. [22] investigated surface roughness and chip formation during milling of stainless steel using the minimum quantity lubrication (MQL) method with the commercial vegetable oil Eraoil KT/2000 as a cooling–lubricating fluid. The highest surface roughness was observed during dry cutting. Surface roughness could be reduced by applying a TiN coating; however, the lowest surface roughness was measured during milling with MQL cooling. When analyzing chip shapes, more deformed and irregular chip forms were observed during dry milling. The TiN coating and MQL method resulted in more regular and acceptable chip shapes. Surface roughness was reduced due to effective lubrication and cooling with the MQL method, as reported in the literature [23,24].

Lv et al. [25] evaluated the effect of different cutting edge geometries on the milling of 316L stainless steel using the finite element method. They analyzed a symmetric cutting edge geometry (K = 1, rounded edge) and asymmetric geometries (K = 0.5, waterfall edge; K = 2, trumpet edge) in solid carbide milling cutters. Their study examined plastic deformation, residual stresses, and temperature distribution. The tool–chip contact length and the effective rake angle (γ) were also assessed.

Grabowski et al. [26] demonstrated that small variations in the cutting edge radius significantly affect cutting forces and surface roughness, with this effect varying depending on the helix angle of the tool's cutting edge. Edge preparation techniques can increase the cutting edge radius and smooth the cutting edge, leading to reduced cutting forces and improved surface finish, particularly when combined with minimum quantity lubrication [27].

Ozcelik et al. [28] investigated the cutting performance in dry and wet conditions during face milling of 316L stainless steel. The experiment aimed to compare tool wear, milling force components, and surface roughness under different operating conditions. The results of the experimental study confirmed that the presence of a semi-synthetic cutting fluid negatively affects the milling of 316L stainless steel. Catastrophic tool failure during wet milling may result from intense thermal stress induced by the application of the semi-synthetic cutting fluid.

Ciftci [29] examined cemented carbide tools with multilayer coatings (TiC/TiCN/TiN and TiCN/TiC/Al_2O_3) applied via chemical vapor deposition (CVD) during dry turning of 304 and 316L stainless steel, focusing on surface roughness and cutting forces. Based on the results, he concluded that TiC/TiCN/TiN-coated cutting tools generate lower cutting forces than TiCN/TiC/Al_2O_3-coated tools. He also found that 316L stainless steel exhibits higher cutting forces than 304 stainless steel, which was attributed to the presence of 2% molybdenum.

Table 1 presents a comparison of coatings used on cutting tools intended for machining stainless steels.

Surface quality after machining is a key parameter in mechanical engineering and materials science. Surface roughness affects the durability and operational performance of components. Factors such as cutting time, tool wear, and machining temperature significantly influence the achieved surface quality.

This study investigated the influence of a PVD coating on the wear of a cutting tool during the milling of 316L stainless steel under air-cooling conditions. The experiment was conducted using a carbide end mill coated with a nanocomposite nACo3 (AlTiSiN) coating. The results indicate that this coating significantly enhances wear resistance, extending tool life while maintaining an acceptable surface quality. The nACo3 coating exhibited high thermal and wear resistance, making it effective for machining hard-to-cut materials.

Table 1. Comparison of selected PVD coatings applied to cutting tools for stainless steel machining: coating type, substrate, mechanical properties, and application performance.

<table>
<tr><th rowspan="2">Coating Type</th><th rowspan="2">Tool Material</th><th rowspan="2">Workpiece Material</th><th rowspan="2">Machining Method</th><th colspan="3">Coating Parameters</th></tr>
<tr><th>Nano Hardness [GPa]</th><th>Dry Friction Coefficient μ</th><th>Maximum Working Temperature [°C]</th></tr>
<tr><td rowspan="2">AlCrN [11]</td><td>Carbide</td><td>316L</td><td>Milling</td><td>32–34</td><td>0.4</td><td>1100</td></tr>
<tr><td colspan="6">The AlCrN coating has excellent resistance to high temperatures and thermal shocks (up to 1100 °C). The presence of chromium increases the resistance to oxidation and also improves the mechanical strength of the coating. AlCrN is an ideal choice for machining with high mechanical and thermal loads, especially in impact cutting operations. In working conditions with or without cooling, it provides exceptional tool life.</td></tr>
<tr><td rowspan="2">TiCN [13]</td><td>Carbide</td><td>304</td><td>Turning</td><td>28–30</td><td>0.3</td><td>500</td></tr>
<tr><td colspan="6">TiCN is a coating that is an extension of classical TiN. The addition of carbon (C) to the TiN structure leads to the formation of a hard, abrasion-resistant titanium carbonitride, which has higher hardness than TiN and better tribological properties. TiCN-coated cutting tools show less damage to the cutting edges of the tools compared to other tested PVD—AlTiN coatings. Therefore, it has been shown that this system provides better and more efficient tool operation in this type of application.</td></tr>
<tr><td rowspan="2">TiAlN [12]</td><td>Carbide</td><td>316L</td><td>Turning</td><td>32–33</td><td>0.5</td><td>800</td></tr>
<tr><td colspan="6">TiAlN coating is a classic and versatile PVD coating, often used on general purpose tools. In this study, the effect of various PVD coatings, including TiAlN and TiN, on the wear of cutting tools during rotation of 316L austenitic stainless steel was evaluated, and it was found that the lowest wear was observed with an 8-layer TiAlN coating at a ratio of 33/67%–TiN/TiAlN. The highest VB values were observed for the uncoated and TiN-coated inserts.</td></tr>
<tr><td rowspan="2">TiN [23,24]</td><td>Carbide</td><td>304 i 420</td><td>Milling</td><td>24–26</td><td>0.4</td><td>600</td></tr>
<tr><td colspan="6">TiN is one of the most commonly used tool coatings due to its favorable cost-to-performance ratio.This paper investigates surface roughness and chip shapes when milling stainless steel (304 and 420) using minimal quantity lubrication (MQL) with a vegetarian fluid on uncoated/coated/TiN-coated WC cutting tools and compares the results to dry machining. The TiN and MQL coatings made the chips more regular and acceptable for both materials being machined. Surface roughness was improved when using TiN-coated tools and using MQL with a vegan fluid.</td></tr>
<tr><td rowspan="2">AlTiN [13]</td><td>Carbide</td><td>316L</td><td>Milling</td><td>34</td><td>0.4</td><td>900</td></tr>
<tr><td colspan="6">The AlTiN coating contains a larger amount of aluminum than TiAlN, which affects the formation of a protective layer of aluminum oxide (Al_2O_3) during operation. This layer acts as a thermal barrier, significantly increasing resistance to oxidation and enabling operation at very high temperatures (up to about 900 °C). AlTiN works great in dry machining, especially in operations with high thermal intensity (e.g., milling or drilling stainless steel).</td></tr>
<tr><td rowspan="2">nACo [30]</td><td>Carbide</td><td>316L</td><td>Milling</td><td>39–40</td><td>0.4</td><td>1200</td></tr>
<tr><td colspan="6">nACo is a coating from the AlTiSiN group. Due to the silicon (Si) content, an amorphous Si_3N_4 layer is created in the structure, which significantly increases the thermal and mechanical resistance of the coating. Thanks to the use of a nanocomposite structure and the addition of silicon, it offers very long tool life when working in the most difficult cutting conditions under high thermal loads—especially where other coatings fail.</td></tr>
</table>

Despite extensive research on the use of PVD coatings in machining, most previous studies have been limited to conventional coatings (e.g., TiN, TiAlN) or focused on general tool life behavior without simultaneously addressing surface quality and wear mechanisms. Furthermore, there is a noticeable gap in studies evaluating the performance of modern nanocomposite coatings like nACo3 (AlTiSiN) in the machining of austenitic stainless steels, especially 316L. This paper aims to fill this gap by providing a comprehensive analysis of tool wear and surface roughness as a function of cutting parameters, offering new insights for industrial applications.

The novelty of this research lies in the evaluation of a less commonly studied nanocomposite PVD coating nACo3—(AlTiSiN) during the dry machining of 316L stainless steel, a material known for its difficult-to-machine properties.

Unlike most prior studies, which focus on more conventional coatings such as TiN, TiAlN, or multilayer systems, this work demonstrates the effectiveness of the nACo (AlTiSiN) nanocomposite in reducing tool wear and improving surface finish.

Furthermore, the integration of surface roughness analysis, flank wear monitoring, and SEM/EDS characterization offers a holistic understanding of the coating performance under various machining conditions.

The outcomes of this research provide actionable insights for industrial applications aiming to enhance machining efficiency and tool durability.

Future studies may explore the use of minimum quantity lubrication (MQL) or cryogenic cooling to further reduce tool wear and improve surface quality.

2. Materials and Methods

This section describes the configuration of the experimental setup, the machining process used, as well as the equipment and measurement methods applied in this study.

2.1. Work Material

In this study, austenitic stainless steel AISI 316L was selected as the workpiece material due to its widespread application across various industries and its poor machinability. The test samples were blocks with dimensions of D300×100 mm. The fundamental mechanical properties of the steel and its chemical composition are presented in Tables 2 and 3, respectively [31].

Table 2. General mechanical properties of AISI 316L stainless steel used as the workpiece material.

Specification	Typical Value
Density (g/cm^3)	7.9
Young's Modulus (GPa)	200
Hardness, Rockwell B	95
Tensile Strength (MPa)	485
Poisson's Ratio	0.3
Elongation (%)	40
Fatigue Strength (MPa)	146

Table 3. Chemical composition of AISI 316L stainless steel in weight percentage (%wt).

C	Mn	Si	P	S	Cr	Ni	N	Mo
0.03	2.00	0.75	0.05	0.03	18.00	14.00	0.10	3.00

2.2. Milling Tools

The tool used in the tests is a carbide end mill with a diameter of 12 mm, a cutting edge length of 26 mm, and an overall length of 82 mm. The geometric details are presented in Table 4. The end mill was manufactured using a 5-axis grinding method from a solid WC–Co sintered carbide rod, grade MK12 (Table 5). Prior to the application of protective coatings in PVD processes, the tool underwent an additional technological procedure involving edge rounding to enhance tool durability. A drag finishing process was performed for the tested tool. The drag finishing process was carried out on the DF-5 machine from OTEC. The applied media with loose abrasive consisted of walnut shell granules impregnated with polishing paste. During this process, the tools rotate around their own axis, for example, for 5 min in the clockwise direction and 5 min in the counterclockwise direction. This ensures a fully repeatable and uniform cutting edge with a rounding accuracy of ±1 µm. To verify this value after the production process, the cutting edge radii were measured using an Alicona Infinite Focus microscope.

Table 4. Geometric parameters of the carbide end mills used in the experimental milling tests.

Parameter	Value	Test Tool
Coating	nACo3	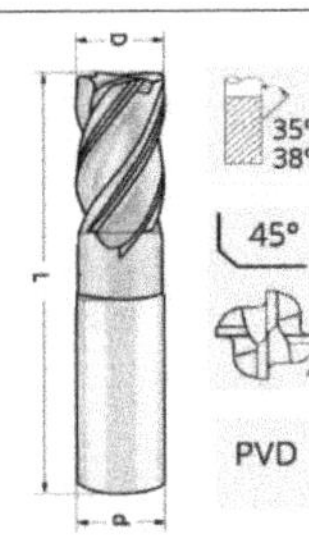
Shank diameter D [mm]	12	
Tool diameter d [mm]	12	
Tool length L [mm]	82	
Number of teeth z	4	
Helix angle β [°]	35–38	
Radial rake angle γ [°]	8	
Radial relief angle α [°]	6	
CER r_ε [µm]	8–10	
Corner	0.25 × 45°	

Table 5. Physical and mechanical properties of the WC–Co carbide substrate (MK12 grade) used in the tool manufacturing.

Grade	Grain Size (µm)	Cobalt Content (wt%)	Density (g/cm^3)	Hardness HV30
MK12	0.18 (ultrafine)	12	14.2	1730

The tool used in the tests was coated with an nACo3 coating produced in a PLATIT π411PLUS device using cathodic arc deposition (ARC-PVD, where PVD stands for Physical Vapor Deposition). The nACo3 coating was applied according to the technology developed by PLATIT. This type of coating is recommended for, among other things, machining stainless steels (Table 6).

2.3. nACo3 Coating Characterization

The nACo3 coating, with a nanocomposite structure based on AlTiSi, was designed for cutting applications such as milling and drilling of hard materials, including hardened tool steels, stainless steels, and heat-resistant alloys. Due to its advanced multilayered structure, the nACo3 coating ensures excellent adhesion to tools and high performance, even in more demanding processes such as milling with ceramic-coated tools and tools made from cubic boron nitride (CBN).

Table 6. Main properties and application range of the nACo3 (AlTiSiN) coating applied on the cutting tools.

Parameters	nACo3	Application
Nano-hardness [GPa]	39–41	
Coefficient of friction [μ]	0.4	
Coating thickness [μm]	3	
Max. Service temperature [°C]	1200	
Coating temperature [°C]	400–500	
Color	blue violet	

The nACo3 coating consists of a multilayer structure of TiN–AlTiN–AlTiN/Si_3N_4, where:

TiN provides high hardness and wear resistance.
AlTiN increases thermal stability and oxidation resistance.
Si_3N_4 minimizes friction and enhances wear resistance under high-temperature conditions.

The nACo3 coating is particularly effective in machining materials that require precise control of heat and friction, making it an ideal solution for advanced cutting processes where both tool durability and high-quality machining are required.

The silicon (Si) content in the coating depends on the deposition rate, which can be precisely controlled during the PVD (Physical Vapor Deposition) process or similar techniques. For the Ti–Si–Al–N coating, Si and Al ions replace Ti in the crystal lattice of c-TiN (cubic TiN), leading to a reduction in the crystal lattice parameters of the coating. This change impacts the mechanical and thermal properties of the coating, such as hardness, wear resistance, and stability at high temperatures [32].

The nucleation process and segregation of the amorphous Si_3N_4 phase in the coating, with an appropriate silicon content, inhibit the grain growth of nanocrystalline nc-(Ti, Al)N. This results in the formation of ultra-small nanocrystalline AlTiN grains, which are deposited in an amorphous Si_3N_4 matrix [30,33].

The grain growth inhibition effect increases the hardness and stability of the coating. Furthermore, the presence of the Si_3N_4 phase, which is chemically stable and resistant to high temperatures, contributes to improved oxidation and wear resistance, even under extreme conditions. Due to this advanced structure, Ti–Si–Al–N coatings are used in cutting tools such as end mills, drills, and turning inserts, particularly in the machining of hard materials like stainless steel, heat-resistant alloys, and hardened tool steels [34,35].

2.4. Milling Process

The machining process involved end milling of a block of 316L stainless steel. Subsequent experimental steps (n) included the removal of a material layer with a width of ae = 1 mm and depth of ap = 18 mm at a constant cutting speed vc = 120 m/min and feed per tooth fz = 0.06 mm/tooth, where z is the number of teeth in the tool. Compressed air at a pressure of 6 bar was used as the cooling medium, supplied through nozzles located on the spindle housing of the milling machine.

2.5. Test Stand, Measurement Methods

The basic test stand was built based on a multi-axis DMU 100 monoBLOCK milling center.

The cutting forces were measured using a piezoelectric force sensor, Kistler 9257B (Winterthur, Switzerland), with a measurement range of ±5 kN, mounted on the machine table (Figure 1). The signal from the sensor is transmitted to a charge amplifier and sent to a computer via a USB connection using a 16-bit analog-to-digital converter with a measurement range of ±10 V. Signal visualization, processing, and recording were performed using a program developed in the LabVIEW environment. The sampling frequency was set to 20 kHz.

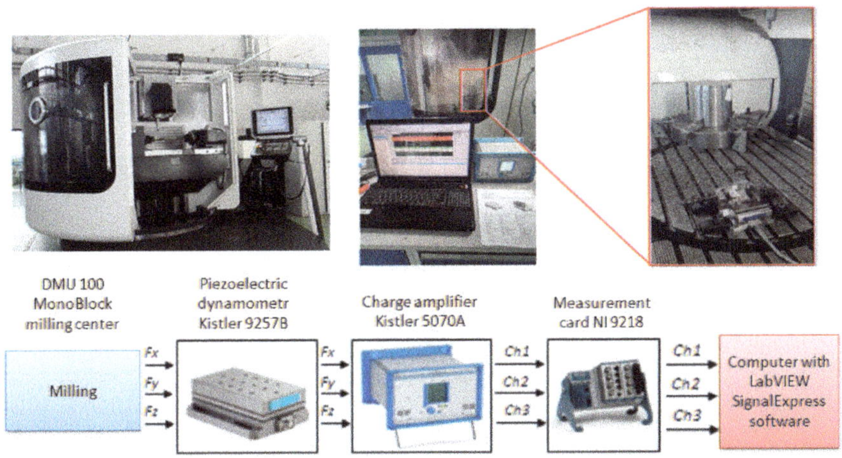

Figure 1. Experimental setup for cutting force measurement using the Kistler 9257B sensor integrated with the data acquisition system.

Tool wear was measured using a DinoLight AM7915MZT microscope (Torrance, CA, USA) combined with a Zoller Genius measuring device (Ann Arbor, MI, USA). The parameter used to evaluate tool wear was the flank wear width VB (mm), which is commonly used to estimate the cutting capability of the tool [36,37]. The method for determining VB and a sample microscopic image are shown in Figure 2.

a)

b)

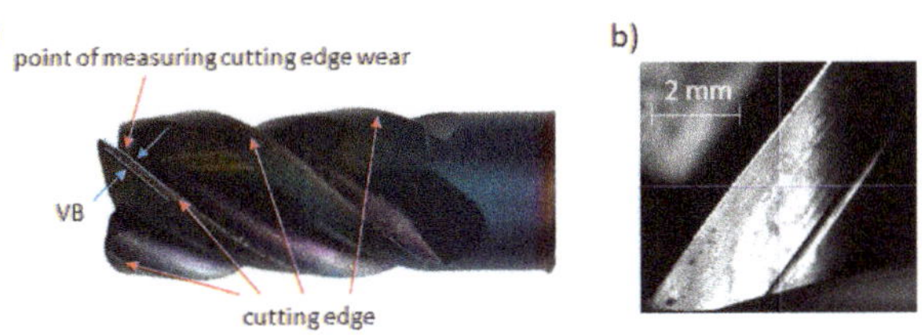

Figure 2. (**a**) Method for determining flank wear width VB on the cutting tool; (**b**) Cutting edge condition at 100× magnification after initial machining.

Surface observations of 316L stainless steel after cutting, with different tool wear times (0, 30, 60, 90, and 120 min), were conducted using a scanning electron microscope (SEM) S-3400N HITACHI (Tokyo, Japan) with an Energy-Dispersive X-ray Spectroscopy (EDS) system for microanalysis (Figure 3a). Surface morphology analysis of the stainless steel after machining was carried out in secondary electron (SE) mode. For the observation of worn tools, backscattered electron (BSE) mode was used, allowing phase and material contrast analysis. The EDS system enabled qualitative elemental analysis at selected points and areas on both the machined steel surface and the worn cutting tool. The combination of SEM observations with EDS microanalysis allowed for the detection of any inclusions, contaminations, or other defects.

Surface roughness measurements after machining included parameters Sa, Ra, and Rz. Additionally, surface maps were made to analyze structural changes. The measurements were performed using an Alicona Infinite Focus G4 microscope (Graz, Austria) (Figure 3b).

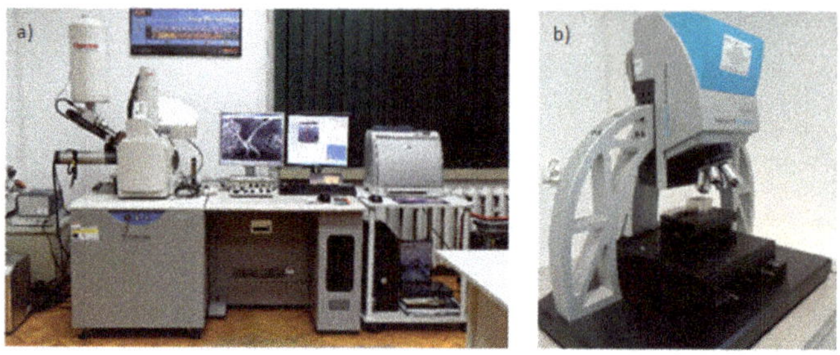

Figure 3. (**a**) Scanning Electron Microscope S-3400N HITACHI (SEM) used for microstructure analysis and (**b**) Alicona Infinite Focus G4 microscope for surface topography and roughness evaluation.

3. Results and Discussion

In the first phase (0 min) of this study, the tool was new. The total cutting force F was about 700 N. After 30 min of operation, the first signs of microscopic wear (abrasive wear) appeared. Minor damage and material adhesion were visible on the coating. Wear began, but its impact on the cutting force was still negligible. Clear signs of wear on the edge were observed after 90 min. The total cutting force increased by about 22% from the start of the tests. Advanced wear of the cutting edge was evident after 120 min. Numerous chippings and traces of adhered material (adhesive wear) were visible. The flank wear VB reached approximately 0.15 mm, and the total cutting force increased to around 885 N. A VB of 0.2 mm was achieved after 150 min, with the total cutting force reaching 972 N (Figure 4). A typical progression of tool wear was observed throughout the tool's life cycle: initially small changes, followed by accelerated wear and eventually critical damage. These phenomena were also observed and described in the literature, where similar tool wear patterns were reported [37,38]. The study was repeated for five tools manufactured within the same production batch. No significant differences were observed. The values of the cutting force components and the condition of the cutting edges were comparable, falling within a 5% measurement uncertainty.

The chemical composition analysis of the coating and the generated build-up after machining 316L stainless steel is shown in Figure 5. At point 1, the presence of nitrogen (N), silicon (Si), and metals such as chromium (Cr), aluminum (Al), and titanium (Ti) was confirmed, which are components of the multilayer coating nACo3 (Figure 5b). In the analysis of the generated build-up (point 2), the presence of elements like carbon (C), silicon (Si), sulfur (S), chromium (Cr), nickel (Ni), and iron (Fe) was confirmed, indicating the presence of the workpiece material 316L stainless steel (Figure 5c). Locally, a dark build-up (point 3) was observed, which may indicate overheating due to high cutting temperatures (Figure 5d). At point 4, traces of adhered material were observed on the coating surface (Figure 5e).

The surface of 316L stainless steel after machining exhibits parallel tool marks—grooves and waves—resulting from the motion of the cutting tool. Surface microstructure analysis was carried out using a scanning electron microscope (SEM) at two different magnifications: 100× and 600×. The lower magnification allowed for observation and evaluation of the general condition of the machined surface over a larger area, while the higher magnification enabled detailed analysis of local damage. SEM images taken at both magnifications for the surface after machining with a new tool (tool life "0") as well as for samples machined for 30, 60, 90, and 120 min are presented in Table 7, illustrating both the overall surface characteristics and the details of damage formed on the machined 316L stainless steel surface. Observations made with a scanning electron microscope (SEM) revealed that the surface quality of the material after machining with a new tool—tool life "0"—is worse compared to the surfaces of samples after 30 and 60 min of tool use. This could be influenced by several factors. In the initial stage of tool operation, this is natural and results from the initial mating of the contact surfaces. The phenomenon occurring here is described in publication [39], where the authors explain various mechanisms of tool run-in. The next period of operation corresponds to low wear intensity, characteristic of the normal tool life stage. The best surface quality was achieved for tool usage times of 30 and 60 min. Extending the machining time to 90 min caused an increase in roughness, deformations, grooves or scratches, and localized discoloration of the material due to excessive friction and high temperature in the cutting zone—burning or overheating. The worst surface quality was observed after the longest cutting time, i.e., 120 min, which may result from wear of the cutting edge (Table 7). In the last period, the intensity of cutting edge wear increases rapidly. Working in this time period (90–120 min) becomes essentially unprofitable, as little is gained from the tool life extension, but a lot is lost due to significant wear and surface quality deterioration after machining. These mechanisms were also described in a similar manner in publication [40], and the authors drew similar conclusions regarding tool wear and its impact on the quality of the machined surface.

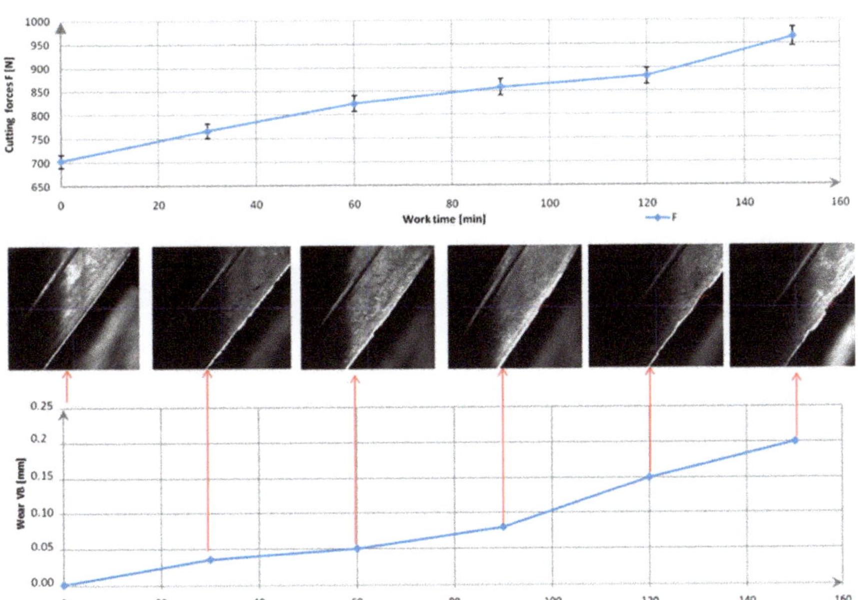

Figure 4. Shows Cutting edge wear development (VB) at 100× magnification and progression of total cutting force (F) over machining time.

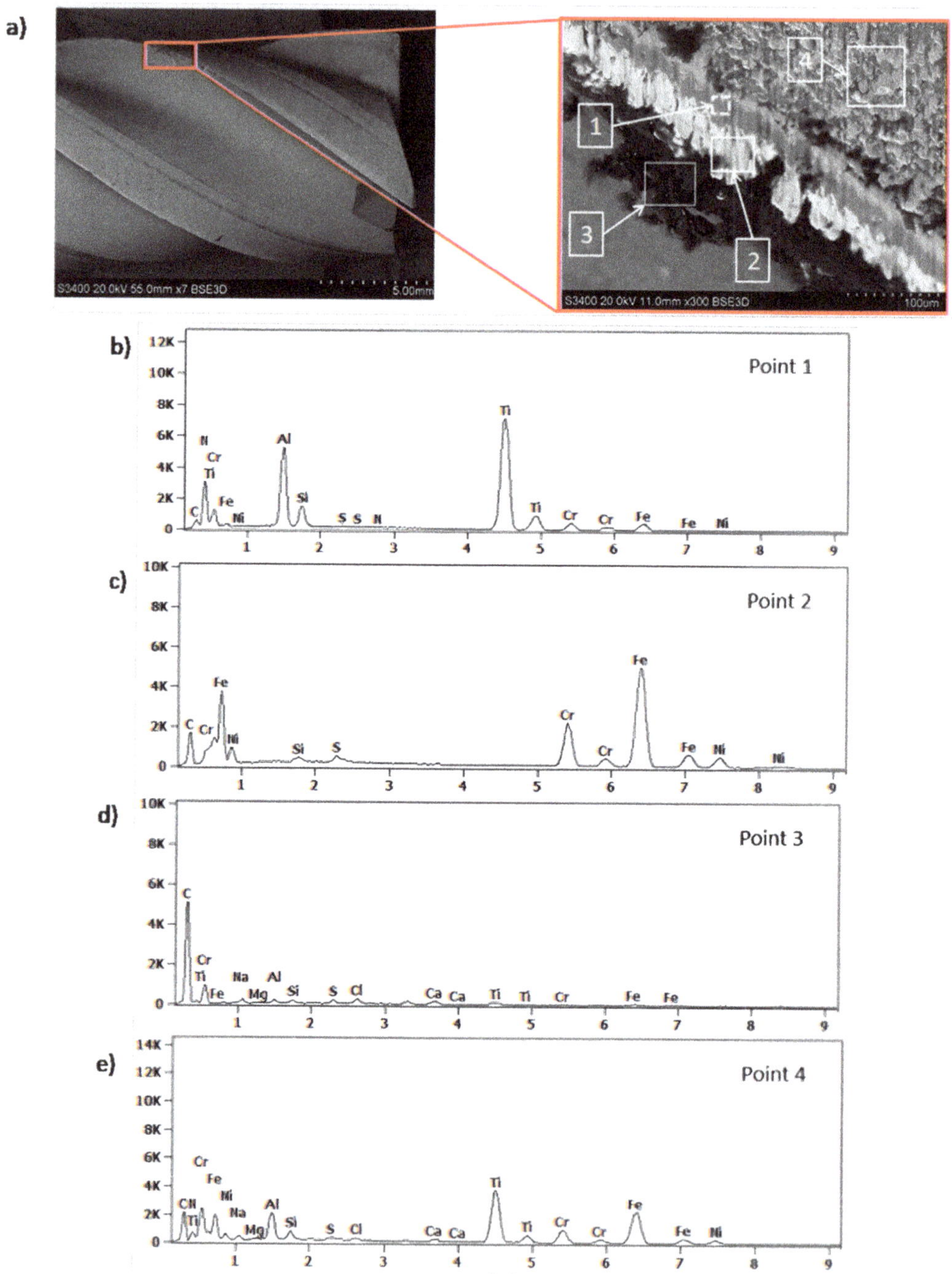

Figure 5. Results of surface and chemical composition analysis of the multilayer nACo3 coating after 120 min of cutting: (**a**) view of the studied surface, (**b**–**e**) EDS spectra at selected characteristic points showing presence of coating components and adhered material from the workpiece.

Table 7. The effect of cutting time on surface quality of 316L—SEM images at 100× (general surface topography) and 600× (local damage characters).

	Surface Photography After Machining at Magnification:	
	Mag. 100×	**Mag. 600×**
Tool life in the machining process, min 0	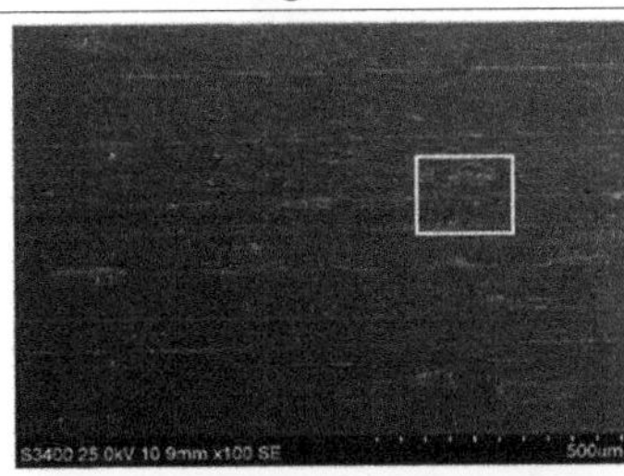	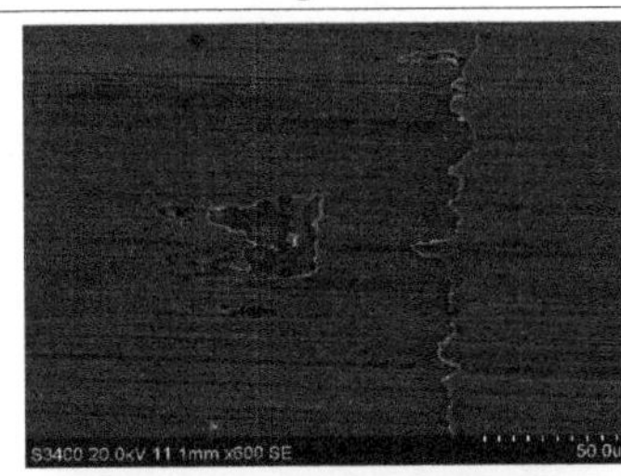
30	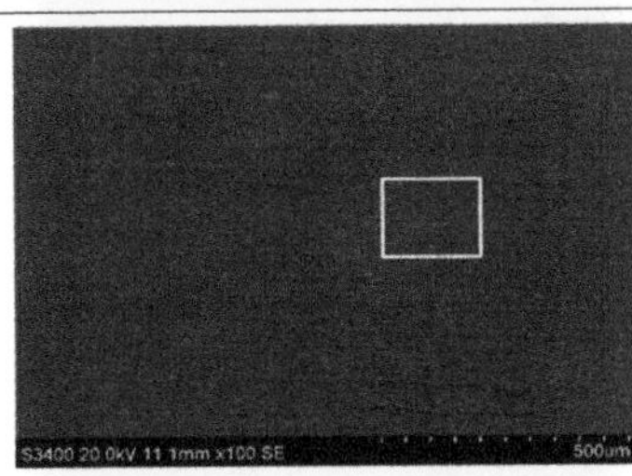	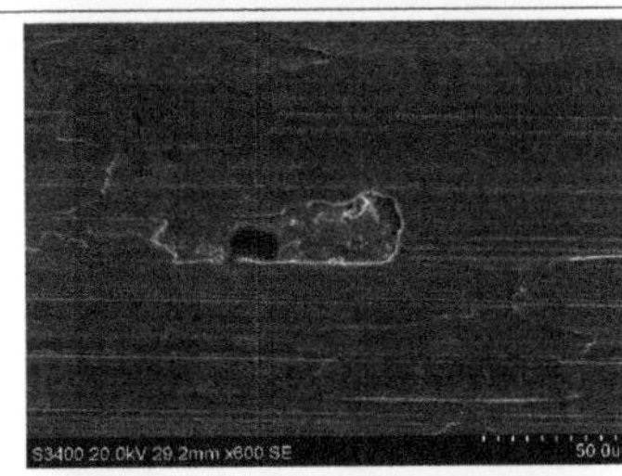
60	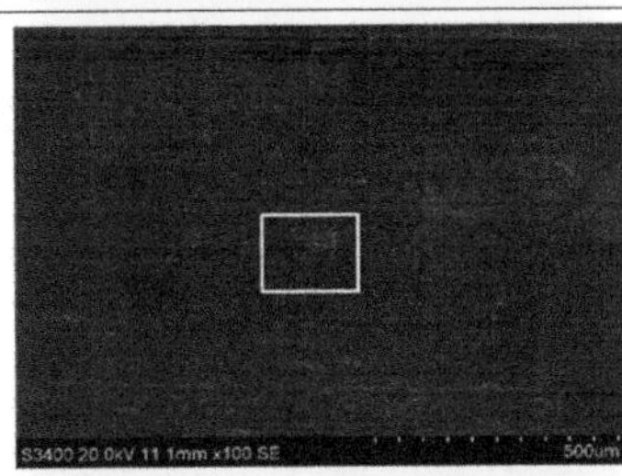	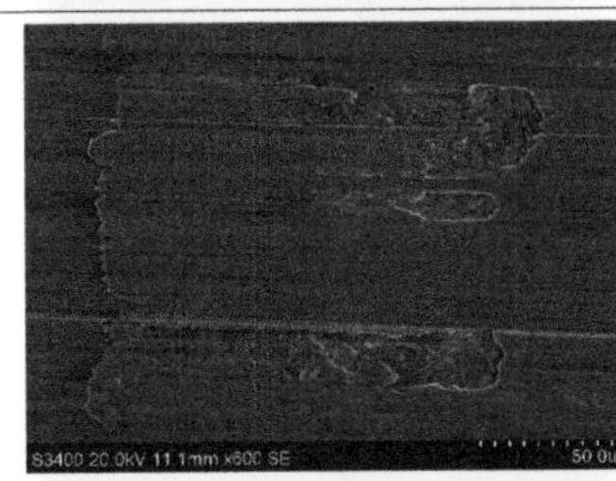
90		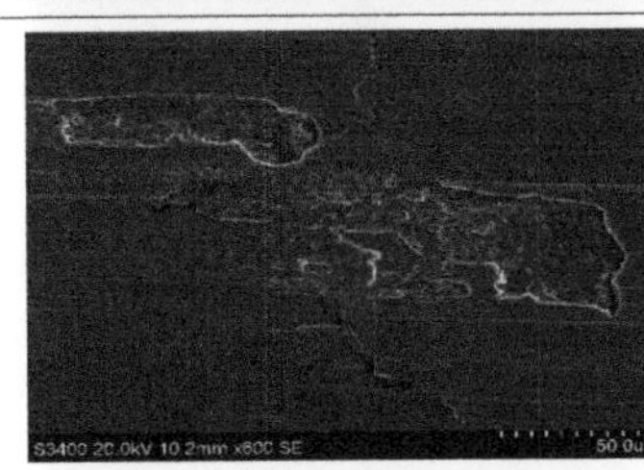
120	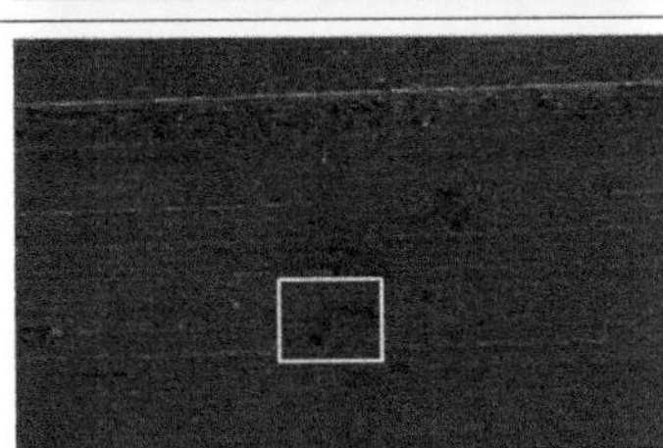	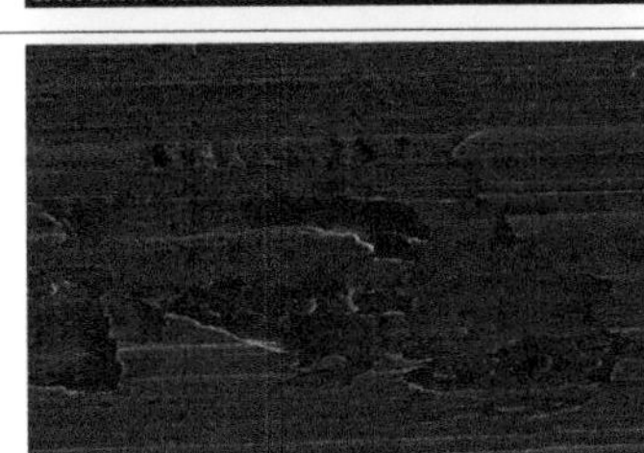

A comprehensive set of surface topography measurements was conducted, with each sample being measured five times under consistent measurement conditions. A statistical analysis was performed for each surface roughness parameter to determine the mean value, standard deviation, and expanded measurement uncertainty. These calculations were conducted assuming a 95% confidence level, in accordance with Student's *t*-test.

Surface images presented in Table 8 show differences in topographical structure at various cutting times. As the cutting time increases, changes on the surface due to tool wear become apparent. At the beginning of machining, the tool is sharp and ensures efficient material removal. The best surface quality was achieved after 60 min of cutting, where Sa reached its lowest value of 0.588 μm, and other parameters also indicated acceptable surface quality. Further cutting resulted in an increase in roughness parameters. The increase in temperature after 90 min can cause changes in microstructure and increased cutting forces, leading to process instability. The cutting edge of the tool starts to round off and wear out. The cutting process begins to generate more heat, causing local deformations and increased cutting forces. There may be the onset of a "pulling" effect, rather than precise cutting, which increases roughness. After 120 min, roughness parameters increased by more than 50%, indicating tool wear and deterioration in machining quality. Severe tool wear can cause material adhesion to the tool (Build-Up Edge, BUE), further worsening surface quality. Authors in the publication [41] presented the process of tool wear during milling of stainless steel. It was shown that the average surface roughness increases with the degree of tool wear, which aligns with the results obtained and described above.

Table 8. Evolution of surface roughness parameters (Sa, Ra, Rz) of machined 316L stainless steel depending on tool usage time.

t [min]	Surface Map	Sa	Ra	Rz
			[μm]	
0	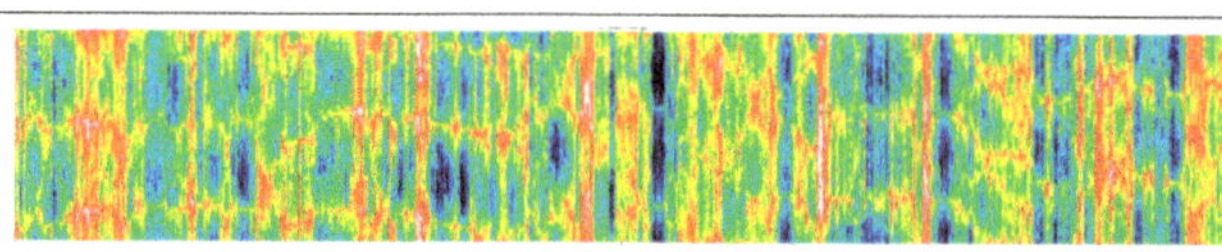	0.831	0.768	0.966
30	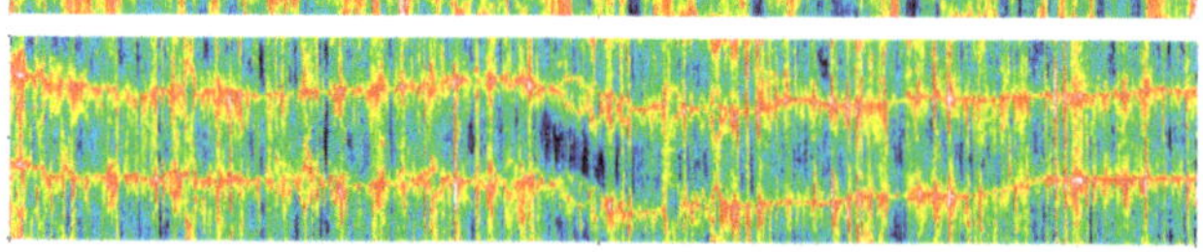	0.629	0.489	0.618
60	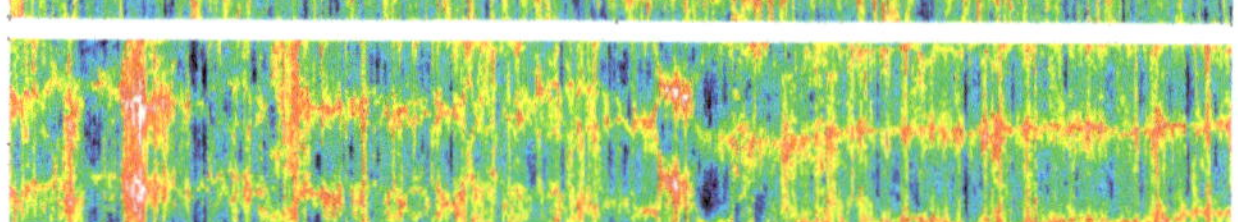	0.588	0.530	0.672
90	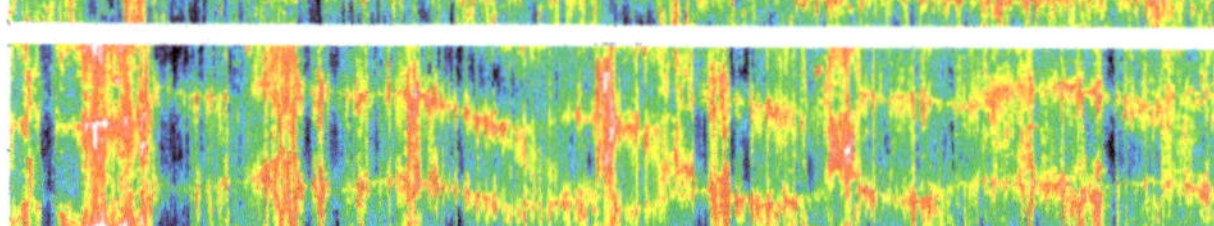	0.760	0.681	0.865
120	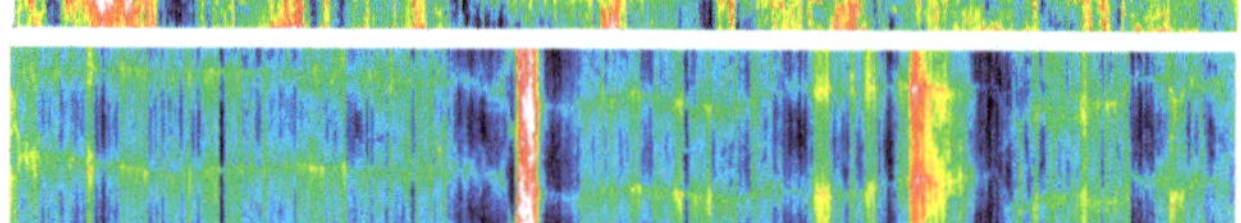	0.990	0.954	1.371

More detailed observations showed that regardless of tool usage time, various defects appeared on the 316L stainless steel surface after machining, with the highest number—as previously mentioned—occurring after the longest tool usage time, i.e., 120 min. In addition to plastic surface deformations, scratches, and nicks, another commonly observed defect was the presence of very hard sulfide inclusions, which cracked under cutting force during contact with the tool, leaving defects on the machined surface (Figure 6).

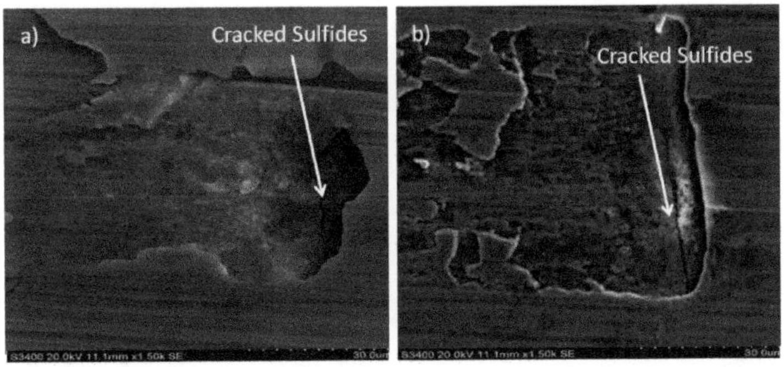

Figure 6. SEM images showing cracked sulfides on the machined 316L stainless steel surface: (**a**) after 30 min and (**b**) after 60 min of machining.

SEM surface observations of 316L stainless steel after machining (Figure 7a), combined with EDS microanalysis, showed that as a result of tool contact with the machined surface, chip adhesion occurs (point 1 Figure 7a,b). Moreover, due to tool contact with hard sulfides (point 2 Figure 7a,c), surface damage occurs, and sulfur is drawn onto the damaged steel surface (area 3 Figure 7a,d). On the damaged surface of 316L stainless steel near the sulfides, black "spots" were also observed (point 4 Figure 7a,e). Black areas on the machined material surface may indicate local burn marks.

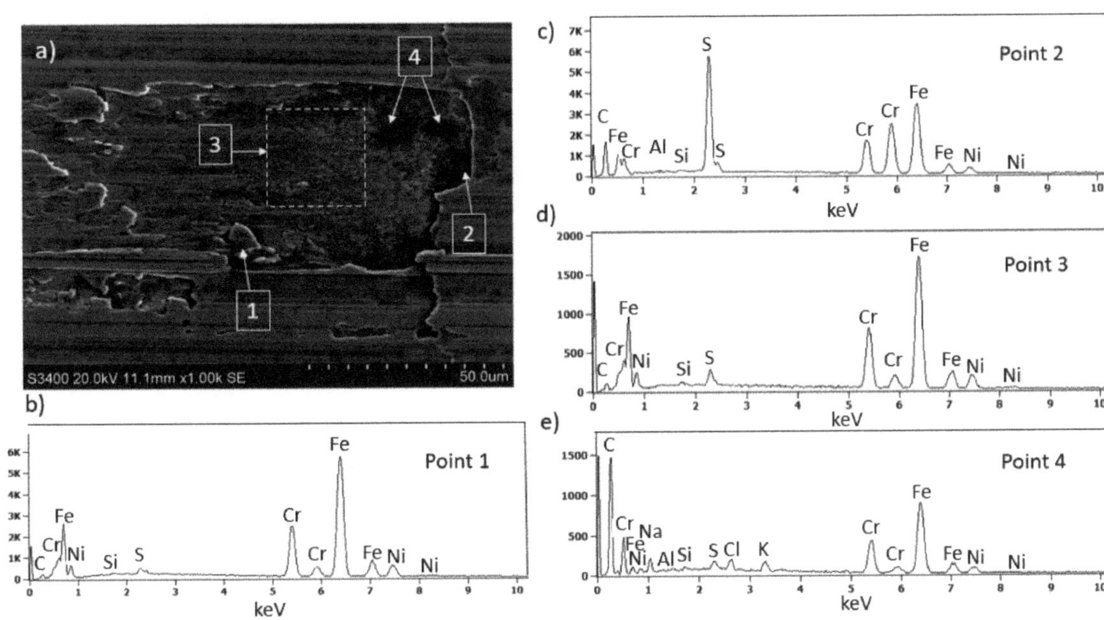

Figure 7. SEM image of the 316L stainless steel surface after 120 min of machining (**a**) and corresponding EDS spectra in marked regions showing chip adhesion (**b**), sulfide damage (**c**), sulfur distribution (**d**), and localized burn mark (**e**).

Observations of 316L stainless steel at the initial stage (Figure 8a) and after 30 min of machining revealed that the local accumulation of sulfides may contribute to burn marks (Figure 8a,b). Similar findings were observed on other surfaces of 316L stainless steel after longer tool operation times. The causes of burn marks on the machined surface of stainless steel could be

- Excessive heating of the steel surface due to inadequate cooling;
- Excessively high cutting speed or large tool force applied to the material surface;
- A dull tool or one with improper geometry, which can increase friction and consequently raise the temperature;
- A high sulfur content or other additives.

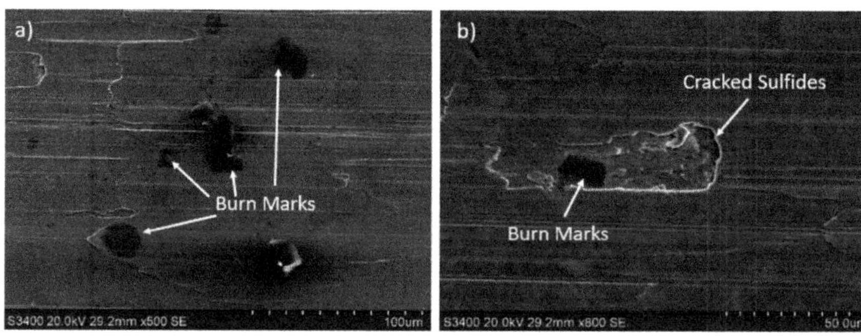

Figure 8. Burn marks on 316L stainless steel surface: (**a**) at the initial stage of machining (tool life 0 min), (**b**) after 30 min of machining.

Other defects identified on the machined material surface include deep wear near the presence of very hard carbides, such as iron and chromium carbides. Carbides often do not fully remove during cutting and can cause local damage, irregularities, and wear on the machined surface as well as tool damage or cracking. Carbides may also be pulled out by the tool, leaving voids or depressions at their location (Figure 9). EDS analysis confirmed that the black area on the milled surface after 60 min of tool operation (Figure 9) is carbide precipitation (Figure 9a,c). Analysis of the voids near the carbide—region 3—revealed the presence of Al, Ti, and Si, elements that are part of the nanocomposite coating nACo3 (Figure 9a,d). This indicates that the tool was damaged due to contact with the hard sulfide. Sulfides were also observed on the machined steel surface (Figure 9e). Similar observations were made by the authors in the publications [42,43], who described the correlation between inclusions on the machined material surface and the occurrence of damage and defects in the machined surface.

The machined material surface also exhibited small, irregular indentations, often near inclusions or material imperfections (Figure 10a), as well as waviness on the machined material surface.

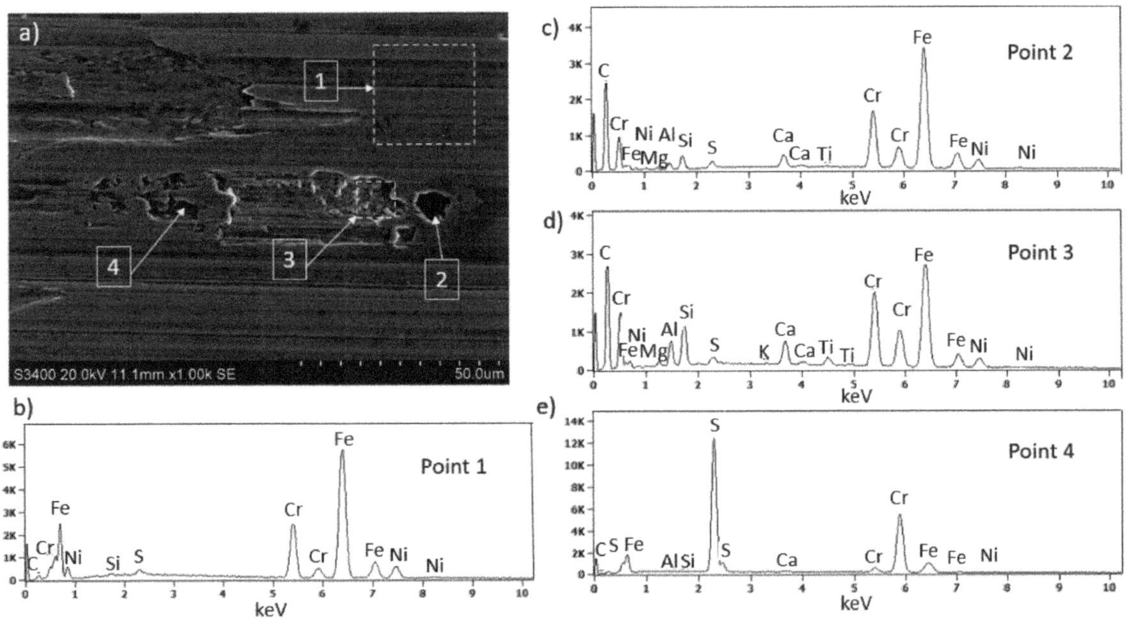

Figure 9. SEM image and EDS analysis of the machined surface after 120 min: (**a**) general view, (**b**–**e**) EDS spectra showing carbide precipitations, coating elements, and sulfide-induced damage.

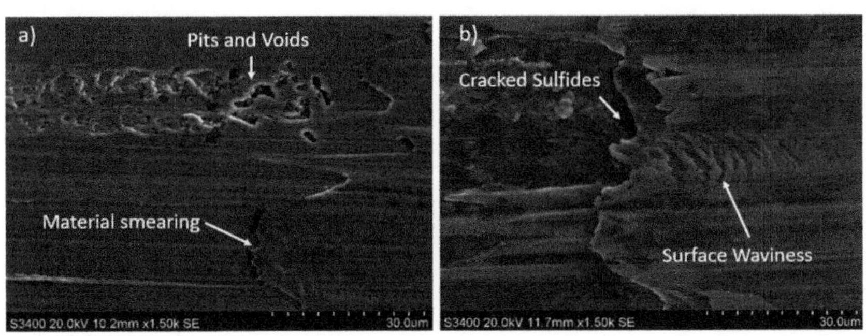

Figure 10. SEM images of 316L stainless steel surface showing (**a**) pitting and depressions after 15 min and (**b**) visible surface waviness after 20 min of tool operation.

4. Comparison with Previous Studies

The results obtained in this study show a clear improvement in both tool life and surface quality when using the nanocomposite nACo3 (AlTiSiN) coating. These findings are consistent with previous research by He et al. [13], who reported enhanced performance of AlTiN-coated tools during dry machining of stainless steel. However, our results demonstrate an even more pronounced reduction in flank wear, likely due to the presence of silicon in the nACo3 structure, which contributes to grain refinement and improved oxidation resistance.

A comparison with the study by Maruda et al. [12] reveals similar trends in the relationship between feed rate and surface roughness. While their work focused on TiAlN coatings, they also observed increased roughness with higher feed rates. Our results confirm this trend but show slightly lower roughness values at comparable feed rates, which may be attributed to the superior tribological properties of the AlTiSiN coating used in our study.

Furthermore, the work of Carvalho et al. [34] emphasized the influence of coating composition on tool wear behavior during the machining of austenitic stainless steels. Their observations regarding oxidation and adhesion wear mechanisms are consistent with our SEM and EDS analyses, which confirm diffusion-related wear processes at elevated cutting speeds.

Compared to the multilayer coatings investigated by Gaurav [11], the single-layer nanocomposite structure evaluated in our research appears to offer more stable performance across a wider range of cutting parameters, particularly in terms of maintaining surface integrity and reducing built-up edge formation.

In summary, while our findings align with existing research in terms of general trends, they also provide novel insights by evaluating a less commonly studied coating type (AlTiSiN nanocomposite) and combining surface roughness and wear analysis in the context of machining 316L stainless steel under varied conditions.

5. Conclusions

The conducted studies showed that the application of PVD coatings, particularly the nanocomposite nACo3 coating (AlTiSiN), significantly improves the tool life and surface quality after milling of 316L stainless steel under air cooling conditions. The nACo3 coating effectively reduces the intensity of tool wear, extending tool life and improving cutting process stability.

The best results were achieved for cutting times up to 90 min, after which a sharp increase in tool wear and deterioration of surface quality occurred. After 90 min, clear signs of wear were observed, and after 120 min, the cutting process became inefficient due to increased cutting forces, intensive wear of the tool's cutting edge, and significant degradation of the machined surface. Roughness measurements confirmed that the lowest values of Sa, Ra, and Rz parameters were achieved during the initial cutting phase; however, after exceeding the optimal time, the tool began to lose its cutting properties, negatively impacting surface quality.

Additional SEM and EDS analyses revealed that a critical issue in machining 316L stainless steel is the material adhering to the tool, which can lead to Build-Up Edge (BUE) and increased friction, causing higher temperatures and intensified wear. Moreover, the presence of sulfide inclusions in the steel's microstructure contributed to local damage on both the machined surface and the tool, further shortening its lifespan. It was found that during the initial phase of machining, the surface quality improved due to tool break-in, but extended cutting time led to excessive heating and increased cutting forces, resulting in surface deformations and burns.

The findings underscore the importance of selecting appropriate cutting parameters, tool geometry, and cooling strategies for optimizing the machining process. Properly chosen PVD coatings, such as nACo3, can significantly enhance tool wear resistance, but controlling cutting conditions is still necessary to minimize negative effects such as excessive heating and friction.

Further research should focus on analyzing the impact of different cutting speeds, lubrication strategies, and new generations of PVD coatings on improving the efficiency of machining 316L stainless steel. It is also worthwhile to investigate alternative cooling methods, such as minimum quantity lubrication (MQL) or cryogenic cooling, which may further reduce tool wear and improve the quality of the machined surface.

316L stainless steel is austenitic and does not undergo martensitic transformation under normal conditions. The cryogenic process can be applied in some specialized applications, such as surgical tools, aerospace components, or precision mechanical parts, where minimizing deformation and improving dimensional stability are crucial.

In conclusion, the results clearly indicate that the application of PVD coatings, such as nACo3, is an effective method for improving the durability of cutting tools and the quality of the machined surface, but it requires proper optimization of process parameters to maximize both technological and economic benefits.

Author Contributions: Conceptualization J.T. and A.S.; methodology, J.T. and A.S.; validation, J.T. and A.S.; formal analysis, J.T., A.S. and G.M.-N.; investigation, J.T., A.S. and G.M.-N.; data curation, J.T. writing—original draft preparation, J.T. writing—review and editing, A.S. and G.M.-N.; visualization, J.T.; supervision, J.T. and A.S.; project administration, J.T. All authors have read and agreed to the published version of the manuscript.

Funding: This research received no external funding.

Institutional Review Board Statement: Not applicable.

Informed Consent Statement: Not applicable.

Data Availability Statement: The original contributions presented in this study are included in the article. Further inquiries can be directed to the corresponding author.

Conflicts of Interest: The authors declare no conflicts of interest.

References

1. Pusavec, F.; Kramar, D.; Krajnik, P.; Kopac, J. Transitioning to Sustainable Production—Part II: Evaluation of Sustainable Machining Technologies. *J. Clean. Prod.* **2010**, *18*, 1211–1221. [CrossRef]
2. Biermann, D.; Iovkov, I. Modelling, Simulation and Compensation of Thermal Effects for Complex Machining Processes. *Prod. Eng.* **2015**, *9*, 433–435. [CrossRef]
3. Zębala, W.; Kowalczyk, R. Estimating the Effect of Cutting Data on Surface Roughness and Cutting Force during WC-Co Turning with PCD Tool Using Taguchi Design and ANOVA Analysis. *Int. J. Adv. Manuf. Technol.* **2015**, *77*, 2241–2256. [CrossRef]
4. Liu, J.; Ma, C.; Tu, G.; Long, Y. Cutting Performance and Wear Mechanism of Sialon Ceramic Cutting Inserts with TiCN Coating. *Surf. Coat. Technol.* **2016**, *307*, 146–150. [CrossRef]
5. Tu, G.; Wu, S.; Liu, J.; Long, Y.; Wang, B. Cutting Performance and Wear Mechanisms of Sialon Ceramic Cutting Tools at High Speed Dry Turning of Gray Cast Iron. *Int. J. Refract. Met. Hard Mater.* **2016**, *54*, 330–334. [CrossRef]
6. Kupczyk, M.; Jozwiak, K.; Cieszkowski, P.; Libuda, P. Influence of Laser Heating on Adhesion of CVD Coatings to Cutting Edges. *Surf. Coat. Technol.* **2007**, *201*, 5153–5156. [CrossRef]
7. Mikołajczyk, T.; Nowicki, K.; Kłodowski, A.; Pimenov, D.Y. Neural Network Approach for Automatic Image Analysis of Cutting Edge Wear. *Mech. Syst. Signal Process.* **2017**, *88*, 100–110. [CrossRef]
8. Mikołajczyk, T.; Nowicki, K.; Bustillo, A.; Pimenov, D.Y. Predicting Tool Life in Turning Operations Using Neural Networks and Image Processing. *Mech. Syst. Signal Process.* **2018**, *104*, 503–513. [CrossRef]
9. Jackson, M.J.; Morrell, J.S. *Machining with Nanomaterials*; Springer Science & Business Media: Berlin, Germany, 2015. [CrossRef]
10. Grzesik, W. *Podstawy Skrawania Materiałów Metalowych*; Wydawnictwo Naukowo-Techniczne: Warszawa, Poland, 1998.
11. Gaurav, S.; Sargade, V.G. Comparative Performance Evaluation of Uncoated and Coated Carbide Inserts in Dry End Milling of Stainless Steel (SS 316L). *Int. Conf. Comput. Intell.* **2011**, *1*, 7–15.
12. Maruda, R.W.; Szczotkarz, N. The Influence of the Type of Coating on the Cutting Tool Wear During Turning of 316L Austenitic Stainless Steel. *Arch. Mech. Technol. Mater.* **2016**, *38*, 45–49. [CrossRef]
13. He, Q.; Paiva, J.M.; Kohlscheen, J.; Beake, B.D.; Veldhuis, S.C. An Integrative Approach to Coating/Carbide Substrate Design of CVD and PVD Coated Cutting Tools during the Machining of Austenitic Stainless Steel. *Ceram. Int.* **2019**, *46*, 5149–5158. [CrossRef]
14. Zlámal, T.; Mrkvica, I.; Szotkowski, T.; Malotova, S. The Influence of Surface Treatment of PVD Coating on Its Quality and Wear Resistant. *Coatings* **2019**, *7*, 439. [CrossRef]

15. Saini, M.; Sharma, R.; Abhinav; Singh, G.; Mangla, P. Optimizations of Machining Parameter in Wire EDM for 316L Stainless Steel by Using Taguchi Method, ANOVA, and Grey Analysis. *Int. J. Mech. Eng. Technol. (IJMET)* **2016**, *7*, 307–320.
16. Szczotkarz, N.; Mrugalski, R.; Maruda, R.; Królczyk, G.; Legutko, S.; Leksycki, K.; Dębowski, D.; Pruncu, C. Cutting Tool Wear in Turning 316L Stainless Steel in the Conditions of Minimized Lubrication. *Tribol. Int.* **2021**, *156*, 106813. [CrossRef]
17. Franczyk, E.; Małek, M. Empirical Study on the Effect of Tungsten Carbide Grain Size on Wear Resistance, Cutting Temperature, Cutting Forces and Surface Finish in the Milling Process of 316L Stainless Steel. *Adv. Sci. Technol. Res. J.* **2023**, *17*, 367–377. [CrossRef]
18. Maranhão, C.; Davim, J. Finite Element Modelling of Machining of AISI 316 Steel: Numerical Simulation and Experimental Validation. *Simul. Model. Pract. Theory* **2010**, *18*, 139–156. [CrossRef]
19. Arrazola, P.J.; Arriola, I.; Davies, M.A. Analysis of the Influence of Tool Type, Coatings, and Machinability on the Thermal Fields in Orthogonal Machining of AISI 4140 Steels. *CIRP Ann. Manuf. Technol.* **2009**, *58*, 85–88. [CrossRef]
20. Umbrello, D.; M'Saoubi, R.; Outeiro, J. The Influence of Johnson–Cook Material on Finite Element Simulation of Machining of AISI 316L Steel. *Int. J. Mach. Tools Manuf.* **2007**, *47*, 462–470. [CrossRef]
21. Arrazola, P.J.; Ugarte, D.; Domínguez, X. A New Approach for Friction Identification during Machining through the Use of Finite Element Modelling. *Int. J. Mach. Tools Manuf.* **2008**, *48*, 173–183. [CrossRef]
22. Uysal, A.; Demiren, F.; Altan, E. Investigation of Surface Roughness and Chip Forms in Milling of Stainless Steel by MQL Method. *Acta Phys. Pol. A* **2016**, *129*, 439–441. [CrossRef]
23. Khana, M.M.A.; Mithua, M.A.H.; Dharb, N.R. Effects of Minimum Quantity Lubrication on Turning AISI 9310 Alloy Steel Using Vegetable Oil-Based Cutting Fluid. *J. Mater. Process. Technol.* **2009**, *209*, 5573–5583. [CrossRef]
24. Sharma, J.; Sidhu, B.S. Investigation of Effects of Dry and Near Dry Machining on AISI D2 Steel Using Vegetable Oil. *J. Clean. Prod.* **2014**, *66*, 619–623. [CrossRef]
25. Lv, D.; Yu, X.; Wang, Y. Evaluation of Cutting Edge K Form Factor in Milling of 316L Stainless Steel: A Study Based on FEM. *Int. J. Adv. Manuf. Technol.* **2023**, *128*, 5223–5236. [CrossRef]
26. Grabowski, M.; Małek, M.; Teimouri, R.; Skoczypiec, S. Effect of Cutting-Edge Geometry on the Machinability of 316L Austenitic Steel. *Adv. Sci. Technol. Res. J.* **2024**, *18*, 110–117. [CrossRef]
27. Ibrahim, R.; Long, C.Y.; Ng, Y.; Rafai, N.H.; Mustapa, M.S.; Ismail, A.E. Experimental Study on the Degree of Surface Generation by Edge Preparation Tools in Milling 316L. *Int. J. Eng. Technol.* **2021**, *8*, 80–88. [CrossRef]
28. Ozcelik, B.; Kuram, E.; Simsek, B.T. Comparison of Dry and Wet End Milling of AISI 316 Stainless Steel. *Mater. Manuf. Process.* **2011**, *26*, 1041–1049. [CrossRef]
29. Ciftci, I. Machining of Austenitic Stainless Steels Using CVD Multi-Layer Coated Cemented Carbide Tools. *Tribol. Int.* **2006**, *39*, 565–569. [CrossRef]
30. Polcar, T.; Cavaleiro, A. High-Temperature Tribological Properties of CrAlN, CrAlSiN and AlCrSiN Coatings. *Surf. Coat. Technol.* **2011**, *206*, 1244–1251. [CrossRef]
31. Natesh, C.P.; Shashidhara, Y.M.; Amarendra, H.J.; Shetty, R.; Harisha, S.R.; Shenoy, P.V.; Nayak, M.; Hegde, A.; Shetty, D.; Umesh, U. Tribological and Morphological Study of AISI 316L Stainless Steel during Turning under Different Lubrication Conditions. *Lubricants* **2023**, *11*, 52. [CrossRef]
32. Haršáni, M.; Sahul, M.; Zacková, P.; Caplovič, L'. Study of Cathode Current Effect on the Properties of CrAlSiN Hard Coatings Deposited by LARC. *Vacuum* **2017**, *139*, 1–8. [CrossRef]
33. Ding, X.; Zeng, X.T.; Liu, Y.C. Structure and Properties of CrAlSiN Nanocomposite Coatings Deposited by Lateral Rotating Cathode Arc. *Thin Solid Films* **2011**, *531*, 1894–3100. [CrossRef]
34. Carvalho, S.; Ribeiro, E.; Rebouta, L.; Pacaud, J.; Goudeau, P.; Renault, P.O.; Rivière, J.P.; Tavares, C.J. PVD Grown (Ti,Si,Al)N Nanocomposite Coatings and (Ti,Al)N/(Ti,Si)N Multilayers: Structural and Mechanical Properties. *Surf. Coat. Technol.* **2003**, *172*, 109–116. [CrossRef]
35. Cselle, T.; Coddet, O.; Galamand, C.; Holubar, P.; Jilek, M.; Jilek, J.; Luemkemann, A.; Morstein, M. TripleCoatings3®–New Generation of PVD-Coatings for Cutting Tools. *J. Mach. Manuf.* **2009**, *49*, 19–25.
36. Zhang, Z.; Liu, Z.; Ren, X.; Zhao, J. Prediction of Tool Wear Rate and Tool Wear during Dry Orthogonal Cutting of Inconel 718. *Metals* **2023**, *13*, 1225. [CrossRef]
37. Zawada-Michalowska, M.; Pieśko, P.; Józwik, J. Tribological Aspects of Cutting Tool Wear during the Turning of Stainless Steels. *Materials* **2020**, *13*, 123. [CrossRef] [PubMed]
38. Caldeirani Filho, J. Influence of Cutting Conditions on Tool Life, Tool Wear and Surface Finish in the Face Milling Process. *J. Braz. Soc. Mech. Sci.* **2002**, *24*, 10–14. [CrossRef]
39. Byrne, G.; Dornfeld, D.; Denkena, B. Wear Mechanisms in Cutting Tools, Advancing Cutting Technology. *CIRP Ann.* **2003**, *52*, 483–507. [CrossRef]

40. Astakhov, V. Surface Integrity in Machining Stainless Steel. In *Tribology of Metal Cutting*; Elsevier: Amsterdam, The Netherlands, 2006; pp. 319–358.
41. Liu, G.; Zou, B.; Huang, C.; Wang, X.; Wang, J.; Liu, Z. Tool Damage and Its Effect on the Machined Surface Roughness in High-Speed Face Milling the 17-4PH Stainless Steel. *Int. J. Adv. Manuf. Technol.* **2016**, *83*, 257–264. [CrossRef]
42. Maciejewski, J. The Effects of Sulfide Inclusions on Mechanical Properties and Failures of Steel Components. *J. Fail. Anal. Preven.* **2015**, *15*, 169–178. [CrossRef]
43. Rieders, N.; Nandasiri, M.; Mogk, D.; Avci, R. New Insights into Sulfide Inclusions in 1018 Carbon Steels. *Metals* **2021**, *11*, 428. [CrossRef]

materials

MDPI

Article

Impact Damage of Aluminum-Composite Sandwich Panels and Constituents

Shun-Fa Hwang * and Ming-Yi Wu

Department of Mechanical Engineering, National Yunlin University of Science and Technology, 123 University Road, Sec. 3, Yunlin 64002, Taiwan; m11011032@yuntech.edu.tw
* Correspondence: hwangsf@yuntech.edu.tw; Tel.: +886-5-534-2601 (ext. 4143)

Abstract: This study investigates the impact failure behavior of aluminum-composite sandwich panels, aluminum sheets, and composite laminates. Aluminum alloy sheets possess excellent ductility and plasticity, while carbon fiber composite sheets exhibit high strength, high rigidity, and superior heat resistance. The sandwich panel structure, composed of two layers of aluminum alloy sheets and a central carbon fiber composite sheet, offers the advantages of being lightweight and high strength. These three types of specimens were subjected to impact energies of 15 J, 25 J, and 50 J. The numerical simulations employ LS-DYNA finite element software, with additional investigations into the energy absorption characteristics, which were employed and compared with the experiment. The experimental results indicate that aluminum alloy sheets only exhibit indentation under all three impact energies. Carbon fiber composite sheets sustain damage without penetration at 15 J but experience penetration failure at 25 J and 50 J. Aluminum-composite sandwich panels exhibit greater resistance to failure as compared to carbon fiber composites. At 15 J and 25 J, the top aluminum layer shows indentation, while the bottom aluminum layer develops cracks. At 50 J, complete penetration occurs. A comparison of damage morphology and force–time curves shows good agreement between the experimental and simulation results. While the carbon fiber composite plate exhibits the highest SEA, it also has the largest damage diameter, indicating more severe damage. In contrast, the aluminum alloy panel has the lowest specific energy absorption (SEA) due to its high weight. The aluminum-composite sandwich panel demonstrates intermediate performance in both damage diameter and SEA, striking a balance between the other two specimens.

Keywords: metal-composite sandwich panel; low-velocity impact; finite element analysis; aluminum alloy; composite material

1. Introduction

With advancements in technology, industries such as aerospace, automotive, and marine engineering have increasingly pursued lightweight materials to enhance performance and efficiency. Carbon fiber-reinforced composites have emerged as a widely used solution due to their low density and high rigidity. These materials exhibit exceptional mechanical properties, being composed of carbon fibers embedded within a polymer matrix to form a strong composite structure. However, most composite materials are inherently brittle and susceptible to impact damage. Unlike metals, which absorb energy through plastic deformation, composites dissipate impact energy through internal damage mechanisms, such as fiber breakage, matrix cracking, and delamination. To improve the impact resistance of composite materials, hybrid structures integrating metals and composites have been developed. One such solution is the metal-composite sandwich panel or fiber metal laminate (FML), a three-layer structure consisting of metal sheets enclosing a composite core. This design leverages the ductility and corrosion resistance of metals alongside the strength and stiffness of composites, creating a structurally efficient and impact-resistant material. Aluminum, titanium, and magnesium are three common types of metals for FML, and the most popular one is aluminum. Aluminum has been combined with different composite materials into some famous FMLs, for example, GLARE (glass fibers), CARALL (carbon fibers), and ARALL (aramid fibers). Among them, GLARE has been studied and applied in airframe construction, fuselage frames, and cargo bay floors [1]. On the other hand, CARALL (carbon fibers and aluminum) is seldom studied.

For AA5083-H116 aluminum alloy plates, an analytical perforation model and a nonlinear finite element simulation were established to predict their perforation behavior under ballistic limit velocities [2]. The same alloy plates were experimentally and numerically investigated by Grytten et al. [3] on the quasi-static perforation with varying plate thickness, boundary conditions, punch diameters, and nose shapes. The static impact behavior of aluminum helideck structures to access offshore installations was an experimental examination [4]. Morin et al. [5] discussed the behavior and failure of stiffened plates made of aluminum alloy AA6082-T6 by both the quasi-static and low-velocity impact loadings and found that the strain rate and inertia effects were negligible. For recent works, Liu et al. [6] presented a material failure criterion, which was the combination of the revised Bressan–Williams–Hill local instability criterion and damage evolution model, to predict their fracture behaviors under different nose shapes of the impactor. As compared to experimental results, the failure criterion had good agreement on the onset of fracture with even coarse meshes. Shi et al. [7] proposed analytical algebraic expressions for predicting the amount of permanent indentation of aluminum panels under impact by a spherical object. For polyurea-coated aluminum plates, Xia et al. [8] used hyperelastic and viscoelastic constitutive models for polyurea and a modified Johnson–Cook model for aluminum to simulate their impact responses. Their results indicated that the simulation results of impact peak force, energy absorption, and deformation were within 5% of the experimental results.

Similarly, for fiber-reinforced composite laminates, extensive research has been conducted on the low-velocity impact behavior [9–16]. For unidirectional fiber-reinforced composite laminates, the main topic was the effect of the stacking sequence on impact behavior [9,10,12], and the quasi-isotropic laminates could provide better resistance to impact. Caminero et al. [9] also discussed the thickness effect, and Zhou et al. [12] considered the residual tensile strength of CFRP laminates after impact. Ouyang et al. [11] focused on the effect of matrix cracks on delamination during impact and indicated that the delamination was easily created around the matrix cracks. For braided composite panels under impact, the effect of the stacking sequence was experimentally studied by Wu et al. [14], and the effect of thickness was investigated by Ge et al. [13]. The temperature effect on the impact behavior of braided composite laminates was considered by Khashaba et al. [15]. Krollmann et al. [16] investigated the impact and post-impact compression properties of hybrid-matrix laminates, in which carbon fiber-reinforced elastomer layers were combined with conventional carbon fiber-reinforced epoxy layers.

For aluminum-composite sandwich panels, Bienias and Jakubczak [17] proposed five damage states to describe the impact behavior, and they included internal degradation and plastic deformation, initiation of new cracks at a fiber/matrix interface and delamination, the initiation and growth of cracks in the metal layers, penetration of the structure and, finally, the withdrawal of the impactor. Drozdziel et al. [18] demonstrated that the damage mechanism of fiber metal laminates with thin-ply composites did not significantly differ from that with a conventional composite ply thickness and confirmed the use of thin-ply fiber metal laminates. Yu et al. [19] revealed that the carbon fiber aluminum laminates had better impact resistance than the glass fiber counterparts. Furthermore, the increase in the yield strength of aluminum alloy could improve the impact resistance of carbon fiber aluminum laminates. Jakubczak et al. [20] used both experimental tests and numerical simulations to find that the main damage modes of fiber metal laminates under low-velocity impact were matrix fracture, fiber cracking, and delamination. Li et al. [21] proposed a novel hybrid laminate by inserting elastomer layers into conventional fiber metal laminates, and these novel hybrid laminates could efficiently absorb more impact energy. Yao et al. [22] discussed the influence of impactor shape on the low-velocity impact behavior of fiber metal laminates by both experiment and simulation.

Aluminum-composite sandwich panels are a relatively new type of fiber metal laminates, which combine the good characteristics of metals and composite materials. This sandwich panel possesses lightweight, better fatigue behavior, and corrosion resistance from composite materials and obtains higher ductility and bearing strength from metals. They are a good substitute for traditional metals used in aircraft and the aerospace industry. Therefore, it is necessary to evaluate their damage condition from collision during the process of maintenance and operation. Especially, the advantage of aluminum-composite sandwich panels needs to be identified as compared to the two original materials. This study addresses this gap by systematically investigating the failure modes, force–time response, and energy absorption characteristics of these three material configurations under varying impact energies. Furthermore, the numerical simulation is also proposed to compare with the experiment and verify its application for low-velocity impact.

2. Thermoplastic Composite Material and Thermoforming Process

The materials utilized in this study include 5052-H32 aluminum alloy sheets, unidirectional carbon fiber composites, and aluminum-composite sandwich panels. The 5052-H32 aluminum alloy, provided by Fabow Industrial Co., Ltd (Taiwan), is a strain-hardened alloy with moderate strength, excellent corrosion resistance, and high fatigue strength. The material's stress–strain curve and mechanical properties are presented in Figure 1 and Table 1, respectively.

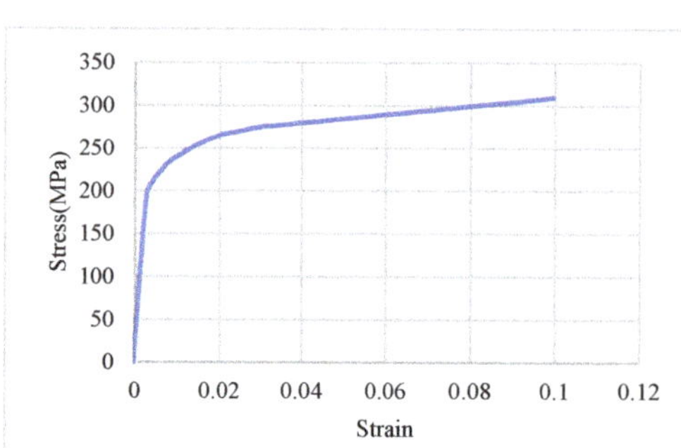

Figure 1. The stress–strain curve of 5052-H32 aluminum alloy [23].

Table 1. Mechanical properties of 5052-H32 aluminum alloy [23].

Properties	Value	Unit
Density	2.68	g/cm^3
Young's modulus	70.3	GPa
Poisson's ratio	0.33	-
Yield stress	193	MPa

The unidirectional carbon fiber prepreg used in this study was manufactured by Formosa Plastics Corporation, with the fiber type designated as TC-36P (12 K). The fiber areal weight (FAW) is 150 g/m^2, and the measured fiber weight fraction is 62%. The detailed material properties of the fabricated composite are provided in Table 2.

Table 2. Mechanical properties of unidirectional composite material [24].

Property	Value	Unit
Density, ρ	1.5	g/cm^3
Longitudinal Young's modulus, E_{11}	128	GPa
Transverse Young's modulus, E_{22}	8.5	GPa
Shear modulus, G_{12}	5.69	GPa
Shear modulus, G_{23}	3.44	GPa
Shear modulus, G_{31}	5.69	GPa
Poisson's ratio, ν_{12}	0.2	-
Longitudinal tensile strength, X_T	1340	MPa
Longitudinal compressive strength, X_C	492	MPa
Transverse tensile strength, Y_T	40	MPa
Transverse compressive strength, Y_C	192	MPa
Shear strength, S	76	MPa

Three types of specimens were prepared: aluminum alloy sheets, carbon fiber composite laminates, and aluminum-composite sandwich panels, each with dimensions of 110 × 110 mm. Due to manufacturing constraints, the exact thicknesses varied slightly, with aluminum sheets at 2 mm, carbon fiber composites at 1.92 mm, and sandwich panels at 1.84 mm.

The aluminum alloy specimens were pre-cut by Fabow Industrial Co., Ltd., while the composite laminates and sandwich panels were fabricated using a hot-press technique. Carbon fiber composite laminates were prepared by stacking twelve layers of unidirectional prepreg sheets in a $[(0°/90°)_3]_s$ configuration. During stacking, a cylindrical roller was used to remove trapped air and prevent void formation. The laminates were then enclosed in release films and placed between aluminum plates coated with a mold release agent before undergoing hot pressing at 140 °C for 30 min without pressure, followed by an additional 1 h curing stage under 50 kgf pressure. The specimens were cooled to room temperature over three hours before final trimming with a diamond saw.

Aluminum-composite sandwich panels were fabricated similarly, with aluminum alloy face sheets (0.6 mm thick) bonded to a four-layer carbon fiber core ($[0°/90°]_s$, 0.64 mm thick). Before bonding, the aluminum sheets were sanded to enhance adhesion.

Impact tests were conducted using an impact testing machine equipped with a spectrum analyzer. The force exerted during impact was calculated using an accelerometer mounted on the impactor, while the damage diameter was measured using a vernier caliper. During the impact process, the accelerometer generated acceleration signals, which were transmitted to the spectrum analyzer. The acceleration–time curve was derived from these signals, and the force–time curve was obtained by multiplying the acceleration by the impactor mass. The impact energy levels tested were 15 J, 25 J, and 50 J, and the impactor had a mass of 7.918 kg and a diameter of 16 mm. By adjusting the drop height, one can obtain the impact velocities of 1.946, 2.513, and 3.554 m/s for the three impact energies, respectively. The reason for choosing these three impact energies was to have different damage phenomena for the three types of specimens.

3. Finite Element Simulation

In this study, LS-DYNA R15 software was employed to simulate low-velocity impact behavior. The simulation model included the impactor, the fixture, and the specimen. The mesh utilized two element types: eight-node solid elements for the impactor, fixture, and aluminum sheets, and four-node shell elements for the composite part. The element sizes were $1 \times 1 \times 1$ mm for the solid element and 1×1 mm for the shell element. The mesh is shown in Figure 2. In the carbon fiber composite specimen, two shell elements represented the composite laminate, with six integration points per shell, totaling twelve points. Each integration point had a thickness of 0.16 mm, matching the experimental $[(0°/90°)_3]_s$ stacking sequence. In the aluminum-composite sandwich panel model, aluminum sheets were represented by solid elements, while the composite core was modeled using a single shell element with four integration points, each 0.16 mm thick, reflecting the $[0°/90°]_s$ stacking configuration.

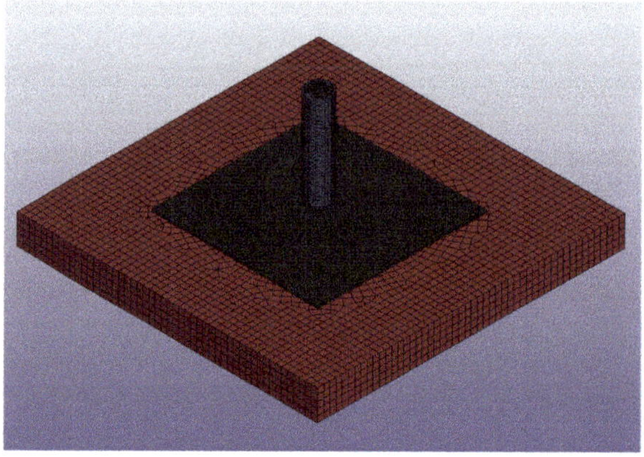

Figure 2. Mesh of the impact model.

To match the experiment and reduce the simulation time, the impactor with the corresponding velocities for different impact energies was released just above the specimen. In the impact experiment, rubber clamps were used at the four corners of the specimen to constrain its movement. In the simulation, this effect was replicated by constraining the corner nodes. Furthermore, contact interactions were enforced between different parts of the model to prevent penetration.

The impactor and fixture were modeled as rigid bodies, as their deformation was not considered. The impactor was set to move vertically while restricting rotational motion, whereas the fixture was completely fixed. The material properties used for these two parts had a density of 6.045 g/cm^3, a Young's modulus of 210 GPa, and a Poisson's ratio of 0.3. Since impact energy is influenced by the impactor mass, the density of the impactor was adjusted based on its volume to ensure that its total mass remained at 7.918 kg, consistent with the experimental setup.

The aluminum alloy sheets were modeled using the *MAT_24_PIECEWISE_LINEAR_PLASTICITY material card, which defines elastoplastic behavior. The material properties assigned to this model are shown in Table 1 and Figure 1. In addition to these properties, the failure strain for the aluminum alloy sheets was set to 0.6 in the simulation. When the maximum plastic strain reached 0.6, the element was considered to have failed and was removed from the calculation. For the composite sheets, the *MAT_54_ENHANCED_COMPOSITE_DAMAGE material card was used. This model was specifically designed for unidirectional fiber-reinforced composites. The material card requires definitions for material properties, failure criteria, element orientation, and failure strain, with the material properties detailed in Table 2. The failure criterion employed in *MAT_54_ENHANCED_COMPOSITE_DAMAGE follows the Chang–Chang failure model [25]. This failure criterion is categorized into fiber direction tensile failure, fiber direction compressive failure, transverse direction matrix tensile-shear failure, and transverse direction matrix compressive-shear failure. In the simulation, if any element exceeded all four failure criteria, it was considered failed and was no longer included in further calculations. If just one failure criterion was fulfilled, parts of the mechanical properties were set to zero, and the element was still active.

In the aluminum-composite sandwich panel simulation, the model consisted of an upper aluminum alloy sheet, a central composite sheet, and a lower aluminum alloy sheet. To simulate the bonding and failure behavior between these sheets, the tie-and-break condition was judged according to the interfacial stresses as the following equation.

$$\left(\frac{|\sigma_n|}{NFLS}\right)^2 + \left(\frac{|\sigma_s|}{SFLS}\right)^2 \geq 1 \quad (1)$$

where σ_n is the interfacial normal stress, σ_s is the interfacial shear stress, and *NFLS* and *SFLS* are the corresponding interfacial strengths. When the computed value reached or exceeded 1, interfacial failure occurred. The normal and shear strengths were set to 8.8 MPa and 54 MPa, respectively.

4. Results and Discussion

In this study, the experimental study involved conducting impact tests on aluminum alloy sheets, carbon fiber composite sheets, and aluminum-composite sandwich panels under three impact energy levels: 15 J, 25 J, and 50 J. A total of nine types of experiments were conducted. To ensure consistency and repeatability, each type of experiment was performed three times. The damage modes observed in the experiments included indentation, damage without full penetration, and complete penetration, depending on the impact energy level.

Figure 3 illustrates the damage conditions of aluminum alloy sheets from both the simulation and repeated experiment under different impact energies, showing both top and bottom views with a full size of 110 × 110 mm. The damage results indicate that under impact energies of 15 J, 25 J, and 50 J, the aluminum alloy sheets only exhibit indentation without cracking or penetration. Due to their high ductility and high thickness, the aluminum alloy sheets were indented without failure, even under the impact of 50 J. The simulation also exhibits similar damage conditions, and the comparison of the predicted damage diameters with the average measured results is listed in Table 3, in which the standard deviation is also displayed in some average values. As shown, the maximum difference between the simulation and the experiment is only 3.6%. Furthermore, both results indicate that the damage diameter increases with the increase in impact energy. Figure 4 presents the corresponding force–time curves. The force–time curves for all three impact energy levels exhibit smooth and symmetrical profiles, indicating the absence of severe damage. As the impact energy increases, both the peak force and damage diameter increase accordingly, as listed in Table 3. The experimental results show that the average peak forces are 4.68 kN at 15 J, 6.26 kN at 25 J, and 9.46 kN at 50 J, with which the simulation has very good agreement. As listed in Table 3, the comparison of the impact time between the simulation and the experiment shows only minor discrepancies. It is interesting to note that at 25 J, the impact duration is longer. This is likely due to the indentation shape closely matching the impactor profile, leading to prolonged contact duration. All these results verify the impact simulation for the aluminum specimens.

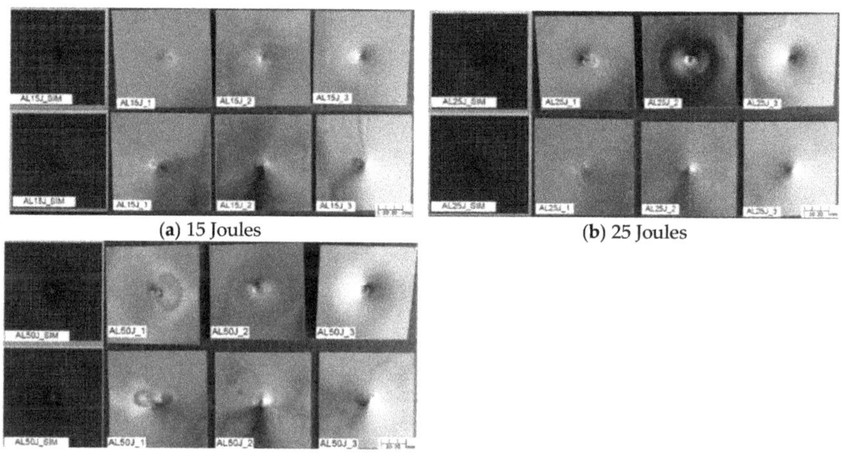

(**a**) 15 Joules (**b**) 25 Joules (**c**) 50 Joules

Figure 3. Simulation and experiment damage of aluminum specimens under 15 J, 25 J, and 50 J.

Table 3. Result comparison of aluminum specimens under 15 J, 25 J, and 50 J.

Specimen	Damage Diameter (mm)	Peak Force (kN)	Impact Time (ms)
AL15J_1	6.9	4.68	6.04
AL15J_2	6.8	4.67	6.04
AL15J_3	7.1	4.71	6.04
Average	6.93 ± 0.15	4.68 ± 0.02	6.04
Simulation	6.8 (−1.87%)	4.69 (0.21%)	6.29 (4.13%)
AL25J_1	8.8	6.26	7.02
AL25J_2	8.6	6.26	7.02
AL25J_3	8.9	6.27	7.02
Average	8.76 ± 0.15	6.26 ± 0.01	7.02
Simulation	8.8 (0.45%)	6.37 (1.75%)	6.8 (−3.13%)
AL50J_1	11.0	9.32	6.24
AL50J_2	11.1	9.36	6.24
AL50J_3	11.2	9.70	6.24
Average	11.1 ± 0.1	9.46 ± 0.21	6.24
Simulation	10.7 (−3.6%)	9.38 (−0.84%)	6.2 (0.64%)

Figure 5 demonstrates the damage profile of carbon fiber composite specimens from both the simulation and repeated experiment under different impact energy levels, showing the top and bottom views with a full size of 110 × 110 mm. The results reveal that under a

15 J impact, the composite sheet sustains a fracture but does not experience full penetration. However, at 25 J and 50 J impact energies, penetration failure occurs. Compared to aluminum alloy sheets, carbon fiber composite sheets exhibit a higher susceptibility to damage. Although carbon fiber composites possess superior strength and stiffness, they also exhibit greater brittleness. When subjected to impact loading, this material is prone to damage, such as fiber breakage, matrix cracking, and delamination. Delamination between different layers is particularly evident on the backside of the impacted specimens, further exacerbating structural failure. The average damage diameters of the experiments under different impact energies are shown in Table 4 and compared with the simulation results. At 25 J and 50 J, the specimens were penetrated and had close damage diameters that were larger than those at 15 J. The diameters from the simulation only have 1.55% to 4.6% errors with the experiment values. A comparison of the failure morphology reveals slight differences between the experimental and simulation results. In the experimental specimens, significant delamination is observed on the backside of the impacted area. However, due to computational efficiency considerations, the simulation model utilizes only two shell elements to represent the composite sheet, preventing the accurate depiction of single-layer delamination.

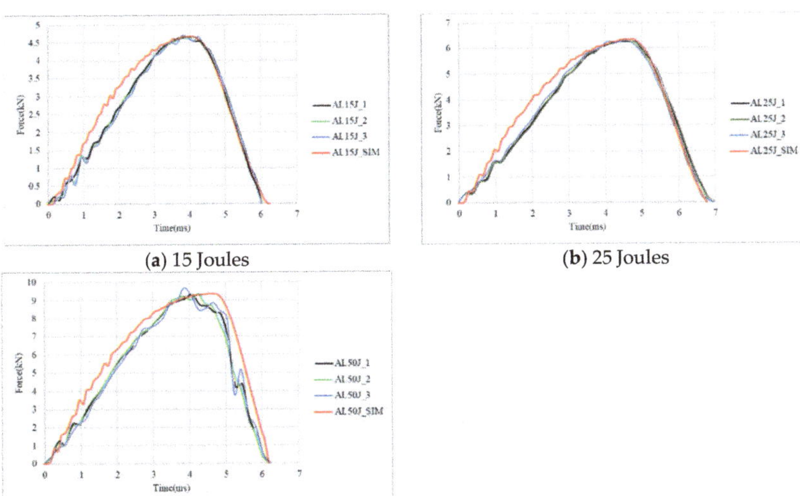

(**a**) 15 Joules (**b**) 25 Joules (**c**) 50 Joules

Figure 4. Simulation and experiment force–time curves of aluminum specimens under 15 J, 25 J, and 50 J.

Figure 6 presents the corresponding force–time curves from both the simulation and experiment. The force–time curves exhibit a common trend across all three impact energy levels: after reaching the peak force, a sharp decline is observed. This phenomenon indicates the initiation of severe internal damage, including fiber fracture, matrix failure, and interfacial delamination, which significantly reduces the specimen's ability to withstand additional loading from the impactor. At 15 J, the composite specimen was not penetrated, such that the impact duration time is the longest. With the increase in impact energy, the impact time becomes shorter, as shown in Table 4. It is observed that beyond a certain impact energy threshold, the increase in peak force and damage diameter becomes less significant. This phenomenon occurs because, at 25 J, the composite sheet has already experienced full penetration, limiting its ability to further resist impact forces. Consequently, even when the impact energy is increased to 50 J, the peak force and damage diameter remain relatively unchanged. Additionally, as the impact energy increases, the impact duration decreases correspondingly. From Figure 6 and Table 4, the peak force from the simulation is less than 1.92% of the experiment, and the impact time has a maximum error of 9.2%. Therefore, the finite element simulation for the impact behavior of composite specimens has reliable results.

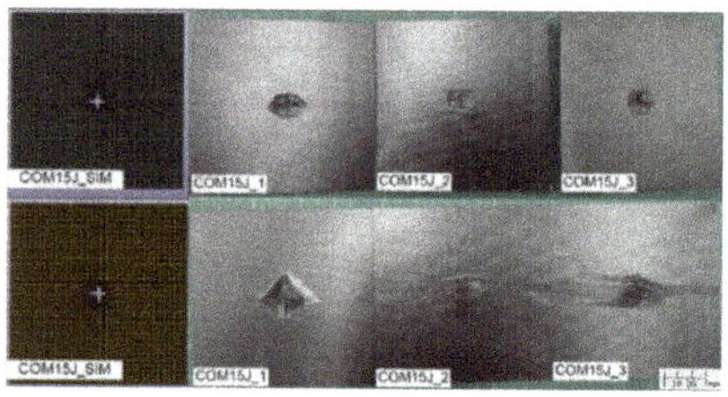

(**a**) 15 Joules

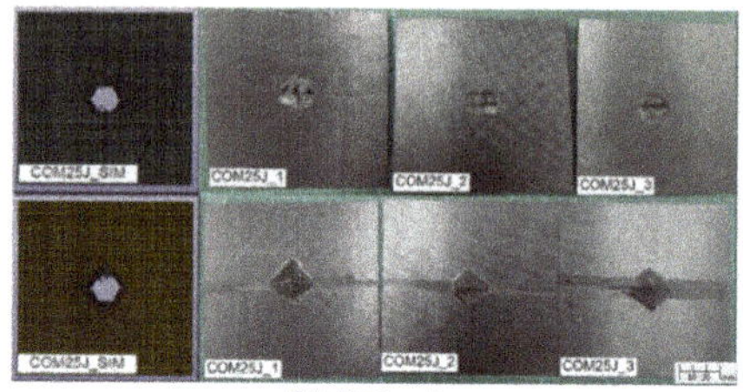

(**b**) 25 Joules

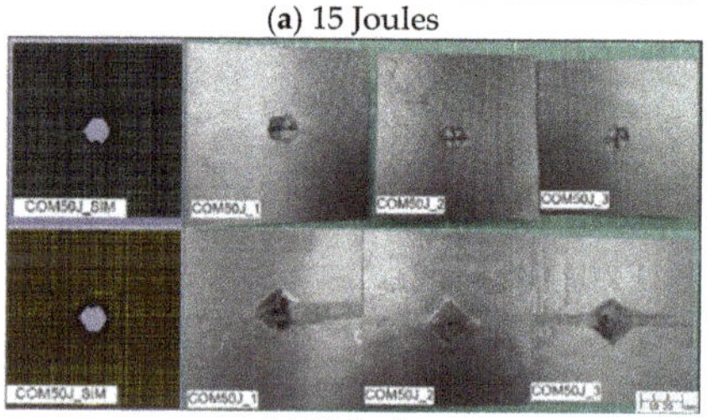

(**c**) 50 Joules

Figure 5. Simulation and experiment damage of carbon fiber composite specimens under 15 J, 25 J, and 50 J.

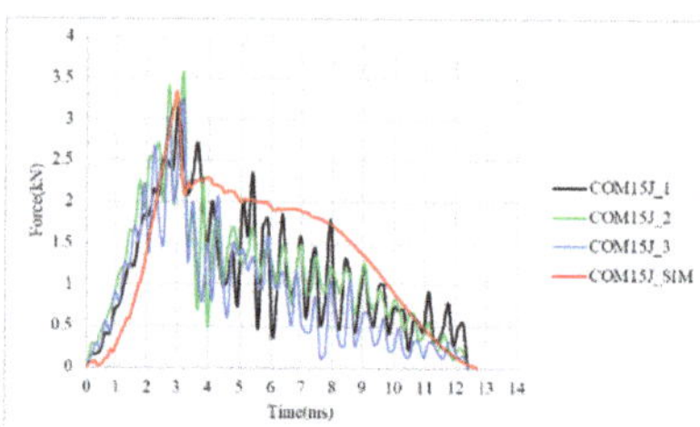

(**a**) 15 Joules

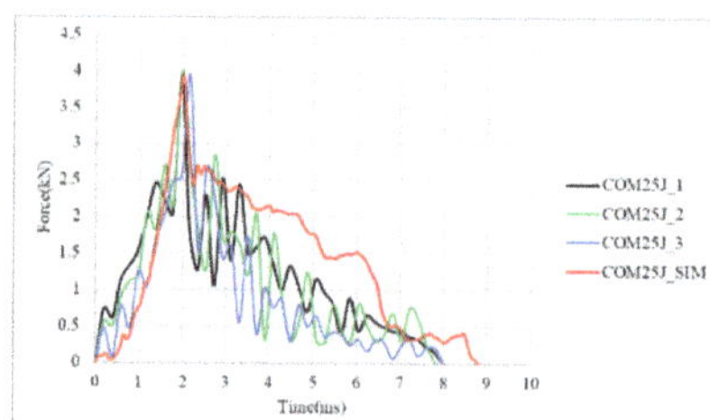

(**b**) 25 Joules

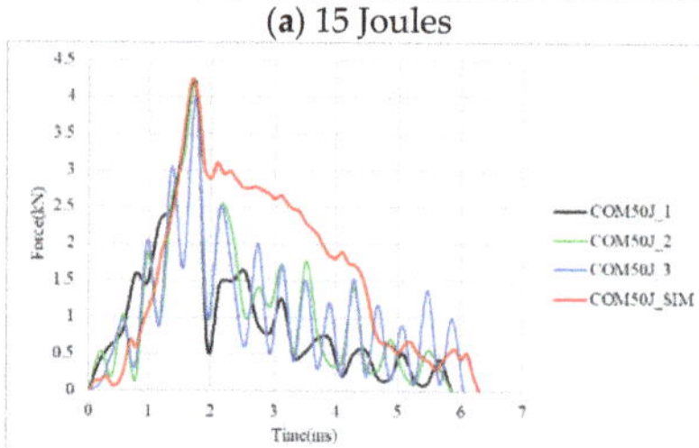

(**c**) 50 Joules

Figure 6. Simulation and experiment force–time curves of carbon fiber composite specimens under 15 J, 25 J, and 50 J.

Table 4. Result comparison of carbon fiber composite specimens under 15 J, 25 J, and 50 J.

Specimen	Damage Diameter (mm)	Peak Force (kN)	Impact Time (ms)
COM15J_1	13.3	3.21	12.4
COM15J_2	12.5	3.55	12.4
COM15J_3	12.9	3.23	12.4
Average	12.9 ± 0.4	3.33 ± 0.19	12.4
Simulation	13.1 (1.55%)	3.38 (1.5%)	12.68 (2.25%)
COM25J_1	15.1	3.90	8
COM25J_2	15.3	4.02	7.9
COM25J_3	15.2	3.95	8
Average	15.2 ± 0.1	3.95 ± 0.06	7.96
Simulation	15.9 (4.6%)	3.93 (−0.5%)	8.7 (9.2%)
COM50J_1	15.5	4.15	5.85
COM50J_2	15.4	4.10	5.85
COM50J_3	15.8	4.23	6.04
Average	15.56 ± 0.21	4.16 ± 0.07	5.91
Simulation	16 (2.82%)	4.24 (1.92%)	6.29 (6.4%)

As for the aluminum-composite sandwich specimens, the damage morphology under different impact energies, with the top and bottom views displayed separately with the full size of 110 × 110 mm, is shown in Figure 7. Figure 8 presents the corresponding force–time curves. The results indicate that under a 15 J impact, the top aluminum alloy layer exhibits indentation while the bottom aluminum layer develops minor cracks. At 25 J, the top aluminum layer continues to indent while the bottom aluminum layer develops more significant cracks. At 50 J, complete penetration occurs. Compared to carbon fiber composite specimens, aluminum-composite sandwich panels demonstrate greater resistance to penetration failure due to the plasticity of the aluminum alloy layers, which absorb impact energy through deformation. The force–time curves for 15 J and 25 J impacts exhibit smoother and more symmetrical profiles, reflecting the high plasticity of aluminum alloy and the absence of penetration failure. However, at 50 J, penetration failure occurs, resulting in a sharp force drop after reaching the peak load, as the specimen can no longer sustain the impact energy.

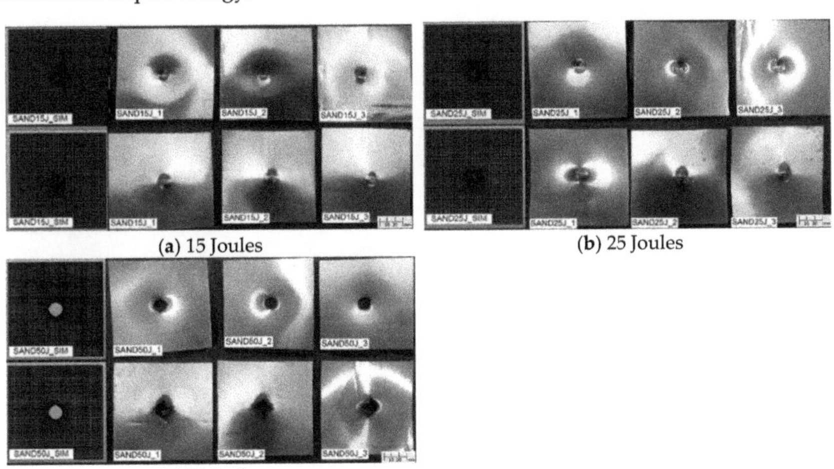

(a) 15 Joules (b) 25 Joules (c) 50 Joules

Figure 7. Simulation and experiment damage of sandwich specimens under 15 J, 25 J, and 50 J.

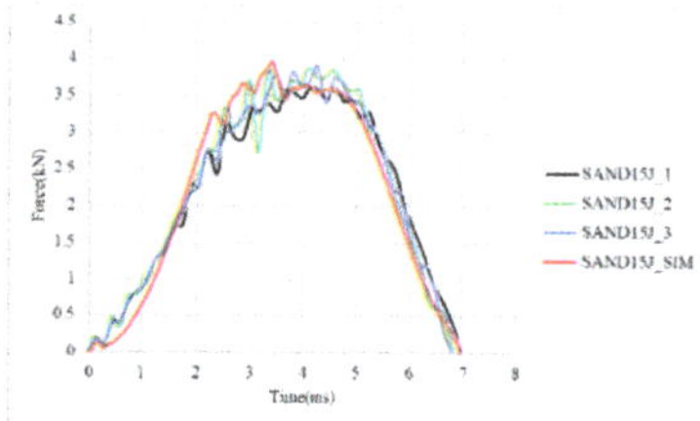

(**a**) 15 Joules

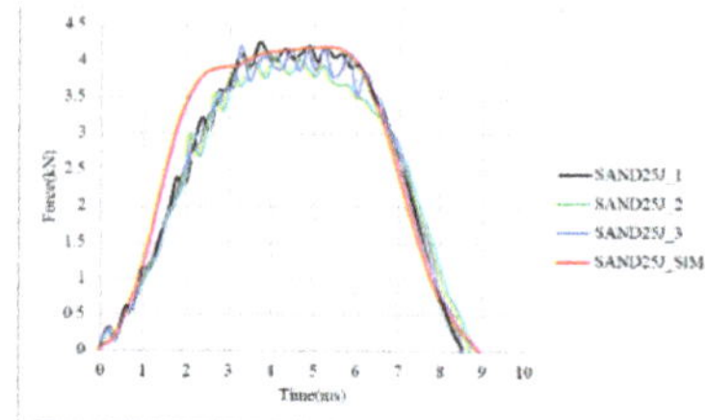

(**b**) 25 Joules

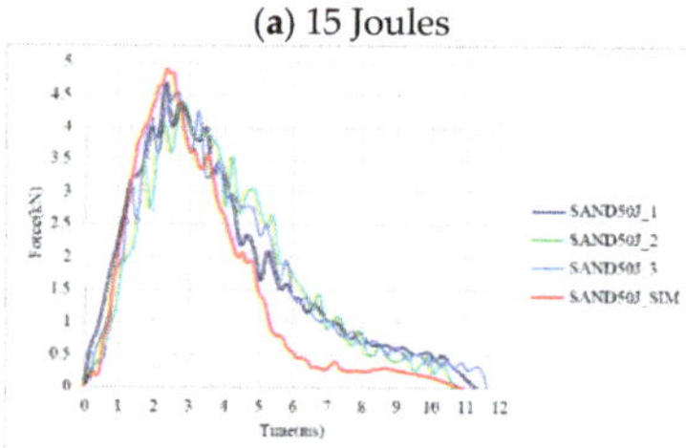

(**c**) 50 Joules

Figure 8. Simulation and experiment force–time curves of sandwich specimens under 15 J, 25 J, and 50 J.

As listed in Table 5, as the impact energy increases, both peak force and impact duration time increase proportionally. For the damage diameter, there is a least value at 15 J, while the two diameters at 25 J and 50 J are very close. The close value on damage diameter may indicate the whole indentation of the impactor, even though there is no penetration at 25 J. Observing the failure deformation reveals a strong agreement between experimental and simulation results. Regarding the force–time curves, the discrepancies between the experimental and simulated curves are also minimal. The maximum error is just 8.6% for the peak force at 50 J.

Table 5. Result comparison of sandwich specimens under 15 J, 25 J, and 50 J.

Specimen	Damage Diameter (mm)	Peak Force (kN)	Impact Time (ms)
SAND15J_1	9	3.62	6.99
SAND15J_2	9.5	3.86	6.99
SAND15J_3	8.9	3.91	6.83
Average	9.13 ± 0.32	3.79 ± 0.16	6.88
Simulation	8.7 (−4.71%)	3.96 (4.48%)	7 (1.7%)
SAND25J_1	14.5	4.26	8.58
SAND25J_2	14	4.09	8.77
SAND25J_3	14.2	4.22	8.58
Average	14.23 ± 0.25	4.19 ± 0.09	8.64
Simulation	14.5 (1.89%)	4.19 (0%)	8.9 (3.0%)
SAND50J_1	14.8	4.7	11.31
SAND50J_2	15.1	4.37	10.72
SAND50J_3	14.9	4.54	11.7
Average	14.93 ± 0.15	4.53 ± 0.17	11.24
Simulation	15.1 (1.13%)	4.92 (8.6%)	11.02 (−1.9)

Figure 9 presents a comparison of the peak force values obtained from the experiments and simulations for the three specimen types under different impact energy levels. The figure demonstrates a strong consistency between the simulation value and the experimental result, with error bars. Additionally, it can be observed that for all impact energies, the aluminum alloy sheet exhibits the highest peak force, and this force significantly increases with the impact energy. The peak force of the aluminum-composite sandwich panel is slightly higher than that of the carbon fiber composite sheet. In addition, the increase in impact energy does not increase the peak force of these two types of specimens much due to the damage of the composite part. Figure 10 compares the damage diameters obtained from the experiments and simulations for the three specimen types under different impact energy levels. The figure shows a strong correlation between the experimental and simulated values. Additionally, it can be observed that the carbon fiber composite sheet exhibits the largest damage diameter, followed by the aluminum-composite sandwich panel, with the aluminum alloy sheet having the smallest damage diameter. Furthermore, the damage diameter of the aluminum specimen keeps increasing with the impact energy. For both the sandwich specimens and the composite specimens, the damage diameter significantly increases from 15 J to 25 J, while the damage diameters at 25 J and 50 J have very close values that are also close to the impactor diameter. The reason should be from the easy damage of the composite part.

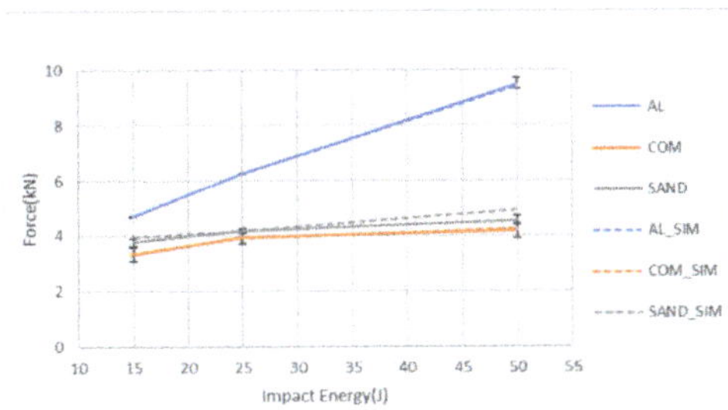

Figure 9. Comparison of the peak forces for the types of specimens under three impact energies from both the simulation and the experiment.

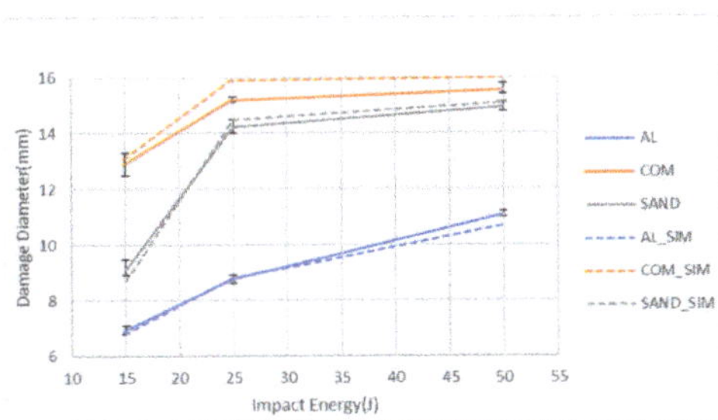

Figure 10. Comparison of the damage diameters for the types of specimens under three impact energies from both the simulation and the experiment.

Figure 11 presents the energy absorption from the simulation for the three specimen types at three impact energies. At 15 J, the carbon fiber composite sheet exhibits the highest energy absorption, reaching 11 J, followed by the aluminum alloy sheet at 10.7 J and the aluminum-composite sandwich panel at 10.4 J. At 25 J, the carbon fiber composite sheet exhibits the highest energy absorption at 24.2 J, followed by the aluminum-composite sandwich panel at 21.2 J and the aluminum alloy sheet at 20.4 J. At 50 J, the aluminum alloy sheet exhibits the highest energy absorption at 44.9 J, followed by the aluminum-composite sandwich panel at 40 J and the carbon fiber composite sheet at 30.1 J. From the above results, at impact energies of 15 J and 25 J, the easy damage of carbon fiber sheets absorbs the highest energy, while at 50 J, they absorb the least impact energy because of less further damage. For the other two types of specimens, the energy absorption maintains a linear increase with the impact energy, and at 50 J, the sandwich panel absorbs lower energy than the aluminum alloy sheet because the former is penetrated. All these suggest that while the carbon fiber composite sheet absorbs the most energy, its brittle nature makes it more prone to damage compared to the other materials, whereas the aluminum-composite sandwich panel leverages both the ductility of aluminum and the strength of the composite layer to balance impact resistance and energy absorption.

The total absorbed energy presented in Figure 11 was further normalized by dividing it by the specimen mass to determine the specific energy absorption (SEA), which represents the energy absorbed per unit mass. Figure 12 illustrates the specific energy absorption values for the three specimen types under different impact energy levels. The carbon fiber composite sheet exhibits the highest specific energy absorption, followed by the aluminum-composite sandwich panel, with the aluminum alloy sheet having the lowest value. Overall, while the carbon fiber composite sheet has the highest specific energy absorption, it also sustains the most severe damage and the lowest peak force, as shown in Figures 9 and 10. The aluminum alloy sheet, due to its highest mass, exhibits the lowest specific energy absorption. However, due to its high ductility, it has the highest peak force and the lowest damage diameter. The aluminum-composite sandwich panel falls between the two, balancing both damage resistance and energy absorption efficiency. Although the sandwich panel's damage diameter is larger than that of the aluminum alloy sheet, it also exhibits higher specific energy absorption. Conversely, while its specific energy absorption is lower than that of the carbon fiber composite sheet, its damage diameter is significantly smaller. In some applications, if one wants to absorb more impact energy with less mass, transfer less impact force, and maintain smaller damage or no penetration, the aluminum-composite sandwich (or FMLs) should be a good candidate.

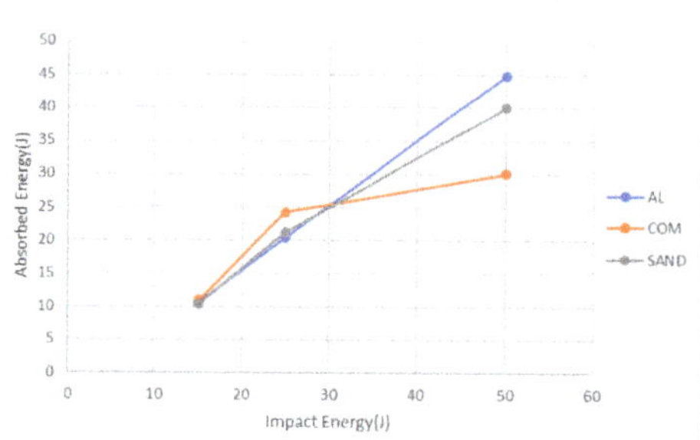

Figure 11. Predicted energy absorbed for the three types of specimens at the three impact energies.

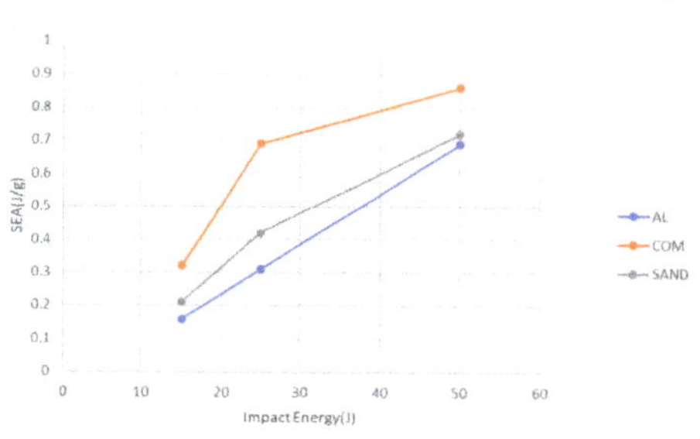

Figure 12. Predicted specific energy absorption for the three types of specimens at the three impact energies.

5. Conclusions

This study conducted low-velocity impact experiments on three types of specimens: 5052-H32 aluminum alloy sheets, carbon fiber composite sheets made from unidirectional prepreg stacking, and aluminum-composite sandwich panels. The impact energies were set at 15 J, 25 J, and 50 J to investigate the failure morphology, damage diameter, and force–time response of each specimen under different impact conditions. Finite element analysis using LS-DYNA was performed to compare the experimental and simulation results, validating their consistency. The aluminum alloy sheets exhibited only indentation under all three impact energy levels (15 J, 25 J, and 50 J). Under an impact energy of 15 J, carbon fiber composite plates exhibit damage without complete penetration. However, at impact energies of 25 J and 50 J, the specimens undergo full penetration failure. Carbon fiber composite plates are more susceptible to damage. Under an impact energy of 15 J, the top aluminum alloy layer of the aluminum-composite sandwich panel exhibits indentation, while the bottom aluminum alloy layer develops minor cracks. At an impact energy of 25 J, the top layer remains indented, but the bottom layer experiences larger cracks. When subjected to 50 J impact energy, the specimen undergoes complete penetration failure.

A comparison of the three types of specimens reveals that, in terms of peak force, the aluminum alloy panel exhibits the highest peak force. Regarding the damage diameter, the carbon fiber composite plate exhibits the largest damage diameter. The aluminum-composite sandwich panel falls between the other two specimens in both peak force and damage diameter. By comparing the experimental results with LS-DYNA finite element analysis simulations, a strong consistency is observed in terms of damage diameter, failure morphology, peak force, and impact duration. This indicates that the simulation model demonstrates a high level of reliability and can effectively predict the outcomes observed in actual experiments. Furthermore, by analyzing the damage diameter and specific energy absorption (SEA), it is evident that while the carbon fiber composite plate exhibits the highest SEA, it also has the largest damage diameter. In contrast, the aluminum alloy panel has the lowest SEA. The aluminum-composite sandwich panel demonstrates intermediate performance in both damage diameter and SEA, striking a balance between the other two specimens.

Author Contributions: Conceptualization, S.-F.H.; methodology, S.-F.H. and M.-Y.W.; software, M.-Y.W.; validation, S.-F.H. and M.-Y.W.; formal analysis, S.-F.H. and M.-Y.W.; writing—original draft preparation, S.-F.H.; writing—review and editing, S.-F.H.; visualization, S.-F.H. and M.-Y.W. All authors have read and agreed to the published version of the manuscript.

Funding: This research was funded by the Ministry of Science and Technology, Taiwan, through MOST 108-2221-E-224-032-MY3.

Institutional Review Board Statement: Not applicable.

Informed Consent Statement: Not applicable.

Data Availability Statement: Data are available on request from the corresponding author.

Conflicts of Interest: The authors declare no conflicts of interest.

References

1. Muniyan, V.; Kumar, V.V.; Suyambulingam, I.; Priyadharshini, S.; Divakaran, D.; Rangappa, S.M.; Siengchin, S. A review of recent advancements in the impact response of fiber metal laminates. *Heliyon* **2025**, *11*, e41756. [CrossRef] [PubMed]
2. Børvik, T.; Forrestal, M.J.; Hopperstad, O.S.; Warren, T.L.; Langseth, M. Perforation of AA5083-H116 aluminium plates with conical-nose steel projectiles—Calculations. *Int. J. Impact Eng.* **2009**, *36*, 426–437. [CrossRef]

3. Grytten, F.; Børvik, T.; Hopperstad, O.S.; Langseth, M. Quasi-static perforation of thin aluminium plates. *Int. J. Impact Eng.* **2009**, *36*, 486–497. [CrossRef]
4. Seo, J.K.; Park, D.K.; Jo, S.W.; Kim, B.J.; Park, J.S. Experimental assessment of the structural behaviour of aluminium helideck structures under static impact loads. *Ships Offshore Struct.* **2018**, *13*, S348–S363. [CrossRef]
5. Morin, D.; Kaarstad, B.L.; Skajaa, B.; Hopperstad, O.S.; Langseth, M. Testing and modelling of stiffened aluminium panels subjected to quasi-static and low-velocity impact loading. *Int. J. Impact Eng.* **2017**, *110*, 97–111. [CrossRef]
6. Liu, B.; Chen, C.T.; Garbatov, Y. Material failure criterion in the finite element analysis of aluminium alloy plates under low-velocity impact. *Ocean Eng.* **2022**, *266*, 113260. [CrossRef]
7. Shi, S.; Lam, N.; Cui, Y.W.; Zhang, L.H.; Lu, G.X.; Gad, E. Indentation into an aluminium panel by the impact of a rigid spherical object. *Thin-Walled Struct.* **2022**, *180*, 109935. [CrossRef]
8. Xia, Y.; Shi, Z.T.; Zhou, Q.; Ao, W.H. Numerical investigation on polyurea coated aluminum plate subjected to low velocity impact. *Int. J. Impact Eng.* **2023**, *177*, 104516. [CrossRef]
9. Caminero, M.A.; García-Moreno, I.; Rodríguez, G.P. Damage resistance of carbon fibre reinforced epoxy laminates subjected to low velocity impact: Effects of laminate thickness and ply-stacking sequence. *Polym. Test.* **2017**, *63*, 530–541. [CrossRef]
10. Long, S.C.; Chen, C.; Wang, H.; Yao, X.H.; Zhang, X.Q. Distribution and propagation of matrix cracks within composite laminates under impact. *Compos. Struct.* **2022**, *281*, 115005. [CrossRef]
11. Ouyang, T.; Sun, W.; Bao, R.; Tan, R.M. Effects of matrix cracks on delamination of composite laminates subjected to low-velocity impact. *Compos. Struct.* **2021**, *262*, 113354. [CrossRef]
12. Zhou, J.W.; Liao, B.; Shi, Y.; Zuo, Y.G.; Tuo, H.I.; Jia, L.Y. Low-velocity impact behavior and residual tensile strength of CFRP laminates. *Compos. Part B Eng.* **2019**, *161*, 300–313. [CrossRef]
13. Ge, X.; Zhang, P.; Zhao, F.; Liu, M.; Liu, J.; Cheng, Y. Experimental and numerical investigations on the dynamic response of woven carbon fiber reinforced thick composite laminates under low-velocity impact. *Compos. Struct.* **2022**, *279*, 114792. [CrossRef]
14. Wu, Z.Y.; Wu, C.J.; Liu, Y.S.; Cheng, X.Y.; Hu, X.D. Experimental study on the low-velocity impact response of braided composite panel: Effect of stacking sequence. *Compos. Struct.* **2020**, *252*, 112691. [CrossRef]
15. Khashaba, U.A.; Othman, R. Low-velocity impact of woven CFRE composites under different temperature levels. *Int. J. Impact Eng.* **2017**, *108*, 191–204. [CrossRef]
16. Krollmann, J.; Schreyer, T.; Veidt, M.; Drechsler, K. Impact and post-impact properties of hybrid-matrix laminates based on carbon fiber-reinforced epoxy and elastomer subjected to low-velocity impacts. *Compos. Struct.* **2019**, *208*, 535–545. [CrossRef]
17. Bieniaś, J.; Jakubczak, P. Impact damage growth in carbon fibre aluminium laminates. *Compos. Struct.* **2017**, *172*, 147–154. [CrossRef]
18. Droździel, M.; Jakubczak, P.; Bieniaś, J. Low-velocity impact resistance of thin-ply in comparison with conventional aluminium-carbon laminates. *Compos. Struct.* **2021**, *256*, 113083. [CrossRef]
19. Yu, G.C.; Wu, L.Z.; Ma, L.; Xiong, J. Low velocity impact of carbon fiber aluminum laminates. *Compos. Struct.* **2015**, *119*, 757–766. [CrossRef]
20. Jakubczak, P.; Bieniaś, J.; Dadej, K. Experimental and numerical investigation into the impact resistance of aluminium carbon laminates. *Compos. Struct.* **2020**, *244*, 112319. [CrossRef]
21. Li, Z.Y.; Zhang, J.Y.; Jackstadt, A.; Kärger, L. Low-velocity impact behavior of hybrid CFRP-elastomer-metal laminates in comparison with conventional fiber-metal laminates. *Compos. Struct.* **2022**, *287*, 115340. [CrossRef]
22. Yao, L.; Wang, C.Z.; He, W.T.; Lu, S.J.; Xie, D. Influence of impactor shape on low-velocity impact behavior of fiber metal laminates combined numerical and experimental approaches. *Thin-Walled Struct.* **2019**, *145*, 106399. [CrossRef]
23. ASM Aerospace Specification Metals Inc. *ASM Material Data Sheet*; ASM Aerospace Specification Metals Inc.: Pompano Beach, FL, USA, 2025.
24. Wu, C.Y. Study of Impact Energy Absorption on the Aluminum-Composite Hybrid Tubes. Master's Thesis, National Yunlin University of Science and Technology, Douliu, Taiwan, June 2019. (In Chinese).
25. Chang, F.K.; Chang, K.Y. A progressive damage model for laminated composites containing stress concentration. *J. Compos. Mater.* **1987**, *21*, 834–855. [CrossRef]

materials

MDPI

Article

Properties of Selected Additive Materials Used to Increase the Lifetime of Tools for Crushing Unwanted Growths Using Hardfacing by Welding Technology

Miroslava Ťavodová [1], Monika Vargová [1], Dana Stančeková [2], Anna Rudawska [3] and Arkadiusz Gola [3,*]

1 Faculty of Technology, Technical University in Zvolen, Študentská 26, 960 01 Zvolen, Slovakia; tavodova@tuzvo.sk (M.Ť.); monika.vargova@tuzvo.sk (M.V.)
2 Faculty of Mechanical Engineering, University of Žilina, Univerzitná 1, 010 26 Žilina, Slovakia; dana.stancekova@fstroj.uniza.sk
3 Faculty of Mechanical Engineering, Lublin University of Technology, Nadbystrzycka 36, 20-618 Lublin, Poland; a.rudawska@pollub.pl
* Correspondence: a.gola@pollub.pl; Tel.: +48-81-538-45-35

Abstract

This article focuses on the possibilities of increasing the service life of tools for crushing unwanted growths. One way to increase their service life is to increase the hardness and resistance to abrasive wear of exposed surfaces of the tool, which are their face and back. At the same time, however, care must be taken to ensure that the shape and weight of the tool is not altered after the additive has been hardfaced on. Thus, the tool was first modified by removing the material by milling from the face and back. Subsequently, two surfacing materials, namely UTP 690 and OK WearTrode 55, were chosen and hardfaced by welding onto the pre-prepared surfaces. After hardfacing by welding, the tools were ground to their original shape and their weight was measured. Subsequently, the tool was sawn, and specimens were created for Rockwell hardness evaluation, material microstructure and for abrasive wear resistance testing as per ASTM G133-95. The OK WearTrode 55 electrode is a hardfacing electrode that produces weld metal with a high-volume fraction of fine carbides in a martensitic matrix. Better results were achieved by the UTP 690 hardfacing material. The hardness was 3.1 times higher compared to the base tool material 16MnCr5 and 1.2 times higher than the OK WearTrode 55 material. The abrasive wear resistance was 2.76 times higher compared to 16MnCr5, and 1.14 times higher compared to the OK WearTrode 55 material. The choice of a suitable pre-treatment for the tool and the selection and application of such additional material, which with its complex properties better resists the effects of the working environment, is a prerequisite for increasing the service life of tools working in forestry.

Keywords: mulching forestry tools; hardfacing; wear-resistance; microstructure; hardness

1. Introduction

As part of the global trend towards environmental protection in forestry and sustainable development of the forest environment, it has become an important issue to address the efficient use of natural resources through the use of modern, highly efficient and safe equipment in forest maintenance [1].

Crushing of unwanted growths is a certain method of mulching the woody mass of the growth of different types and thicknesses [2]. Mulching is widely recognized as an effective soil and water conservation measure all over the world. It is a mechanized type of work that destroys and crushes above-ground parts of vegetation. Globally, numerous laboratory and field studies have demonstrated that mulching is beneficial in reducing soil and water loss across a variety of environmental conditions, including agricultural lands [2–4]. Working bodies of agricultural machinery and equipment operate in harsh conditions. Mulching cutters shred various types of vegetation, including stumps, trees, bushes and their roots, and mix wood residues with soil, which acts as an abrasive [4]. Mulching devices are used as adapters that are clamped to the forest wheeled tractor. The main part of the adapter structure is a rotating cylinder with a rotation speed of $n = 1000\ min^{-1}$, which is attached to the base machine. Tools are placed around the circumference of the rotary cylinder [5,6]. These machines can work even in broken terrains, with large slopes. They are often used in forestry to process unwanted growth, especially in inaccessible terrain under high voltage lines, etc.

The wear and tear of mechanisms and machine parts is a phenomenon constantly accompanying the operation of any machinery, and it causes measurable financial losses [7,8]. Deterioration processes usually begin in the superficial layer of a given part. The consequence of wear is a progressive loss of mass and initial, nominal dimensions [9,10]. The fastest wearing part of the assembly is the tools. Tools for crushing unwanted growth operate in a strongly heterogeneous environment. Due to the impact of the working environment, in which there are various types of wood, soil, minerals and rocks, they are subject to high wear. This wear is caused by the impact and abrasion of hard abrasive particles. Abrasive wear is believed to be responsible for up to 50% of failures, or the replacement of parts, components and tools [11]. After the loss of the wolfram–carbide WC tip, there is a rapid loss of material precisely on these surfaces, which leads to a total loss of its function, namely the crushing of unwanted growth. These tools are also subject to impact load. The results of long-term research by the authors [5,12–14] show that the observed wear of the material is caused by cyclic plastic deformation. Strengthening of the material before increased stress, which is a manifestation of cold plastic deformation, increased its hardness, and therefore also its brittleness [15]. Cyclic repetition of the plastic deformation of the tool under abrasion conditions leads to a loss of material, mainly on the back surface of the tool, over time (after the loss of WC tips).

The working bodies of the mulching cutter are made of different materials depending on the requirements for strength, wear resistance, corrosion resistance and other properties. One of the most common materials is high-quality carbon steel. Steels with the addition of chromium or tungsten can also be used to improve performance. Another popular material for the production of mulching cutter workpieces is carbide materials such as tungsten carbide, which have high hardness and wear resistance. Materials based on ceramics or composites can also be used [4,12,16].

There are several methods for increasing the service life of tools—heat, chemical heat treatment, hardfacing by welding, etc [17]. Based on long-term research in the field of increasing the service life of tools for crushing unwanted growths, the tools were modified through heat treatment. However, the results showed that the hardness of the modified tool (with 16MnCr5 material) was 1.75 times higher than the hardness of the unmodified tool, but at the same time, the performed heat treatment had almost no effect on the resistance to abrasive wear [5]. One of the very effective measures for increasing wear resistance is the application of a suitable additional material on functional surfaces [18]. Despite being used for over 100 years, there still is much to discover, as modern metallurgy provides more and more sophisticated alloys, which then have to be studied to find the best technological parameters in order to fully utilize complex material properties [7,18–20]. Hardfacing is a commonly used method to improve the surface properties of agricultural tools, components for mining operations, soil treatment equipment and others [21–23]. However, it is still not widely used for wood processing applications in forestry. The alloy is homogeneously applied to the surface of the base material (usually low-carbon or medium-carbon steels) by hardfacing in order to increase hardness and wear resistance without significant loss of ductility and toughness of the base material [13,23–25]. According to the author Kiswel [3], K-700HT flux-cored wire or welding flux-cored wire VELTEK-N565, which is used for hardfacing of mining equipment, impact drilling bits, screens, augers and parts of high-manganese austenitic steels, providing a hardness of 54 to 60 HRC [4].

The main components of Fe-based alloys are chromium, molybdenum and boron, which make Fe-based alloys resistant to wear. Wang et al. [25] studied the microstructure, hardness and wear resistance of Fe-based alloys by adding the elements ferrotitanium (Fe-Ti), ferromolybdenum (Fe-Mo), ferrovanadium (Fe-V) and graphite, which were applied by the arc welding process [26]. The results showed that the hardness and wear resistance of the hardfacing metal layer increased with an increasing proportion of Fe-Ti, Fe-V, Fe-Mo and graphite. Their amounts must be controlled in the range of 8–10% graphite, 12–15% Fe-Ti, 10–12% Fe-V and 2–4% Fe-Mo. However, cracks begin to form in the layer of hardfacing metal if the amount of Fe-Mo is more than 5% [27,28]. Important factors that determine the resistance to abrasive wear include the hardness, size, shape and intensity of the particles, and the hardness, shape, size and number of hard phases and their distribution in the base metal. Some authors [25,28,29] state that wear resistance increases with the increasing hardness of hard structural elements (carbides, borides, etc.) and their increasing share in the structure. During impact wear, the material is exposed to impact and high pressure, due to which it deforms or locally breaks—it can even crack.

Previous research has not considered that by hardfacing through welding the additive material to the surface of the tool, the weight of the tool will be increased. As several of these tools are mounted on the rotor of the base machine adapter (depending on the rotor dimensions, approx. 30 to 50 tools), this change in tool weight will also change the rotor loading dynamics, which may also affect the mulching process itself. In order to avoid an increase in tool weight, it is necessary to remove material from the tool, which will then be replaced in the next step by being welded on an additional material. At the same time, the shape of the tool must also be preserved, so it is necessary to grind the tool to its original shape after hardfacing by welding. To improve the surface, new machining methods have been developed in recent years. As the authors of [30] state in their research, a new body-armor-like abrasive tool (BAAT) with a soft and hard combined substrate (metal–elastomer substrate) to achieve high-shear and low-pressure grinding, it is possible to almost completely remove surface residual defects such as pits and deep scratches under optimal conditions. Also, the surface quality was significantly improved after grinding [30]. In their research, to prepare a flexible grinding body-armor-like abrasive tool (BAAT) with an increased tangential force-to-normal-force ratio, Wei. et al. found that smaller grinding grain sizes generated lower surface roughness values in zirconia ceramics after grinding [31].

The aim of this research is to increase the resistance to abrasive wear as well as the hardness of the tool body, because after a very short time, these tools need to be changed, which causes frequent machine downtime. Also, the efficiency and quality of the mulching process is reduced when using tools are worn out. At the same time, these tools are quite expensive—the price ranges up to EUR 100. This causes both technical and economic problems for companies operating in forestry. For this reason, it is necessary to look for solutions to increase the service life of these tools.

2. Materials and Methods

The two areas on the tool that are most stressed during its operation were identified, namely the face and back of the tool (Figure 1a). Since the tool operates in a highly heterogeneous environment, after a short time, the WC tip is lost and there is a significant and rapid loss of material on the face and back of the tool (Figure 1b). Figure 1c shows a plastically deformed surface with traces of pressed-in abrasive elements from the working environment of the tool. The ferritic–pearlitic structure cannot withstand such a significant abrasive load as the tool body is subjected to.

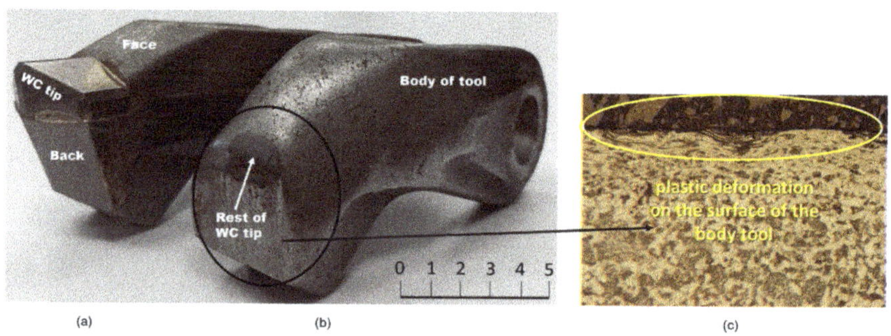

Figure 1. New, unused tool (**a**); damaged, non-functional tool (**b**) and microstructure of material with plastic surface deformation (**c**) [2,14].

The body of the tools for crushing unwanted growths is made of 16MnCr5 (according to STN EN ISO 683-3:2022) [5,32], which is a stainless structural manganese–chromium steel designed for cementing. The steel is well ductile in hot annealing and after soft and cold annealing, and is easy to machine and weld. It is ideal for machine components that are subsequently hardened to a diameter of 35 mm and is suitable for high core strength cementing. It is used for the manufacturing of components such as shafts, gears, camshafts, valve lifters, piston pins and gear couplings.

The most exposed areas of the tool need to be strengthened, which is supposed to increase the resistance of the tool to the loss of the WC tip and also to increase the life of the tool even after the loss of the WC tip. The novelty lies in the modification of the (new) tool by removing material from the face and back. This removed material is then replaced by a new material—an additional hardfacing material which, due to its chemical composition and structure, is better able to withstand the effects of the working environment. At the same time, this pre-preparation ensures that the weight and shape of the tool are maintained.

To verify the suitability of the choice of hardfacing materials, it is necessary to carry out laboratory tests, which include the following:

- Rockwell hardness measurement,
- Microstructure assessment by SEM microscopy,
- Abrasive wear resistance test—in principle ball-on-flat.

2.1. *Tool Modification Before Hardfacing by Welding*

When the additive material is hardfaced onto the exposed surfaces of the functional parts of the tools without prior preparation, the overall weight of the tool increases and the shape of the tool changes. This process will also affect the loading dynamics of the rotor to which the tools are mounted. Such a change can become one of the causes of faster tool wear. One way to prevent this is to pre-prepare the exposed surfaces of the tools by removing material from the tool body. Modifications were carried out on the face and back of the tool (Figure 2).

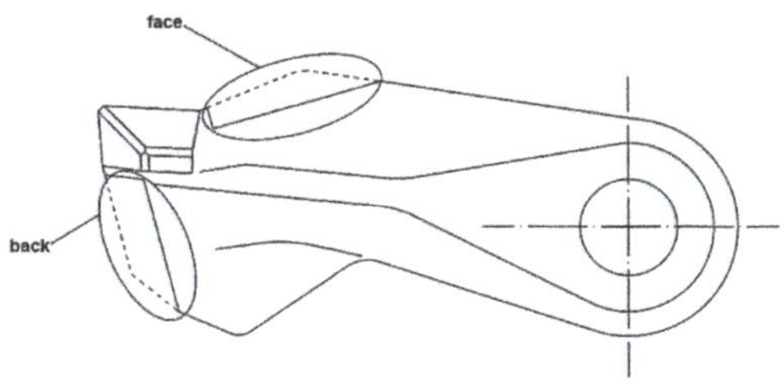

Figure 2. Modified tool surfaces.

The tool surfaces were prepared by chip machining—milling with a face mill with a diameter of 18 mm. A 20 × 30 mm surface was milled on the face of the tool (Figure 3a) and, 20 × 15 mm two surfaces were milled on the back (Figure 3b) to maintain the shape of the tool. The milling was carried out to a depth of 5 mm.

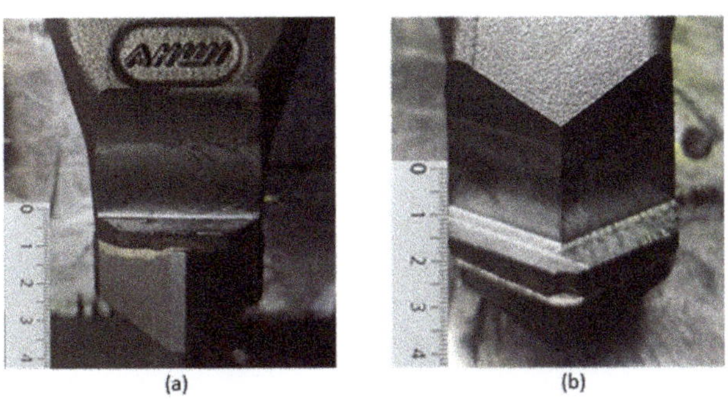

Figure 3. Modified face of tool (**a**); modified back of tool (**b**).

2.2. Hardfacing Materials and the Hardfacing Process

The selected hardfacing materials were subsequently hardfaced onto the pre-prepared surfaces. These hardfacing materials should provide better wear resistance due to their chemical composition and structure. The hardfacing material was applied to each tool in two layers. After the first layer of hardfacing material was applied, the hardfaced surfaces were cleaned and the second layer of hardfacing material was applied in a direction rotated by 90° with respect to the direction of the first layer of hardfacing material.

The manual arc hardfacing (MMA) method was used for the experiment. Two hardfacing electrode materials were chosen, namely UTP 690 (STN EN 14700: E Fe4 [33], Bohler) and OK WearTrode 55 (STN EN 14700: E Z Fe6 [33], Esab). The chemical composition of these electrodes is in Table 1. The characteristics of hardfacing electrodes are as follows:

- The UTP 690 is used for the repair and production of cutting tools, especially for the restoration of cutting edges and working surfaces. The hardfacing metal is highly resistant to friction, compression and impacts, even at elevated temperatures of up to 550 °C. Using this electrode, it is also possible to manufacture new tools by hardfacing on unalloyed and low-alloy base steels. After hardfacing, it forms a martensitic structure. The weld deposit is equivalent to a high-speed steel with increased Mo content. The type of electrode coating is rutile. The hardness of the coating in the second layer is 62 HRC [34].

- The OK WearTrode 55 is a high-stress electrode for welding wear-resistant functional surfaces under simultaneous impact stresses with the necessary partial corrosion resistance. Machining of the coating is possible by grinding. After welding, it forms a martensitic structure [19,35]. It is used as a coating for parts of agricultural and forestry machinery, mixers, transport equipment, etc. The type of electrode coating is lime-basic. The hardness of the coating in the second layer is 52–59 HRC [35].

Table 1. Standard chemical composition (mass.%) of the base material and electrodes [34–36].

	C	Si	Mn	Cr	Mo	P	S	V	W	Ti	Fe
16MnCr5	0.14–0.19	0.1–0.37	1.1–1.4	0.8–1.1	-	max. 0.035	max. 0.035	-	-	-	balance
UTP 690	0.9	0.8	0.5	4.5	8.0	-	-	1.2	2.0	-	balance
OK WearTrode 55	0.7	0.6	0.7	10.0	-	-	-	-	-	-	balance

The tools were hardfaced by certified welders in the company ŽOS, a.s. in Zvolen, Slovakia. Electrodes with a diameter of 2.5 mm was used for hardfacing by welding. The hardfacing surfaces were cleaned manually with a wire brush and degraded with an industrial solvent. When hardfacing by welding with UTP 690 electrodes, the electrodes were dried at 300 °C for 2 h in a Zepacomp drying oven. The tool was preheated to a temperature of 250 °C. The welding current was set to I = 90 A and the voltage U = 24.5 V. During the hardfacing by welding with OK WearTrode 55 electrodes, the electrodes were dried at 200 °C for 2 h in a Zepacomp drying oven. The tool was preheated to 200 °C. The set welding current was I = 90 A and voltage U = 24.5 V.

It was important to minimize the heat input during the hardfacing by welding process. After each 20 mm section, it was necessary to measure the interpass temperature using a touch thermometer. The interpass temperature was prescribed by the manufacturer at 200 °C. If the specimen exceeded this temperature, it was necessary to wait until the specimen temperature dropped below the specified interpass temperature to proceed with the surfacing. The tool was always cleaned during the hardfacing process to prevent inclusions from impurities in the weld metal. After steaming, the samples were allowed to cool freely in air to room temperature and then cleaned. Figure 4a shows the tool with UTP 690 and Figure 4b shows the tool with OK WearTrode 55.

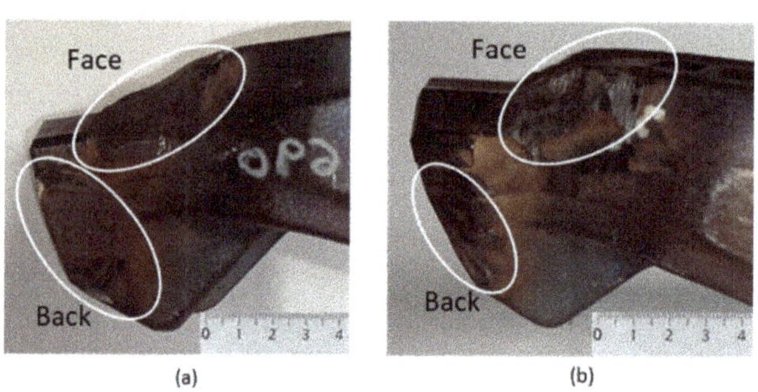

Figure 4. Tool after grounding of hardfacing: UTP 690 hardfacing material (**a**); OK WearTrode 55 hardfacing material (**b**).

After the additional materials were hardfaced onto the pre-prepared tool, the hardfacing material was ground to maintain the shape of the tool. Figure 4a,b show the tool after modification—the material was removed from the face and back of the tool with the subsequently hardfaced and grounded additive material UTP 690 (Figure 4a) and OK WearTrode 55 (Figure 4b).

2.3. Evaluation Methods

In order to evaluate the suitability of the selected hardfacing electrodes, laboratory analyzes were performed on the samples:

- Measurement of the weight of the tools before and after modification on the AG2000C digital scale (Axis 4 Sp. z o.o., Gdańsk, Poland).
- Rockwell hardness measurement according to ISO 6508-1 on universal hardness tester with max. load 250 kg, model UH250 (Bauhler, Boston, MA, USA) [37].
- Evaluation of the microstructure of the tool body material and additional hardfacing materials on an Olympus GX71 metallographic microscope with an Olympus DP12 camera (Olympus, Tokyo, Japan). Samples for microstructure evaluation were prepared in a standard method. To develop the structure, 2% Nital etchant (2% HNO_3 solution in ethyl alcohol) was used for the base material and Cor etchant (120 mL CH_3COOH, 20 mL HCl, 3 g picric acid, 144 mL CH_3OH) was used for the hardfacing materials.
- Test of resistance to wear, performed according to ASTM G133-95 with macroscopic evaluation of traces after the test on a Zeiss Stemi 2000 stereomicroscope (Carl Zeiss AG, Oberkochen, Germany) [38].
- Wear of a flat sample with determination of the volume V_f, evaluated with a 3D profilometer TalySurf CLI 1000 with a confocal and touch induction sensor (Taylor Hobson, Leicester, UK).

The wear resistance test was performed according to the ASTM G133-95 Standard Test Method for Linearly Reciprocating Ball-on-Flat Sliding Wear. This test method covers laboratory procedures for determining the sliding wear of ceramics, metals, and other candidate wear-resistant materials using a linear, reciprocating ball-on-flat plane geometry. The direction of the relative motion between sliding surfaces reverses in a periodic fashion such that the sliding occurs back and forth and in a straight line. The principal quantities of interest are the wear volumes of the contacting ball and flat specimen materials. The load is vertically downward through the spherical specimen against a horizontally mounted flat-shaped specimen. Dimensional changes for both ball and flat specimens are used to calculate wear volume and wear rate. The tangential force can be measured continuously during the oscillating contact and serves to obtain data on the coefficient of friction [38].

In Figure 5a, the Bruker TriboLab instrument (Billerica, MA, USA) on which the test of resistance to abrasive wear was performed, with a detail of the working area of the instrument (Figure 5b) is shown. In Figure 5c is a schematic representation of the device [38]. In Figure 5d, the sample on which the test was performed is shown.

Flat sample wear—flat sample wear volume V_f is calculated from the stroke length and the average cross-sectional area of the wear track. The wear volume of a flat sample V_f (mm^3) is calculated from Equation (1) [38]:

$$V_f = A \cdot L \tag{1}$$

where

A—average track cross-sectional area [mm^2],
L—stroke length [mm].

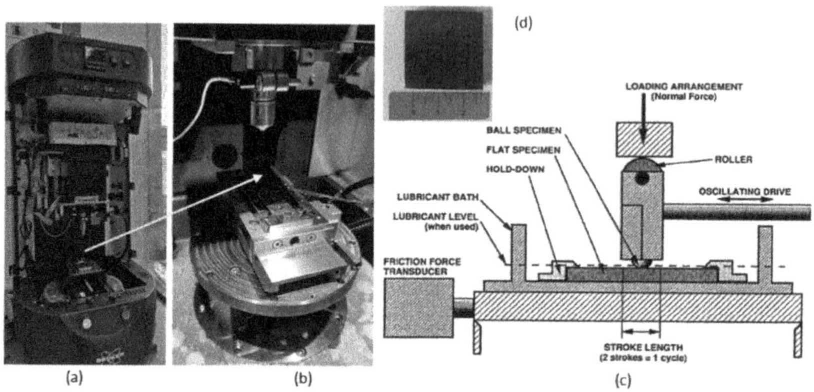

Figure 5. Bruker TriboLab instrument (**a**); detail of the workspace (**b**); schematic representation of the wear test apparatus [38] (**c**); tested sample (**d**).

The samples for the tests were prepared by abrasive water jet cutting technology. The dimensions of the samples were 30 mm × 30 mm × 5 mm (Figure 5d). The prepared sample was inserted into the device, and the time, load force and path of the ball movement along the flat sample were set. After the set time, a trace of the ball was observed on the sample.

The test conditions were set as follows:

- Loading force: 120 N,
- Frequency: 5 Hz,
- Test duration: 15,000 s,
- Track length: 10 mm,
- Ball material: sintered carbide (94% WC + 6% Co),
- Ball diameter: 6.35 mm,
- Hardness of the ball: 56 HRC.

3. Results and Discussion

Next, the results of the performed laboratory analyzes and tests on samples of the base material and hardfacing electrodes are presented for their evaluation of selected properties and determination of suitability for use in order to increase the service life.

3.1. Weight Measurement

In order to determine whether the weight of the tool changed after the modification, measurements of their weight before and after the modification were taken. Table 2 shows the measured weights of the instruments.

Table 2. Weight of tools.

	New Unmodified Tool	Modified Tool
UTP 690	1730 g	1715 g
OK WearTrode 55	1730 g	1714 g

It can be concluded that the correct pre-preparation of the tool, followed by the hardfacing by welding of the additional material, does not lead to a significant change in weight.

3.2. Hardness Measurement

The measured hardness values according to Rockwell according to ISO 6508-1 [37] are in Table 3, five hardness measurements were taken on each sample and the average HRC value was calculated from them.

Based on the results from the hardness measurements, we can conclude that both hardfacing materials achieved significantly higher hardness compared to the base tool body material 16MnCr5. Some authors report that hardness correlates strongly with abrasive wear resistance [25,39]. Thus, based on the results from hardness measurements, we can predict that the welds with the highest hardness will best resist the effects of an abrasive environment. However, other authors, in turn, have reported in their studies that hardness has no effect on wear resistance [13,40]. Therefore, it was necessary to carry out microstructure assessment and wear testing for the base material of the tool body and for the individual hardfacing materials.

Table 3. Hardness values according to Rockwell.

		Values HRC					
		1.	**2.**	**3.**	**4.**	**5.**	**Average**
Sample No. 1	16MnCr5	20	21	20	19	20	20 ± 0.71
Sample No. 2	UTP 690	60	61	62	62	63	62 ± 1.14
Sample No. 3	OK WearTrode 55	50	52	52	51	53	52 ± 1.14

3.3. Microstructure Evaluation

Figure 6a shows the microstructure of the tool body material 16MnCr5. It is a ferritic–pearlitic steel in a not-heat-treated state. Figure 6b shows the martensitic structure of the hardfacing material UTP 690. Needle-like formations are observed in the hardfacing metal of UTP 690, which indicates that this weld metal could better transfer the loads acting on it. In Figure 6c is a hard deposit layer of hardfacing material UTP 690 on the base material (BM). Their connection is without errors, no defects are observed in the remelting zone, which would predict an insufficient connection of the materials. There are no defects in the hardfacing metal (cracks, voids, inclusions, etc.). Figure 6d shows also martensitic structure of the OK WearTrode 55 hardfacing material. Figure 6e shows a hard deposit layer of the OK WearTrode 55 hardfacing material on the base material. There are also no defects, cracks or pores that would reduce the quality of the mixing of the base material with the hardfacing material.

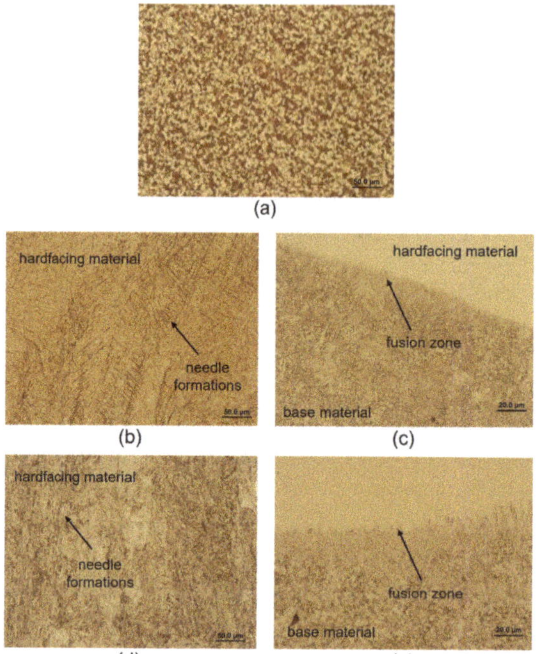

Figure 6. Microstructure of 16MnCr5 (**a**); microstructure of UTP 690 (**b**); fusion zone UTP 690-BM without observation of defects (**c**); microstructure of OK WearTrode 55 (**d**); fusion zone OK WearTrode 55-BM without observation of defects (**e**).

3.4. Wear Resistance Test with Flat Specimen Wear Evaluation

Figure 7 shows the samples after the wear resistance test. Three tracks of straight lines were made on each sample.

Figure 7. Samples after the wear resistance test (no. 1—base material, no. 2.—UTP690, no. 3—OK WearTrode 55).

Macroscopic observation of the traces after the test was performed on a stereomicroscope. In Figure 8a, three tracks can be seen after the passage of the ball on sample no. 1 (tool body material 16MnCr5). From the details in Figure 8b, it can be seen that the track is uniform, the printed volume of material is uniform over the entire length of the track, a straight line.

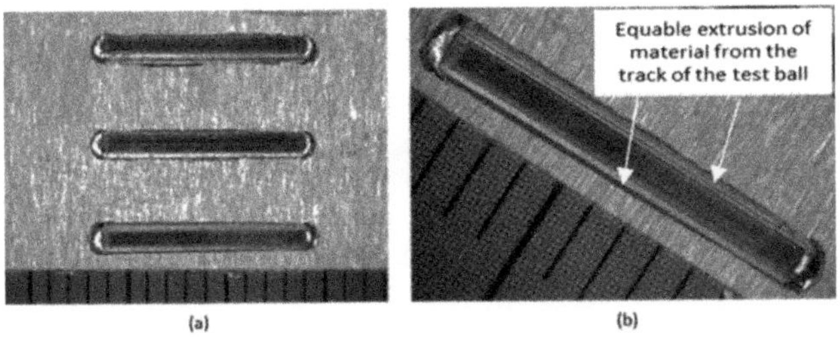

Figure 8. Tracks on sample no. 1—16MnCr5, mag.12.5× (**a**); track detail mag. 25× (**b**).

In Figure 9a, there are three tracks of the ball on sample no. 2 (UTP 690). From the details in Figure 9b, it can be seen that the trace is not uniform. The pushed-out volume of material is larger on one side than on the other.

Figure 9. Tracks on sample no. 2—UTP 690, mag.12.5× (**a**); track detail mag. 25× (**b**).

Three tracks of the ball on sample no. 3 (OK WearTrode 55) are shown in Figure 10a. From the detail in Figure 10b, it can be seen that the track is not uniform. The pushed-out volume of material is larger on one side than on the other side of the track. However, we can conclude that the pushed-out material is more uniform than in sample no. 2.

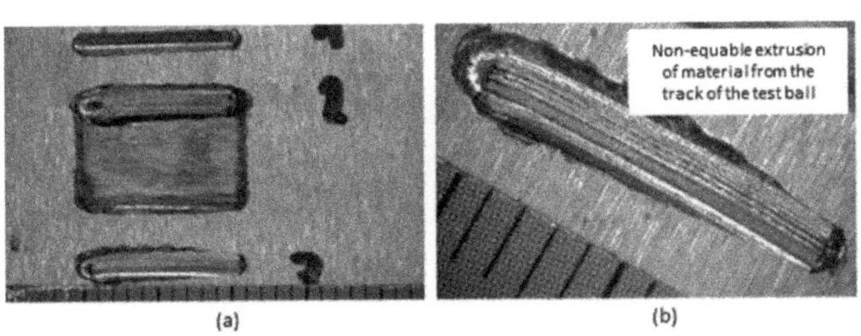

Figure 10. Tracks on sample no. 3—OK WearTrode 55, mag.12.5× (**a**); track detail mag. 25× (**b**).

Figure 11 shows the average cross-sectional area of the trace for material 16MnCr5. For the hardfacing material UTP 690, the average cross-sectional area is shown in Figure 12, and for the hardfacing material OK WearTrode 55, that is shown in Figure 13. Here, it is possible to observe a more pronounced pushing-out of the material on both sides of the track (green areas).

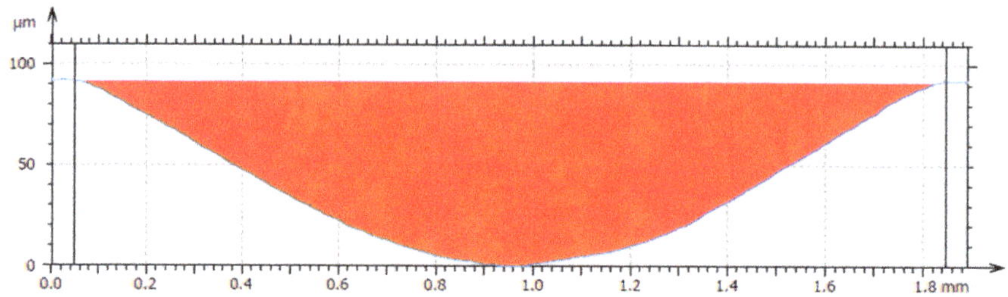

Figure 11. Average cross-sectional area of the 16MnCr5 track.

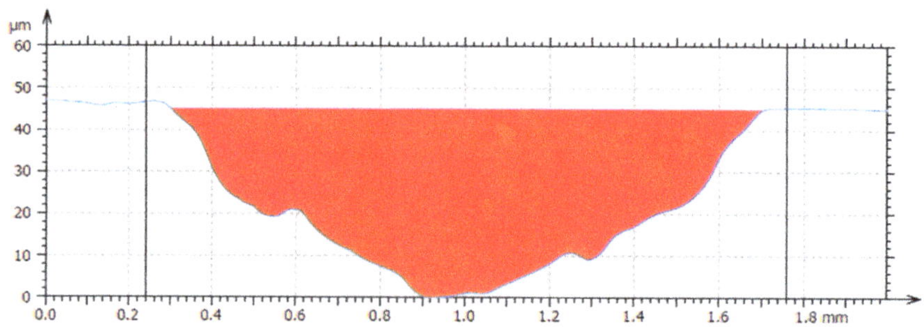

Figure 12. Average cross-sectional area of the UTP 690 track.

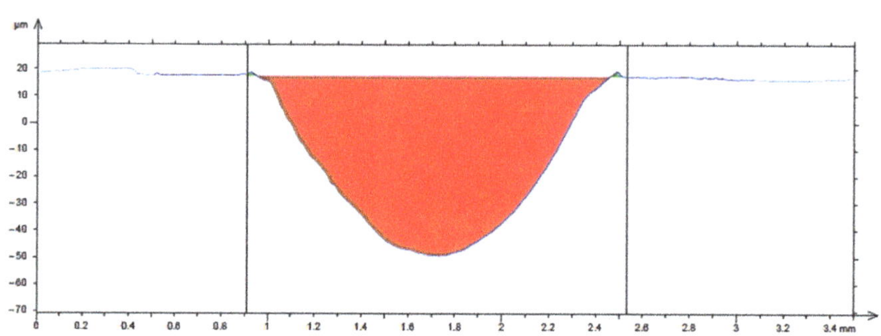

Figure 13. Average cross-sectional area of the OK WearTrode 55 track.

Based on the measured data from individual measurements, the wear volumes of the flat sample were calculated for each material according to Equation (1) (Table 4).

Table 4. Data obtained from the test of resistance to wear.

Sample		Max. Track Width (mm)	Max. Track Depth (mm)	Track Cross-Sectional Area Content (mm^2)	Wear Volume of Flat Sample (mm^3)
Sample No. 1	16MnCr5	1.790	0.0910	0.0968	0.968
Sample No. 2	UTP 690	1.450	0.0441	0.0350	0.350
Sample No. 3	OK WearTrode 55	1.507	0.0440	0.0350	0.400

Figure 14a shows a graph comparing the hardness of individual materials. The base material of the tool for crushing unwanted growths, 16MnCr5, reached an average hardness of 20 HRC. The hardfacing UTP 690 material reached an average hardness of 62HRC, and the OK WearTrode 55 reached an average hardness of 52 HRC. Compared to the base material, the hardfaced materials achieved significantly higher hardness, by more than three times. Some authors report that hardness is strongly correlated with resistance to abrasive wear [11,20,21,25,26,28]. However, the authors of [11,13,25] used different test methods than those used in this experiment to evaluate the resistance to abrasive wear. Figure 14b shows a graph comparing the values of the wear volume of the flat sample V_f for each material tested. The lowest volume loss was achieved by UTP 690, namely 0.350 mm^3. Compared to OK WearTrode 55, this is only 0.05 mm^3 more, even though the hardness of UTP 690 was 10 HRC higher. The base material 16MnCr5 reached the highest volume loss value, namely 0.968 mm^3.

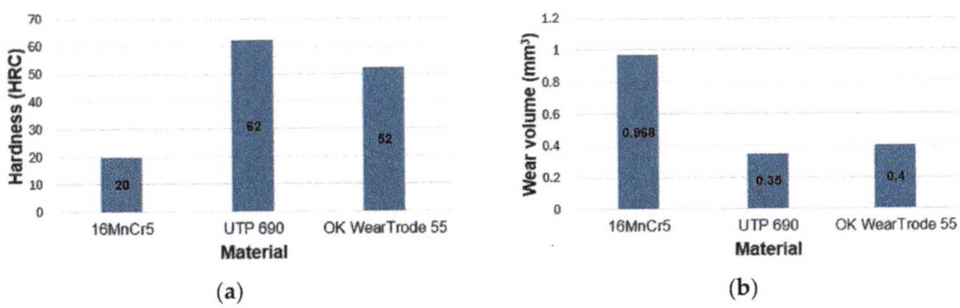

Figure 14. Graphs for comparing the values obtained in the experiment—hardness according to Rockwell (**a**); volume loss of material V_f (**b**).

The hardfacing material UTP 690, with its chemical composition represented by chemical elements Cr and W, also increased in hardness and wear resistance. The elements V and W refined the grain. These elements contribute to increasing abrasion resistance. Vanadium increased the toughness needed to balance the increased hardness with the carbides C, Cr and W. The chemical composition of the OK WearTrode 55 hardfacing material is richer in Cr, but elements such as Mo, V and W are absent. Figure 6 shows images from a light microscope, showing the structures of individual materials. Both hardfacing materials have a martensitic structure with the presence of needle-like formations, which could make these hardfacing materials better able to transfer the loads acting on it; therefore, it is assumed that they will better withstand the effects of an abrasive environment. As stated by some authors [9,19,21], the microstructure of the material also has a great influence on wear resistance.

4. Conclusions

Based on the results of experiments carried out in laboratory conditions and presented in the article on two selected hardfacing materials UTP 690 and OK WearTrode 55 applied to the base material 16MnCr5, we can conclude that:

- Appropriate pre-treatment of the tool can ensure that its weight is maintained after the additional material has been hardfaced by welding.
- The HRC hardness of both hardfacing materials reached a higher average value than the hardness of the base material 16MnCr5.
- Both hardfacing materials achieved significantly better results compared to the base material of the 16MnCr5 tool—the UTP 690 coating achieved 2.76 times higher and

the OK WearTrode 55 coating achieved 2.42 times higher abrasion wear resistance than the 16MnCr5 tool body material.

- Comparing the hardfacing materials with each other, better results were achieved by the UTP 690 hardfacing material: resistance to abrasive wear was 1.14 times higher compared to OK WearTrode; also, its hardness reached 1.19 times higher value.

In order to confirm the results of the laboratory measurements and tests, efforts will be made to make the given modifications to the tools that will be tested under operational conditions.

Author Contributions: Conceptualization, M.Ť. and D.S.; methodology, M.Ť.; software, M.V.; validation, M.V., A.R., A.G. and M.Ť.; formal analysis, A.R. and D.S.; investigation, M.V. and A.G.; resources, M.Ť.; data curation, A.G. and M.V.; writing—original draft preparation, M.Ť., M.V. and A.G.; writing—review and editing, M.Ť.; visualization, D.S.; supervision, M.Ť.; project administration, A.R. and D.S.; funding acquisition, A.R. and D.S. All authors have read and agreed to the published version of the manuscript.

Funding: This research was funded by "Kultúrna a edukačná grantová agentúra MŠVVaŠ SR under the grant number KEGA 015ŽU-4/2023 Modernization of teaching chip technologies with elements of information technologies based on networked virtual laboratories and by "Vedecká grantová agentúra MŠVVaŠ SR a SAV" VEGA 1/0073/24 Experimental analysis of shapes and materials of working tools of additional equipment of forestry mechanization technology".

Institutional Review Board Statement: Not applicable.

Informed Consent Statement: Not applicable.

Data Availability Statement: The original contributions presented in this study are included in the article. Further inquiries can be directed to the corresponding author.

Acknowledgments: This work was supported by "Vedecká grantová agentúra MŠVVaŠ SR a SAV" under the grants numbers VEGA 1/0073/24 and by "Kultúrna a edukačná grantová agentúra MŠVVaŠ SR KEGA 015ŽU-4/2023.

Conflicts of Interest: The authors declare no conflicts of interest.

References

1. Grebner, D.L.; Bettinger, P.; Siry, J.P. *Introduction to Forestry and Natural Resources*; Academic Press: Cambridge, MA, USA, 2014; p. 496, ISBN 978-0-12-386901-2.
2. Hnilica, R.; Kotus, M.; Jankovský, M.; Hnilicová, M.; Dado, M. Qualitative classification of mulchers. *Agron. Res.* **2017**, *15*, 1890–1896. [CrossRef]
3. Dengxing, F.; Guodong, J.; Yangyang, W.; Xinxiao, Y. The effectiveness of mulching practices on water erosion control: A global meta-analysis. *Geoderma* **2023**, *438*, 116643. [CrossRef]
4. Kolomeichenko, A.; Gribov, I.; Soloviev, R.; Maximov, E.; Karyakin, S.; Dudareva, N.; Deev, V.; Prusov, E. Design evaluation and material selection of a mulch milling cutter working bodies. *E3S Web Conf.* **2023**, *402*, 8. [CrossRef]
5. Ťavodová, M.; Kalincová, D.; Kotus, M.; Pavlík, Ľ. The possibility of increasing the wearing resistance of mulcher tools. *Acta Technol. Agric.* **2018**, *21*, 87–93. [CrossRef]
6. Abhishek, K.; Nithin, B.; Nithin, N.; Punith, K.; Samapth, P. Fabrication of Agricultural Waste Shredder Machine. *Int. J. Sci. Eng. Technol. Res.* **2018**, *9*, 112–116.
7. Świć, A.; Gola, A. Economic analysis of casing parts production in a flexible manufacturing system. *Actual Probl. Econ.* **2013**, *141*, 526–533.
8. Salehi, S. Fuzzy multiple criteria group decision-making in performance evaluation of manufacturing companies. *Appl. Comput. Sci.* **2023**, *19*, 28–46. [CrossRef]
9. Pawlik, J.; Bembenek, M.; Góral, T.; Cieślik, J.; Krawczyk, J.; Łukaszek-Sołek, A.; Śleboda, T.; Frocisz, Ł. On the Influence of Heat Input on Ni-WC GMAW Hardfaced Coating Properties. *Materials* **2023**, *16*, 3960. [CrossRef]

10. Mendez, P.F.; Barnes, N.; Bell, K.; Borle, S.D.; Gajapathi, S.S.; Guest, S.D.; Izadi, H.; Gol, A.K.; Wood, G. Welding Processes for Wear Resistant Overlays. *J. Manuf. Process.* **2014**, *16*, 4–25. [CrossRef]
11. Blau, P.J.; Dehoff, R.R. Development of a two-body wet abrasion test method with attention to the effects of reused abradant. *Wear* **2013**, *302*, 1035–1039. [CrossRef]
12. Ťavodová, M.; Hnilicová, M.; Miturska, I. Analysis of Hard Facing Material Designed for Application on the Forestry Tools. *Adv. Sci. Technol. Res. J.* **2019**, *13*, 111–119. [CrossRef]
13. Falat, L.; Džupon, M.; Ťavodová, M.; Hnilica, R.; Ľuptáčiková, V.; Čiripová, L.; Homolová, V.; Ďurišinová, K. Microstructure and abrasive wear resistance of various alloy hardfacings for application on heavy-duty chipper tools in forestry shredding and mulching operations. *Materials* **2019**, *12*, 2212. [CrossRef] [PubMed]
14. Vargova, M.; Tavodova, M.; Monkova, K.; Dzupon, M. Research of Resistance of Selected Materials to Abrasive Wear to Increase the Ploughshare Lifetime. *Metals* **2022**, *12*, 940. [CrossRef]
15. Świć, A.; Wołos, D.; Zubrzycki, J.; Opielak, M.; Gola, A.; Taranenko, V. Accuracy control in the machining of low rigidity schafts. *Appl. Mech. Mater.* **2014**, *630*, 357–367. [CrossRef]
16. Nahak, S.; Dewangan, S.; Chattopadhyaya, S. Discussion on Wear Phenomenain Cemented Carbide. *Procedia Earth Planet.* **2015**, *11*, 284–293. [CrossRef]
17. Chrobot, J. A concept for a production flow control system toolset for discrete manufacturing of mechanical products. *Appl. Comput. Sci.* **2025**, *21*, 143–153. [CrossRef]
18. Šebek, M.; Falat, L.; Kováč, F.; Petryshynets, I.; Horňak, P.; Girman, V. The effects of laser surface hardening on microstructural characteristics and wear resistance of AISI H11 hot work tool steel. *Arch. Metall. Mater.* **2017**, *62*, 1721–1726. [CrossRef]
19. Müller, M.; Hrabě, P. Overlay materials used for increasing lifetime of machine parts working under conditions of intensive abrasion. *Res. Agric. Eng.* **2013**, *59*, 16–22. [CrossRef]
20. Viňáš, J.; Brezinová, J.; Brezina, J.; Hermel, P. Innovation of Biomass Crusher by Application of Hardfacing Layers. *Metals* **2021**, *11*, 1283. [CrossRef]
21. Brezina, J.; Brezinová, J.; Viňáš, J. Possibilities for Using New Types of Additive Materials for Hardfacing. *Defect Diffus. Forum* **2020**, *405*, 171–178. [CrossRef]
22. García-Orenes, F.; Cerdà, A.; Mataix-Solera, J.; Guerrero, C.; Bodí, M.B.; Arcenegui, V.; Zornoza, R.; Sempere, J.G. Effects of agricultural management on surface soil properties and soil-water losses in eastern Spain. *Soil Tillage Res.* **2009**, *106*, 117–123. [CrossRef]
23. Krauze, K.; Bołoz, Ł.; Wydro, T. Parametric Factors for the Tangential-Rotary Picks Quality Assessment. *Arch. Min.* **2015**, *60*, 265–281. [CrossRef]
24. Winczek, J.; Gucwa, M.; Mičian, M.; Koňár, R.; Parzych, S. The evaluation of the wear mechanism of high-carbon hardfacinglayers. *Arch. Metall. Mater.* **2019**, *64*, 1111–1115. [CrossRef]
25. Kováč, I.; Mikuš, R.; Žarnovský, J.; Drlička, R.; Harničárová, M.; Valíček, J.; Kadnár, M. Increasing the wear resistance of surface layers of selected steels by TIG electric arc surface remelting process using a powder based on $CaCN_2$. *Int. J. Adv. Manuf. Technol.* **2022**, *123*, 1985–1997. [CrossRef]
26. Wang, X.; Han, F.; Liu, X.; Qu, S.Y.; Zou, Z.D. Microstructure and wear properties of the Fe–Ti–V–Mo–C hardfacing alloy. *Wear* **2008**, *265*, 583–589. [CrossRef]
27. Wang, X.H.; Han, F.; Liu, X.M.; Qu, S.Y.; Zou, Z.D. Effect of molybdenum on the microstructure and wear resistance of Fe-based hardfacing coatings. *Mater. Sci. Eng.* **2008**, *489*, 193–200. [CrossRef]
28. Ťavodová, M.; Vargová, M.; Falat, L. Possibilities of modification of ploughshares used for winter maintenance of forest roads. *Manuf. Technol.* **2020**, *20*, 834–844. [CrossRef]
29. Selech, J.; Majchrzycki, W.; Ulbrich, D. Field and Laboratory Wear Tests of Machine Components Used for Renovation of Dirt Roads—A Case Study. *Materials* **2023**, *16*, 6180. [CrossRef]
30. Tian, Y.; Zhang, G.; Liu, B.; Ahmad, S.; Liu, S.; Zhao, S. Study on high-shear and low-pressure grinding using a new BAAT with soft-hard combined substrate for single-crystal silicon. *Int. J. Adv. Manuf. Technol.* **2024**, *131*, 467–482. [CrossRef]
31. Wei, C.; Tian, Y.; Ogunbiyi, O.F.; Han, J.; Fan, X.; Gu, Z. Effects of abrasive grain size of flexible body-armor-like abrasive tool (BAAT) on high-shear and low-pressure grinding for zirconia ceramics. *J. Manuf. Process.* **2024**, *120*, 827–836. [CrossRef]
32. *STN EN ISO 683-3*; Heat-treatable steels, alloy steels and free-cutting steels—Part 3: Case-hardening steels. Slovak Office of Standards, Metrology and Testing: Bratislava, Slovakia, 2022.
33. *STN EN 14700*; Welding consumables—Welding consumables for hard-facing. Slovak Office of Standards, Metrology and Testing: Bratislava, Slovakia, 2023.
34. UTP 690. Available online: https://www.valtec.sk/navarove-elektrody/elektrody-bohler-utp-690-25x350mm4kg142ks-37637/ (accessed on 20 January 2025).

35. OK WearTrode 55. Available online: https://www.gobako.sk/item/121/ok-8458-pr-25mm-elektroda-ok-weartrode-55hd (accessed on 20 January 2025).
36. Steel 16MnCr5. Available online: https://www.jkz.cz/cs/produkty/konstrukcni-oceli/csn-14-220-17131-16mncr5/ (accessed on 20 January 2025).
37. *ISO 6508-1*; Metallic Materials, Rockwell Hardness Test, Part 1: Test Method. International Organization for Standardization: Geneva, Switzerland, 2016.
38. *ASTM G133-95*; Standard Test Method for Linearly Reciprocating Ball-on-Flat Sliding Wear. ASTM: West Conshohocken, PA, USA, 2002.
39. Slota, J.; Kubit, A.; Gajdoš, I.; Trzepieciński, T.; Kaščák, Ľ. A comparative study of hardfacing deposits using a modified tribological testing strategy. *Lubricants* **2022**, *10*, 187. [CrossRef]
40. Brziak, P.; Bernasovský, P.; Mráz, Ľ.; Piussi, V.; Grgáč, P.; Mráz, Ľ.; Hrivňák, I.; Kálna, K.; Pecha, J.; Országhová, J.; et al. *Materiály a ich Správanie sa pri Zváraní–2. Kniha Učebných Textov pre Kurzy Zváračských Inžinierov (Materials and Their Behaviour at Surfacing Welding–2nd Study Book for Courses of Welding Engineers)*, 2nd ed.; Welding Research Institute: Bratislava, Slovakia, 2023; ISBN 80-88734-10-X.

MDPI AG
Grosspeteranlage 5
4052 Basel
Switzerland
Tel.: +41 61 683 77 34

Materials Editorial Office
E-mail: materials@mdpi.com
www.mdpi.com/journal/materials

www.ingramcontent.com/pod-product-compliance
Lightning Source LLC
Chambersburg PA
CBHW061509210926
51850CB00008B/1866
9783725866243